The World Outside

The World Outside

Selected Wildlife Studies and Stories

David Stephen

GORDON WRIGHT PUBLISHING
55 MARCHMONT ROAD, EDINBURGH, EH9 1HT
SCOTLAND

Photography by David Stephen

British Library Cataloguing in Publication Data

Stephen, David
The World Outside.
1. Wildlife
I. Title
639'.9   QL82

ISBN 0-903065-38-X

Typeset by Jo Kennedy
Printed and bound by Spectrum Printing Company

In Memoriam

Daisy Ferguson

and may her god value

the stuff she was made of.

'What is man without the beasts? If all the beasts were gone, men would die from great loneliness of spirit; for whatever happens to the beasts also happens to the man. All things are connected. Whatever befalls the earth befalls the sons of the earth.'

Chief Seathl, N. American Suwamish Indian.

Introduction

If someone were to tell you today that on one Highland estate, in four seasons, gamekeepers had killed 48 otters, 246 pine martens and 106 polecats, you wouldn't believe it. Nor would you believe that, on the same place, they had also killed 27 sea eagles, 63 goshawks and 275 kites. And you'd be right not to. But if you make the years 1837-1840, and the place Glen Garry, the figures would be true, because that's what the record says.

Oddly enough they killed only 11 foxes, so foxes couldn't have been as thick on the ground in those days. They managed, however, to knock off 67 badgers, 198 wildcats, 301 stoats and weasels, 98 peregrine falcons, 78 merlins, 18 ospreys, 63 hen harriers, 15 golden eagles and over 600 buzzards (common, rough-legged and honey). That's what they claimed, and recorded for posterity. And you mustn't think that Glen Garry was a special den of iniquity: it was merely down to the standard of its time. So was the Duchess of Sutherland's estate, where 224 eagles, 1155 hawks and kites, and 901 wildcats, polecats and pine martens, were killed between 1831 and 1834.

And you might think—well, things are better now: the killing times are over. But are they? Do we really do better by our wildlife? Well, there are species on that list which we don't kill any more because they are no longer with us to be killed. The ones we still have with us are still killed in many places, Acts of Parliament notwithstanding.

It is true, of course, that in modern times, as in more distant ones, many factors have been at work in changing the face of Scotland and its wildlife complex: the ruthless destruction of the old forest after the failure of the Risings of '15 and '45, the Coming of the Sheep, the spread of towns, the building of roads and railways, changed farming practices, the draining of marshes, river pollution, hydro-electric schemes and reafforestation with great blocks of exotic conifers. But the one constant and active factor in the destruction of wildlife in the nineteenth and twentieth centuries, has been the game preserver. Add to that the increasing disruption and destruction of habitats and you realise how precarious is the future of wildlife—a future that depends on Man.

A good example of that is the otter, belatedly given total protection by law in Scotland. The Government dragged its feet for a year before activating the legislation, and for any kind of offence against the otter you can be fined up to £1000. But does anyone really believe that will stop the boyos in the northwest and the islands from trapping and snaring otters? They're at it now, and I believe they'll still be at it when this book appears. It is a fact of life that those who did not kill otters before the Act will go on not killing them, and that those who killed them before the Act will go on killing them. As they still do with eagles, falcons, harriers and the rest. It has to be emphasised that there are landowners and keepers who have always scrupulously observed the law (they didn't need a law to make them like that) but there are many more, a great many more, who do not.

Long before the days of the game preserver's ascendancy other factors were at work in influencing the disappearance or survival of wildlife species. Since his advent Man has been a main factor in changing and disrupting habitats, and in direct killing-out of wildlife species. One way or another he is still doing it.

Climatic change has had nothing to do with the extermination of any species in the past thousand years. It is interesting to look at the record.

The giant Irish Elk, the pride of the Pleistocene, probably died out before it could meet a human being in Scotland. Climatic change was against its survival but scientists reckon that it was on the way out anyway. Unlike the red deer it failed to adapt to changing conditions. It was a magnificent beast, with a spread of antlers weighing up to ninety pounds.

Other losses in prehistoric times were the lemming and the northern rat-vole, which were unable to survive the warmer Atlantic climate. The northern lynx lasted until Man gave it the final push into oblivion. And from then on Man took over as the star actor in the drama of wildlife.

In historic times we have lost the wild ox, the wild boar, the native pony, the brown bear, the reindeer, the elk, the beaver and the wolf. Then the polecat. The reindeer has since been brought back, and polecats are being planted here and there by private individuals. The extermination of these species was in almost every case related to the destruction of the forest—the great Wood of Caledon, remnants of which can still be seen at Rannoch, Strathspey and elsewhere.

It is perhaps difficult to realise that we still had the elk, 'the deer of shambling gait', when Wallace marched to Stirling. Its last strongholds were in the far north of Scotland, and there are a number of references to it in Gaelic tradition. In the old poem *Bas Dhiarmid,* we read:

Glen Shee, that glen by my side,
Where oft is heard the voice of deer and elk.

The brown bear disappeared earlier, probably in the ninth or tenth century, and Caledonian bears were taken to Rome to appear in the brutal circuses there. The beaver was hunted for its skin and Man was probably the main agent responsible for its extinction.

The wolf outlasted them all, surviving long enough to become British. All out war was waged against the wolf, and the last one is reckoned to have died in 1743 on the Findhorn. It seemed to me to be fitting to open this book with an account of him.

The following pieces, observations and stories are a selection from my writings in *The Scotsman* over a period of twenty-five years.

David Stephen

Contents

Cry Wolf

Folklore has given the Scottish wolf the same kind of reputation that history foisted on John Graham of Claverhouse, and you can still sell rechauffe frightfulnesses about either more easily than you'll gift the truth. The 'pitiless eyes that scare the dark with their green and threatening light' make the wolf a suitable dog for 'Bloody Clavers'.

The Scottish wolf survived long enough to become British, but nobody knows exactly where or when the last one made his exit bow.

There is no doubt that the wolf was still howling when Montrose made his incredible winter foray over the Corrieyarick to surprise Argyll at Inverlochy, still howling when Claverhouse shattered McKay at Killiecrankie; still howling when the clansmen marched to Sheriffmuir. And maybe there was still the echo of a howl somewhere when the last of the Stuarts raised his standard at Glenfinnan.

The records say that the last wolf in the North-east was killed in Kirkmichael, Banffshire, in 1644. The last wolf in Perthshire, we are told, was killed by Sir Ewan Cameron of Lochiel at Killiecrankie nine years before Claverhouse got the bullet not of silver.

The story that may be true, and for many reasons deserves to be true, has Scotland's last wolf still breathing in 1743. The killer of that last wolf, perhaps the last wolf indeed, was MacQueen of Pall-a-chrocain, stalker to the Laird of MacKintosh.

The story is that when 'the black beast' (melanic or evil?) had killed two children the laird called a *tainchel* to round him up, and MacQueen was appointed leader of the expedition. As the day wore on, with no sign of MacQueen, the laird became angry; but his anger turned to elation when the stalker arrived at dusk with the bloody head of the wolf hanging from his belt.

MacQueen told a wonderful story, a masterpiece of restraint and under-statement, which in my view is one of the most telling and graphic animal stories in the literature.

'As I came through the slochk by east the hill there, I foregathered wi' the beast. My long dog there turned him. I buckled wi' him, and dirkit him, and syne whuttled his craig, and brought awa his countenance for fear he might come alive again, for they are precarious creatures.'

Brief, brilliant, honed to the bone. Danger? The man didn't say so. Personal prowess? The man made no claim. He simply buckled wi' the beast and dirkit him. And the wolf let him. MacQueen was not afraid. Nor did he suggest that the wolf made any move to attack him.

When I first read this account many years ago I liked it, but I was dubious about it. Given the legends then masquerading as fact this wasn't surprising.

One could hardly find a story about wolves that weren't trailing, or crowding in on, some lonely backwoodsman, trapper or traveller, to eat him up.

Looking at it now, in the light of thirty years' research on the wolf, especially in North America, culminating in Mech's definitive synthesis, one is brought to realise how acceptable MacQueen's account is in its totality. The surprising thing about it is that it's no surprise. It would find favour with any modern researcher because it fits the facts about the wolf as now known.

Let us measure MacQueen's account against the findings of modern research.

D.F. Parmalee (of Kansas State Teachers' College) and a co-worker chased a dog wolf and four pups on Ellesmere Island, North Canada, caught two of the litter and carried them off. On their way back to camp they shot several ptarmigan, which they hung from their shouldered gun barrels. In the dark they heard something behind them, and found it was the bitch wolf following so close that she was nosing the dead ptarmigan. Several times they had to drive her off with snowballs. She skulked about their tent all night.

Adolph Murie, who was 'the first biologist to conduct an intensive and objective ecological study of the wolf,' and who has spent long periods in Alaska at this work, tells how he crawled into a wolf den and took a pup, just after the bitch had bolted. Dog and bitch did nothing more than howl and bark in the distance until he left.

Lois Crisler's husband took a whole litter of pups from a den while the adult wolves 'bounded around crying.'

Dr Mech tells how he and his associates found a snared wolf who allowed herself to be held by the scruff of the neck when free, while Mech ear-tagged her, fastened a radio collar round her neck, and examined her teeth. She remained as docile as though anaesthetised.

Mech also reports, on the evidence of observers Audubon and Bachman, how an Indiana farmer descended into the wolf pit where three big wolves had been caught, and killed them there.

Then there is the account by a conservation officer-pilot, Robert Hodge, of Minnesota, who has shot hundreds of wolves from aircraft. He landed once, to follow on foot, a wolf whose forelegs he had broken. When he came up to finish the beast off he was met by a meek, docile animal wagging its tail like a frightened dog.

My own insignificant contribution to this list is that when our bitch wolf Magda was seven months old I had to corner her to catch her and carried her in my arms to her new quarters. Her only reaction was to empty her bladder and bowel over my midriff.

Which brings me back to MacQueen. While Murie's strongest impression of Alaskan wolves was their 'friendliness', and Mech says they have a personality that in humans would be called 'agreeable', the record suggests that Scottish wolves were a particularly and peculiarly ferocious breed. But were they?

The preamble to the MacQueen story is that the wolf had killed two children on the hills by the Findhorn. Prof. Ritchie, in his classic *The Influence of Man on Animal Life in Scotland,* deals with the wolf rather less than objectively. He mentions the MacQueen story as a persistent tradition, but apparently takes for granted the killing of two children.

14

Yet you would be hard put to it to find an authentic case of a wild non-rabid wolf attacking anybody, and the consensus of informed opinion in North America and Russia is that wolves do not attack people.

We are all familiar with the stories of wolves in Eastern Europe invading towns and killing babies or priests at the altar. Yet Hanzak and Vasilevsky, with whom I once collaborated on a mammal book, could not authenticate one of them. A leading Soviet mammologist told Robert L. Rausch that one of his associates had tried to document reported attacks by wolves on humans.

'It was impossible to do so, and it was concluded that, with the exception of possible killing of small children wandering alone into remote areas, reports of such attacks had no basis in fact. People like to believe that wolves are dangerous.'

On this the biologists of the two super-Powers clearly agree, which means that attacks on people, where and if they occur, are by *rabid wolves,* which are no different from rabid dogs. Pet smugglers please note . . .

Ritchie, whose classic work is a fine shorthand source for students of Scotland's man/wildlife past, fell too easily into the traditionalist camp with his use of terms like plague, pestilence, scourge, savage marauders and wolfish breed. His delvings were prodigious; his results invaluable. Yet what comes through to me is that the wolf was hunted because of his predation on livestock, with predation on humans as a kind of gloss.

According to Boece, the legendary King Dorvadilla, who reigned two centuries before Christ, ordaint:

'The slayer of ane wolf to have ane ox to his reward. Oure elders persewit this beast with gret hatreut, for the gret *murdir of beistis* done be the samin . . .'

Edeir, contemporary of Julius Caesar, 'delitit in no thing more than in chais of wild beistis, with hounds and rachis, and specially of wolffis, for they are *noisum to tame bestiall.*

In 1527 Boece wrote: 'The wolffis are *richt noisum to the tame bestiall* in all parts of Scotland except ane part theirof namit Glenmores . . .'

The only mention of predation on people reported by Ritchie is from Bishop Leslie, of Ross, towards the end of the sixteenth century, and this is worth quoting:

'Our nychbour Inglande has nocht ane wolf . . . but we now nocht few, ye contrare, verie monie and maist cruel, cheiflie in our North contrey, quhair nocht only invade they scheip, oxne, ye and horse, but evin men, specialie women with barne, outragiouslie and fercelie thay ovirthrows.'

I don't know what the evidence is for it, but Scottish wolves were said to be grave robbers. Mrs D. Ogilvie, lines from whose verses I quoted at the beginning, ends them with:

'And he digs the dead from out the sod,
And gnaws them under the stars.'

In the *Orkneyinga Saga* the skald wrote about the battle of Waterfirth between the Vikings and the Skyemen:

'There I saw the grey wolf gaping
O'er wounded corpse of many a man,'

And Ritchie wrote: 'And so at last the inhabitants of Ederachillis were compelled to carry their dead across the sea to the lonely island of Handa, there to lay the poor bodies in peace far from the reach of the prowlers of the night.' They did indeed, and maybe they had reason to, as people had later with human grave robbers, but Ritchie was really after the wolf in the way he told it.

15

But what of the children the MacQueen's wolf had killed? If they were, it was a tragedy by any standard. Or were they, perhaps, like the two old women drowned by Claverhouse in Galloway who were never drowned? Against that you have the Russian speculation about children wandering into remote areas—a suggestion, nothing more.

And you also have the word of D.H. Pimlott, of Ontario, who wrote the following:

'Perhaps even more powerful testimony is that, in spite of one of the highest wolf populations in the world in that area, thousands of children canoe and camp in the wilderness section of Algonquin Park each year, and there are no reports of any one of them having been attacked or threatened by wolves.'

What about attacks by Scottish wolves on adults? There is no hard evidence that I can find but, according to tradition, wolves were so numerous in the sixteenth-century Highlands that a man was taking his life in his hands to travel through the wilds of Lochaber or Rannoch. Ritchie says that 'such places were almost impassable on account of their savage tenants.' Our wolves must surely have been a very special breed . . .

As a result of this belief hospices or spittals were erected wherein travellers could seek refuge after dark. The Spittal of Glenshee is a present-day reminder of them. No doubt these spittals were used by travellers in fear of wolves but were they really in any danger? And what of those who couldn't reach one in time? Were they killed and eaten? Tradition is silent on this.

But surely, you might say, there must be a grain of truth somewhere in these stories. That is a point, and the rabid wolf is the probable answer. The Russian view is that 'such stories probably are based on attacks by rabid wolves, with the usual exaggeration as they are passed on, eventually becoming established as folklore.'

It is difficult to square this with official Soviet policy, which is to exterminate the wolf. Plotnikov reported about 42,000 wolves killed in 1946: the figure had dropped to 8800 by 1963.

The Americans did their share of killing between the wars when the Federal Government redd them from most of the country. Today the U.S. Department of the Interior has listed the wolf as an endangered species, and in Canada 'control' rather than total destruction is the declared aim.

To all of this I can add only a postscript, which is all I am competent to add. Wolves have fascinated me since the day, forty-eight years ago, when I saw my first one alive and free, and last year a lifelong wish came true when I got a timber wolf male pup five weeks old, followed four months later by a bitch aged five months. We called them Marquis and Magda, and maybe you can guess why . . .

Marquis was four months in the house (two compartments of which he successfully wrecked) and fixed himself first of all on my German Shepherd bitch Lisa, who is very big. Then he graduated to my wife, then to me. Magda was put out with Marquis the day she arrived, and soon accepted the dog.

Both wolves are now a year old, and it has taken Magda seven months to accept completely the two human beings in the pack. This is exactly according to the script evolved after much study by Mech and others.

Lisa remains the pack leader, the hominids are treated with friendly respect. We spend time every day with the wolves because they are social animals and our society is a happy one.

Marquis occasionally tries to strike the dominant male posture, but is quickly given the correct wolf response and immediately accepts his second position. After that Lisa allows him to rough-house with her and Magda.

Marquis is now four inches taller than Lisa. I had a Canadian visitor who said he was the finest timber wolf he had ever seen in captivity.

The other day some men were watching the three of them brulzying around, and one of them said to me:

'That wolf could kill that dog.'

'He could,' I agreed, 'but he won't. She wears the stripes.' Protocol and precedence are strongly built into the tightly-knit wolf society, and Marquis is a good-natured clown in the hands of my wife.

Watching this pair I realise how well the American researchers have done their work on wolf behaviour, protocol and personality, as well as on facial and posture signalling. They might have written the script for Marquis and Magda.

Marquis shuns all strangers: that is he will stand back from them. Because wolves are scared of people, except those who have brought them up. But small children are different, and Marquis shows a keen interest in them. I took in a Little Red Riding Hood one day and he wanted to play with her. I didn't let him because he's a big wolf and plays rough. As badgers do, and big puppies.

One day, when Marquis was lifting his leg to me to have his belly scratched, I said to him:

'If you chaps are the ravening, murderous, man-killing marauders you're supposed to be, why did primitive man choose you to turn into his dog?'

He licked my mouth.

Blackbird of the Mountains

The ring-ouzel is the mountain blackbird—the blackbird of the deer forest and the high grouse moor—its neighbours the red stag and the mountain hare, the red grouse, the golden plover, and the wheatear.

It is a migrant, summering in the Highlands and wintering on both shores of the Mediterranean.

When you're in the mountains you need never be in doubt about the ring-ouzel. It's so obviously a blackbird. And, as in the common blackbird, the cock is black while the hen is brown.

But there's no possibility of confusion between the species. In the first place, you don't expect to see the blackbird where you see the ring-ouzel. In the second place, the ring-ouzel has a white crescent on his breast which is obvious and striking.

In the female the white gorget is narrower, duller, and tinged with brown. Young birds are very like fledgling blackbirds, and have no white gorget of any kind.

During the flight north, in spring, and south, in autumn, ring-ouzels will be seen in most counties in most years, and it is at such times that confusion with the common blackbird is most likely.

There can be no confusion with blackbirds in normal plumage, but, as everyone knows, pied blackbirds are not at all uncommon, and a blackie with white on his chest might just manage to pass himself off as a ring-ouzel.

But the closed wing in the ring-ouzel is paler than the surrounding feathers, due to the fact that the secondaries and coverts have grey edges. This is a useful guide.

Most ring-ouzels in Scotland nest between 1000 and 1500ft., and seldom over 2000ft. They like the steep heathery slopes above mountain burns, where the ground is terraced by sheep walks and deer paths. And they like to get the nest tucked under a clump of heather, and, frequently, against a boulder.

Both sexes work at building the nest, which is of coarse grass with heather twigs and earth in the foundation. The finished lining is done with finer grasses. The eggs usually number four, and are very similar to those of the blackbird.

On the hill, you're far more likely to hear the ring-ouzel before you see it. It may chatter or chuckle.

The chatter is a loud *'tac-tac-tac-tac,'* and this alarm may be sounded before you're anywhere near the nest. The chuckle is a grating cry, but much more subdued than the chatter.

It also has a *'tik-tik'* call, very reminiscent of the blackbird. And, sometimes, you might be forgiven for thinking that the bird churring at your ear is a whitethroat.

Fledglings ready to leave the nest call a great deal to the parent birds when they are coming with food. The old birds use the grating cry frequently at this time, and the call of the young always reminds me of the musical *'chirling'* of young magpies.

The chicks are fed on the typical thrush mixture—worms, larvae, and full-winged insects, but small snails are a frequent item.

Like the blackbird, the ring-ouzel turns to berries and fruit in the autumn—hawthorn, sloe, ivy, rowan, and elder, plums, cherries, gooseberries and currants.

During the nesting season the ouzel is both shy and bold. When disturbed it will fly from the nest for a distance of up to a mile, taking advantage of all cover on the way to keep out of sight.

It is, I should say, the shyest and least approachable of British thrushes, and is most wary about returning to the nest if there's a human being anywhere in sight.

Some birds, with well-grown young, will strike at a human intruder as the mistle thrush does; but I have never met them. The bird I usually see least of when I'm standing by a ring-ouzel's nest is the ring-ouzel.

You can sometimes bring the bird in about your ears if you handle the fledglings and they start to call. But even that doesn't always work. You're far more likely to hear the old birds chattering and chuckling out of sight.

But against other intruders the ring-ouzel is indeed bold in defence of its nest. It will assault hoodie and jackdaw, terrier or stoat.

But the sound and fury achieves little if the nest is known to the intruder. No ring-ouzel can fight off a hoodie. No two ring-ouzels can do it. And most ouzel nests which are robbed on the hill are robbed by hoodie crows.

Some ring-ouzels over-winter in Britain, and birds seen at the end of the year can usually be considered as doing so. Nestlings ringed in this country have been taken again in the South of France, Spain, and Algeria.

I used to see ouzels in the North of Spain in large numbers in November, and I always liked to think as I watched them, that they had come down from the wild places that had known Montrose and the men who marched with Prince Charles Edward.

Bashful Badger Can Become
A Terrible Fighting Machine

Scotland has plenty of badgers. The brock is widely distributed in Highlands and Lowlands, and is often present where not many people imagine him to be. This is because he is largely nocturnal and seldom gets into serious trouble. The past decade or so has been kinder to the badger, perhaps, than any other in the last century; if brock still has his enemies among men, he has also many more friends.

Like so many carnivorous mammals and birds, the brock took his share of hammering from the game preservers. He was killed out not for what he did but for what he might do, a principle which still activates his enemies. It is a sad reflection on twentieth century education that a man can still give, as his reason for killing badgers, nothing better than: 'I don't like the dirty brutes!' Everything else apart, the 'dirty' is a gross libel.

The brock is bear-like; squat, big-clawed, hand-footed. He is strong-jawed, powerful, retiring, tolerant, non-aggressive and a terrible fighting machine. But he is the last only when self-effacement, turning the other cheek, and the peace conference have failed. Once he goes to war he can break a terrier's jaw, cripple a collie, kill a fox, or take the hand off a man. Men, however, don't usually attack with their hands; even the English badger-diggers use tongs.

Bear-like though he may be in conformation, the brock is all weasel: a member of the great Mustelidae family. The flick of the head may be bear, but once he starts moving he is all weasel, with the typical rippling hump-up run of the clan. In action, he looks very like the wolverine.

Everywhere he goes he is heralded by his white face. The night would have to be dark indeed before you could miss that candy-striping—head on or in profile. It is an announcement. The late Harry Mortimer Batten made out a case, many years ago, for the black and white markings being camouflage. But the conditions he described were very special: bright moonlight among broken masonry. In fact, in moonlight or dark, in the half-light of evening or at dawn, the face of the brock is as plain as the stripes on a skunk. The pattern is probably a warning one. Whatever it is, it takes a lot of missing.

The brock is a beast of ancient lineage; his ancestors knew the cave bear. In modern times he has had the experience of being evicted by mechanical earth movers, and buried by bulldozers; he has been built round by city overspills; for weeks at a time you can set his time of emergence by the ten o'clock jet; then suddenly, he finds his runway tarmacked and his home range floodlit. And he quits.

Badgers are stay-at-homes, strongly territorial, thrawn to move. Though they will range far enough in a night, they'll be home in the morning, if not in the main den then at a spare one nearby, unless they've moved in to spend a day with neighbours farther out. But they'll be back next day, or the next, although they may give a sett a few days' rest. A well-established badger sett will hold badgers almost continuously so long as there are badgers in the neighbourhood. The brock's sett is his castle, and his castle is his home.

The earthworks of an established sett—as distinct from temporary scrape-outs or new diggings—are unmistakeable: tons of earth fronting a varying number of deep burrows, so that the whole resembles a series of small quarries. Single holes, with single mounds, are just as easy to identify; size and depth give them away. Badgers have beds, and at the occupied sett there will be, at different times, old bedding stuffed in a hole, or bedding out to air, or trails showing that new bedding has been trundled in. There will also be latrines in use.

'Dirty brock' is a libel on the badger. The beasts keep their home clean. You'll never find it cluttered up with rubbish as you will a fox den, and if it is cluttered outside one or two doors you can be sure there's a fox lodging there. Foxes quite often move in on badgers, and vixens sometimes whelp in a badger sett, keeping to a single end, out of the badgers' way. Badgers usually put up with a fox lodger; but sometimes the brocks take a turavee and throw the lodger out. Badgers can, and at times do, kill fox cubs, which is why so many foxhunters don't like them. But such killings must be rare.

All observers who have watched badgers closely are agreed—at least I know of none who disagrees—that they never carry food into the den, or to the den mouth. It follows from this that if lambs or poultry are found at a badger den the culprit is a fox lodger. I have used the word 'never' here, always a dangerous one to use where animals are concerned, but in this context it appears to be justified.

This isn't to say that badgers never kill poultry. Some of them do; but a poultry-killing badger is exceptional, and most lead blameless lives in this respect. So far as I know there are only a few authentic cases of badgers killing lambs, and I know of none at first hand. Neal puts the figure for Britain as low as one lamb a year. Almost all cases of alleged lamb-killing are based on dead lambs found at a den, and it can usually be taken for granted that a fox carried them there.

Badgers don't appear to touch carrion—that is flesh which is really high or putrefying. But we have to be quite clear that we mean exactly this, because they will eat dead animals—rats, rabbits, mice, small birds—which have been dead for days, but which are still fresh as any Christmas turkey. I've brought wild badgers to a variety of baits—maize, raisins, dehydrated horsemeat, fish, honey, rabbits—but never to real carrion.

The brock's technique with a nest of young rabbits is distinctive; he digs down vertically on to it, and not in from the entrance as a fox does.

Small prey rather than large is the badger's way: rats, mice, voles, rabbits, slugs, earthworms, caterpillars, beetles. He takes many wild fruits, including doghips, and a variety of roots, bulbs and tubers. He will eat grass. Gamekeepers often accuse him of eating gamebirds and their eggs, but one needs more evidence than say-so. One can imagine a brock eating a nestful of eggs if he comes on them, but there is no evidence that he looks for them, and only the scantiest that he attacks adult birds. Evidence from dung and stomach analyses is completely negative.

Then there are wasps. Badgers delight in tearing out bikes and devouring the grubs. They'll wreck every bike they come across. They also like the grubs and honey of the bumble bees.

In the badger there is delayed implantation. The main mating period is spring, but there is a second period in autumn, and coupling may take place at any time between the two extreme dates. Implantation takes place in December/January, after which development is complete in two months. This explains why a captive badger could be a year in captivity and then produce young. Cubs are born in February and March, earlier in the south, later in the north. The cubs begin to appear at the den mouth when they are about two months old, although they are active below ground before then.

Small cubs can be very naif, and will even accept milk from a saucer. At three months of age they will still accept one's foot as a normal part of the scenery, and run over it again and again during play. They are extremely playful at this stage, and can make as much noise as a litter of piglets in the bracken and undergrowth. As June wears on they become more concerned about foraging, and usually leave soon after emerging from the sett. The cubs are often out in good light, and on their way, some time before the adults appear.

Typical emerging times for July in Lowland Scotland are: first cub 9.50p.m., second 9.53p.m., third 9.58p.m. Sow at 10.10p.m. and boar at 10.25p.m. Compare this with 5 September, when times were sow at 8.45p.m. and boar at 9.00p.m.

Badgers aren't often seen about by day, but the spring mating period is a likely time. Cubs are more prone to come out exploring in daylight, in cover. Some Highland badgers have their young above-ground, among rocks or bracken, and such cubs are readily put on foot by day.

A badger's nose is good, very good, and can't be fooled. His hearing is acute, but his eyesight isn't as good as yours or mine in the dark. A sniff of you, however slight, or a peep out of you, will send him rumbling down into his den like a runaway hutch; but he'll look you in the face without flinching so long as you are not actually recognisable for what you are. You'll see a badger when he won't see you.

An emerging badger, exploring with his nose, and listening hard, will bob up and down several times before coming right out. If he hears or smells nothing suspicious he'll stay out, and presently leave for the night. Sounds he understands are taken for granted—aircraft, motor cars, the sneeze of a crow on a branch, the patter of other badger feet nearby, the bark of a roe-buck—but the scrape of a boot, or the sniff to catch the dreep, will send him down or away, or freeze him momentarily. Behaviour in this respect varies with age and place. Young brocks are more tolerant than old ones, and my experience has been that Highland brocks are flightier than Lowland ones.

Like the weasel clan, and the fox, the badger has musk glands with which he sets scent. You can watch badgers doing this. It is certainly an advertisement, probably marks territory, and enables badgers to find their way and each other. Sows set it for cubs, as you can see by the way the cubs can follow the mother's line some time after she has left on her foraging.

Badgers don't hibernate, although they are less active in winter. Snow on the ground doesn't keep them in, not even in the Highlands; but prolonged hard frost they don't like. December is a quiet month. January is often the opposite. A February freeze-up doesn't keep them in much, presumably because they have cubs by then, or are about to have them. A badger once came to a frozen heap of

21

sand in my stackyard to dig out a nest of rats. He failed because the crust was like iron. I dug them out with a pick, the terrier killed them, and the badger came the next night and ate them—on the spot.

One assumes that most badgers which die below ground are buried there by the others: walled off in a blind end. This is borne out by the fact that one finds skulls and other bones in new diggings from time to time, bones long clear of tissue and ingrained with the colour of the local earth. There is, however, one record of a dead brock being dragged from a sett by the others and buried outside. An increasing number of badgers are being killed on the roads at night by cars, so one must now consider modern traffic a significant mortality factor in badgers as it is in hedgehogs.

Badger runways are clearly defined when in nightly use, and could easily be mistaken for footpaths, until they pass under some obstacle where a human being couldn't go, upright or stooped; then the mistake becomes apparent. The brocks use these routes to and from the den as people use pavements.

Weight for size, badgers are far more solid than foxes, especially when they've laid on a lot of fat towards the end of the year. A boar measuring a yard long, including tail, will weigh anything from twenty-five to forty pounds or more. The last two Scottish boars I weighed topped forty pounds. Boars are heavier, thicker, blunter, more braid-faced than sows. The cubs have short faces which lengthen as they grow.

Apart from the snorts, snuffles, grunts, squeaks and yelps one would expect from such an animal, the badger is normally very silent. Until one screams! Nobody knows why a badger should scream as it sometimes does, but the sound is fit to compare with the most eldritch display of any vixen. I must say I have heard it only in the neighbourhood of the sett, within a hundred yards or so, and nearly always when a big brock had moved around into a position where it would be able to own stray whiffs of my scent (I think). But this could be mere coincidence. Anyway it's a gey noise.

Adders—Beware of Man!

When I come across notices like 'Beware of the Bull' or 'Beware Guard Dogs', or 'Fallen Rock', I tend to take them seriously at first reading, because a three-year-old dairy bull could wreck a man as easily as it could a china shop, because a big rock could flatten a man as a road roller flattens a frog, and because a big, fierce guard dog could take a pound off a man faster than any slimming pill.

The sort of notice I don't take seriously is the one that reads 'Beware Adders'. It is a joke, as the notice 'Motorists Beware of Sheep' is a joke, and I would rewrite both to read 'Sheep Beware Motorists', and 'Adders Beware People'.

How many sheep have killed motorists, as against the number of sheep killed by motorists? The same question might be asked about adders. When did you last hear of an adder killing a man? And how many times have you listened to people describing how they killed an adder? Things being as they are the adder has far more to fear from man than man has from adders.

Of course, the adder is poisonous; but man is poison. Nine people out of ten become spontaneously generated St Georges as soon as they see an adder, and go to work on the premise that the only good adder is a dead one. Therefore the logical notice on a hill occupied by adders is 'Adders Beware'.

Having said all this I freely admit that the adder is not the kind of subject that can profitably be discussed on a basis of right and wrong, or sense and nonsense, or good and bad, or useful or useless. The fact is that most people fear adders and are hostile to them, some do not fear them and are neutral, and a few who don't fear them actually like them.

I don't like or dislike adders; I am interested in them, like to see them, and never molest them. I would rather let one go than kill one. Which proves nothing except that that is the kind of person I am.

What annoys me about the 'Beware Adders' notice is that it plays on people's fear of snakes, nourishes it, exploits it, and implies that the adder is hanging about, lying in wait for the first person to come near so that it can strike out and bite him.

The plain truth is that the adder is not in the least an aggressive snake; in fact it is a highly nervous snake which takes evasive action at the first hint of man's approach, and which will not strike until trodden upon or otherwise assaulted. In short, man has to appear to be the aggressor.

You might say that while all this is perhaps true the fact remains that some people have been bitten who didn't know there was an adder present until they were bitten. This is true. Yet the fact remains that most people can go over adder ground without ever seeing one alive, and that the man who is seeking adders, like myself, has to look for them.

I have been bitten twice in my life, when I was in my teens, and in both cases I took the needle while handling the snakes too carelessly. I've handled adders many hundreds of times since, without the thought of danger ever entering my head.

The fear of the adder is irrational. More people are bitten by dogs each year than are bitten by adders, and dog bite can be much more dangerous than adder bite.

One of the unfortunate by-products of this general hostility towards adders is that the harmless little slow-worm is often killed in mistake for one. I have, in fact, received through the post packages marked 'Live adder—with care', only to find inside a dead slow-worm.

Dogs and sheep are sometimes bitten by adders on the hill, and sometimes such bites prove fatal. But this kind of mishap is far from common.

One of the natural enemies of the adder is the hedgehog. Some hedgehogs, as is well established, are great killers of adders. Others show little interest in the snake, while yet others will withdraw at sight of one. The idea that the hedgehog is immune to the venom of the adder has gained currency but is false. If the hedgehog is bitten on the face it is more likely to die than recover.

Night Attack—by Heifer

The whaups (curlews if you like the English of it) had their nest in the big hayfield, off-centre in the south-east quadrant, with the wood to the north, the road on the west, the cornfield on the south, and the big pasture on the east. There were four eggs in the nest.

It was dark and clammy chill an hour before daybreak, and the bird was sitting on her eggs with her long beak pressed into her shoulder, sound asleep. She hadn't budged for the past half hour despite the owl activity around her ears. Before that she had come awake when the owl brushed the top of my hide with its feet, making as if to pitch then changing its mind.

The owl was mousing in the hay. Once in a while I could hear the fruity hoots of him, then he would flap past me like a big bat, on hushed wings, sometimes silent, sometimes with a ringing war whoop. He might as well have been fitted with a silencer for all the attention the whaup paid to him.

With the dour greyness of first light came the swish and thud of hooves, which spelt bullock or heifer beast. The footfalls came to a halt behind me, and were replaced by snorts. That brought the whaup awake, and she long-necked on the nest. The heifer dunted my hide with her big head, and I knew that she would be in on top of me if I didn't do something about her.

I elbowed the cloth at my back, and made contact with her face. That made her jump back snorting, but didn't discourage her for long.

I peeped out at the whaup, and saw she was now standing in a half crouch over her eggs.

The heifer was moving in for another dunt. I had to do something about her. Taking a chance that the whaup would blame it all on the cattle beast I clouted the cloth hard and said in a low fierce voice, or at least what was meant to be a fierce voice: 'Get back over that fence, you bitch. Go on! Scram!'

She thudded away a few yards, and stopped. The whaup lifted from her eggs, a shadowy shape in the ground gloom, and pitched a few yards on the other side. If the heifer would now go I was sure the bird would be back immediately.

In fact, the bird came back while the heifer was still swithering. She straddled her eggs, jabbed her long beak to the ground on the edge of the nest, and lowered herself carefully. She shuffled her wings, and settled, pressing beak to shoulder.

I could hear the black-and-white, steamy breathing nuisance with varnished nose coming up behind me again, and again I punched the back cloth. The whaup raised her head, but settled again at once. The heifer prodded me back.

There is nothing more thrawn, more persistently curious, more aggravating, more stimulating to the production of colourful invective than a dedicated heifer beast with a single track mind. She was determined to wear my hide on her head like a trophy, and I was equally determined she wasn't going to force me into the open and so scare the whaup, maybe for keeps.

So we had a boxing match, five second rounds with intervals of maybe five minutes. And all the while the whaup sat on, lifting her head now and again in query, but not taking fright.

It was an hour before the nuisance shambled off, and in the good light I could see her making for the pasture fence, which she had broken through earlier.

I said to the whaup: 'I'm sorry about all that.'

She raised her head, turned towards me, shuffled on her eggs, tucked her beak into her shoulder, and went to sleep.

She slept till the barking of the farm dog was heard in the field. Then she came awake and crouched low on the nest. Presently her mate was in the air, calling in alarm, demonstrating against the dog. The dog moved in until I could hear its panting. Then the bird on the nest leaped into the air to join her partner in chivvying the dog.

In a couple of minutes the farmer arrived, and I let myself out of my night prison. The birds were flying around.

'You had company,' the farmer said.

'A boxing match,' I answered. 'I had a bitch of a heifer prodding at me for half the night.

I walked back to my car while he went on to mend the fence. From the roadside I spied the whaups, and in a minute or two I saw one of them pitch near the hide. She was on her way back to her eggs.

A week later she hatched four chicks.

The Part-Time Conservationist

A former neighbour of mine, a farmer who no longer farms because his farm no longer exists, had forty-two pairs of house martins nesting with him, and he was fond of them and in a way proud of them, although there's no reason in the world why a man should be fond or proud of house martins any more than he should be of dogs or cats or whatever.

Anyway, he had reroofed the building in which they nested, which meant removing almost all the old nests, and he was a bit worried about how the birds would react. I assured him it wouldn't bother the birds at all, because the new roof was made of wood and no different from the old one except that it was new. When the birds arrived in the spring, they set about nesting from scratch, and everything appeared to be all right.

But the nests kept falling down; they wouldn't stick. The birds would get a crescent of plaster up then, as soon as it dried out, it would come down. 'They must be using a housing scheme mixture,' I said to the farmer. 'It's what they've always used,' he replied. So we decided it must be the new wood, but the farmer wasn't prepared to let it rest there.

He dug a hole in the yard, put a shovelful of clart from a fieldgate into it, then a shovelful of well-rotted dung from the midden. Then he added water, and, finally, he sprinkled in a little cement. The birds used the mixture, and all the nests stuck. They were still sticking when he left the farm.

Those who take the view that the world wouldn't have been held up on its axis if forty-two pairs of house martins had failed to breed, would be right; it wouldn't. But this farmer liked his house martins and when the crunch came he was prepared to make a personal effort to save them. It was a one-man act of conservation, achieved by a little effort and a little time taken off.

The same man opened a door into a loft on the day the swallows arrived so that they could get in to nest. And it remained open until they left in the autumn. But, of course, he was not alone in that kind of behaviour.

Personal efforts and small parings from profits, are more than ever the cost of keeping what pleases you in being, for modern farming (as distinct from farmers) isn't very kind to wildlife generally, and sometimes hits hardest the species the farmer would most like to keep. The partridge and the peewit are two such. The corncrake was another.

The barn owl is another. Modern buildings, and old buildings renovated, rarely suit it for nesting, and combine harvesting has changed the old stackyard from a place of abundant food supply to one like Mother Hubbard's cupboard. The old symbiotic relationship between owls and men is breaking down for want of stacks and mice in between.

Another neighbour of mine had barn owls nesting with him for decades. Then the time came to renovate the building in which they nested. Before their end was pulled down the farmer had a great crate fixed to the wall at the other end, and an entrance to it cut in the masonry. The birds moved in to the new place and nested. If the old site had been pulled down first they would have been homeless and would have left. It was the little forethought and the little effort, at some cost, that saved them.

Of course things have changed since then and the owls have gone. What the farmer had saved by effort, and was prepared to keep by effort, had to go because modern farming left no place for it. The farmer as conservationist faces a great dilemma.

The Trouble With Dogs . . .

Chesterton, that other *Ursa major,* said it all, or nearly all, when he said that a dog was all right so long as you didn't make the mistake of spelling it backwards. It's a common enough mistake. And it can lead to all sorts of trouble.

Like the man who couldn't get out of his own house until his wife told the dog it was all right to let him leave. I had some of it myself, when I paid a visit.

Or like the couple who couldn't go out in their car together if god were coming along; god wouldn't have two at a time in his car. So, if god were going along, the man or his wife had to stay at home. Otherwise it meant leaving the dog at home. And he didn't like that.

Or like the time, many years ago, when I went into a house and was told I couldn't sit on a certain sofa because the dog wouldn't share it with anyone. He bit anyone who tried. You could have the sofa when he wasn't looking, or when he was out, but not when he was there, using it. I cuffed him off, and sat down, whereupon he accepted me and agreed to share. We both owned the sofa after that. He used his teeth on everybody else, but reserved his tongue and tail for me.

Not so long ago, I was in an office doing some business. When I was being shown out by the owner, his guard dog, following, took a piece of cloth from me as a souvenir. 'I'll see it doesn't happen again,' the man said. 'I'll see it doesn't happen again,' I said, 'even although it costs the dog a few teeth.'

In none of these cases was the dog fundamentally to blame, but admitting that doesn't solve any problems in a confrontation. You can place the blame where it belongs—and, of course, it belongs with the people—while regretfully placing your foot in the dog's face. I have never subscribed to the proposition that a dog is entitled to a first bite at me.

There are, when you come right down to it two ways of dealing with a dog. Either you write the script and remain in control, or you opt out and let the dog write its own script as it goes along. Many dogs are born script writers, and succeed in getting their scripts accepted. No dog should be allowed to do this.

The dog rehearsed in your script, the one you make for it, plays out its part without knowing of any other. It never has to unlearn. The scriptwriting dog has a tough time when someone comes along in the end and tears it up.

In any household, in my view, the dog should be quite unequivocally last in the peck order. This doesn't bother it at all because the pattern is innate, the wild dogs having a strict order of precedence (a peck order) and protocol. A house dog that is first, or anywhere but last, in the peck order is a nuisance and a potential menace, and the situation can be ludicrous when people conform. I see nothing wrong with my dog vacating my seat when I appear on the flair heid, as they say.

One can do nothing with any dog unless it is first of all trained in obedience. Any other way is like trying to put a roof on a house before you've built the walls. A disobedient dog is an abomination. It is the simplest thing in the world (although it takes a little time and care) to get dogs, any dogs, to accept anything, including instructions. It is a constant source of amazement to me to meet so many folks who would nail a child's foot to the floor the first time he put it wrong, but who will accept the most outrageous behaviour from a dog.

I expect, and get, obedience from my dogs, collectively and individually, and I've had five at one time. I expect, and get, from them tolerance where tolerance is asked for. My terriers are expected to kill rats at any time, and rabbits when required, but to leave cats, weasels, stoats, polecats, hedgehogs, children, people, sheep, cattle, pigs or anything else severely alone. And I expect all this even when my back is turned.

I expect my Labradors to kill nothing, and to behave in every way exactly like the terriers. I expect my present German Shepherd to do the same, and she does more. She has mothered my fox cubs, and taken charge of my roe deer. She behaves in a civilised manner towards all visitors, bothering no one, and has an especial interest in children.

When my two ganders act up, giving her the fizz treatment she turns and walks away. My greylag goslings, free flying, open up like the Red Sea to let her through; it never occurs to them to be afraid of her.

She plays the game according to the rules: my rules. I don't know what rules she would have made if she had been left to make them. I haven't consulted her, and have no intention of doing so. I wrote the script.

The dog can feel whatever it likes. What matters is what it does. My terrier isn't all that fussy about foxes, except when she's in the mood. When she's not in the mood she stands grumbling and nickering while they lick her face or pat her with a paw, or grapple with her. But she does nothing more. She knows the score. She can no more stop girning than she can stop breathing.

There's nothing in the least special about these dogs. They just happen to be

27

like all the others I've had. And if I have yet others there will be no problems in making them the same. They're not born like that; they're made. And the making is so easy it's almost not true.

What about watch dogs, or guard dogs? The same thing applies: the dog should be a guard, and bark and demonstrate accordingly. But it should do nothing more until told. It should not be savage by its own choice. It should be the opposite. And this is the cause of half the trouble with guard dogs. The biggest mistake anyone could make is to buy a savage dog in the belief that it will be a super guard dog. It will be a super headache.

This point was well brought out in a recent TV programme about working dogs. The RAF German Shepherds were all well mannered dogs and, as the man said, they had to be taught to attack people after being trained to tolerate them. *Taught* is the operative word. Not their will, but your will, be done.

Half the bother with watchdogs, it seems to me, is a failure to understand this principle: that a dog should be totally obedient, tolerant and well mannered, and actively aggressive only when instructed. The choice should never be the dog's. It should not write the script, unless where it is left alone in charge of premises that are locked. During working hours the dog should be friendly to everybody, or withdrawn and neutral, but never aggressive. This is where the teaching comes in.

An aggressive dog is a liability. A peaceful dog that becomes aggressive when told is the ideal. But not everybody can make such a dog. An aggressive dog writing its own script is nothing to be proud of. The owner needs his head examined. Such a dog will land him in serious trouble, even if it doesn't half-kill someone in the meanwhile.

Aggressiveness becomes worse as dog size increases. A big fierce dog is always more dangerous than a small fierce one. It has more powerful weapons. This works the other way, too. The threat is greater as man size goes down, and the small child is at greatest risk, even from a small dog.

It follows that such dogs are a menace when running loose. Dogs shouldn't be running loose at any time, of course, but there's never any scarcity of them. It's a fact of life. The savage dog should never be loose, even when accompanied, unless the man has written the script for it. But, then, if he has written the script the dog won't be savage.

At the end of the day the responsibility is the owner's, and there are more dogs than responsible owners. That, too, is a fact of life.

Paper World of the Wasp

A wasp's bike is a work of art—a paper world, up to the size of a big football, with padded scalloped shell and tier upon tier of hexagonal cells with supporting pillars between, with accommodation for perhaps 2000 wasps and followers, yet as near weightless as such a substantial weighable structure can be. It is the work of a season—a short season; and it lasts a season—no more.

'What happens to wasps' nests when the wasps have finished with them?'
Well they lie derelict underground, or wherever, and flake, and sag, and

discolour, and slowly break down—that is if they are left untouched after all the wasps and their grubs have died. But many of them—how many it is almost impossible to guess—come into use by woodmice, either as storehouses for food or as sleeping places.

The mice rummage about in the interior of the paper citadel, pulling down pillars and pulling aside comb, until they have the room they want for a sleeping nest or a storehouse. The mice will also block up the entrance to the nest with old grass and earth. It is easy to observe this if you keep captive woodmice, and bring an old wasps' bike to them, and put it in their living quarters.

Quite often, over the years, I have visited wasps' bikes many months after the end of the wasp season, to find some of them still empty and others occupied by woodmice. I have found berries and grain in some of them; others were used as sleeping nests.

Some years ago I found one occupied by a small weasel, and from time to time, during the winter and early spring, the evidence of weasel occupancy was there even when the weasel was not.

Nests not occupied by such small mammals become, as they break down, the haunt of many soil creatures, including such things as earwigs, centipedes and snails.

One of the things that happens regularly to wasps' nests, whether occupied or not, but usually when the colony is at its height, is destruction by a badger. The badger is a specialist predator on wasp grubs, and every year one finds nest after nest torn out and wrecked. The scattering of paper shell and comb is the giveaway.

Woodmice will use such scatterings in their burrows, and can sometimes be seen carrying the paper away. With captive mice this is no problem.

The cycle of paper production begins when the queen wasps emerge from hibernation in late spring. Only the young queens from each year's nests survive the winter; the old queen, her workers, and all the remaining grubs, die.

The surviving queens first feed themselves and gain strength; then they search for a nesting place. Some wasps nest in trees; others, perhaps most, in holes in the ground—holes that are often old woodmouse burrows, which makes another connection between wasps and woodmice. But queens will nest in many places, including attics, outhouses, floors, bird nesting-boxes and gutterings.

Once the queen has found a suitable place she begins to make paper. But if she has chosen a hole in the ground, as so many do, she has first of all to carry out some earth to make room for the roof of her nest (she builds from the roof down). She goes in with a mouthful of paper pulp and comes out with some earth so that the hole grows at the same time as the nest.

Sources of paper are often old posts, weathered but not rotten, but the queen may go to work on a bit of the greenhouse from which the paint has flaked, or make do with the stems of dried plants. When a queen is rasping off wood with her jaws she makes a noise that the human ear can easily hear, and as she comes back to the same place time after time it is easy enough to watch her at work.

The wood is pulped by the queen, using saliva, and pasted on to the roof of the hole as a house martin lays on mud under the eaves of a house. She makes the roof, and the first tier of cells, herself; when her first young are on the wing, about a month from egg-laying, they take over the work of building and foraging and the queen confines herself to laying eggs.

The nest grows through the summer, and by August has a busy thriving

29

community. The whole has been the work of only a few months, yet is a considerable achievement. All the work expended produces only wasps for a year. All, except the young queens, die before or with the first frosts. And the great citadel becomes a ghost town with a few shrivelled, mummified wasps entombed.

Any time after that the mice take over.

Even Dead Wasps Sting

A stung adult is like an adult with toothache; he prompts more levity than sympathy. A stung puppy, on the other hand, is like a stung child—a pathetic thing on which we shower our sympathy.

In the case of the child our sympathy is absolute. In the case of the puppy it is tempered with a kind of sorrowful mirth, because there is nothing more droll than the antics of a stung puppy, and it is difficult to avoid mixing a smile with the ministrations.

My old Border Terrier bitch, who died last year, was the perfect canine early-warning system against invading wasps. Just as she knew my car engine from all others, she could tell wasp-buzz from bee-buzz or the revving of a bluebottle's two-stroke. When she made a hurried, strategic withdrawal from her basket, to hide behind my seven league boots on their shelf, we knew there was a vespa in the kitchen, or infiltrating, in search of jam-drams and a mortal drunk. When she was just a slipperful of puppy she had been stung in a foot, by the simple act of treading on a wasp which was dead but whose stiletto hadn't been informed of the fact, which is a common oversight with the breed.

She was a pathetic wee soul, whimpering, whining, crying, and offering me the outraged paw for magic I didn't have. She got over it in an hour or so, but from that day she became an expert on the subject of the Clan Vespa and all its septs, and backed her judgement unreservedly.

All wasp haters, or wasp belters (not always the same thing), should, if they keep dog or cat, take great care about gathering wasp bodies from the floor, because the stiletto can remain active for a minute or two after the vespa is apparently dead. What happened to my old bitch happened later to one of her grandsons.

My old dog became completely demoralised when a wasp entered the house, and would lie trembling violently behind my boots until I had swiped the intruder down. But it seems that dogs have to learn wasp-fear from experience. My Labrador, now well on in years, has chopped wasps throughout her life, and still does so although she can hardly see. She has never been stung.

My present Border puppy chases wasps, and chops at them, as she does at flies and moths, and I cannot break her from the habit. I simply try to get to the wasp ahead of her.

I have become a wasp belter through force of circumstances. I have no quarrel with the yellow-jackets myself, and haven't been stung in over thirty years. But the domestic situation makes neutrality difficult, at least indoors: the dogs chase wasps, my wife runs from them, so I have to swipe them to keep the dogs from catching them and my wife from being accidentally caught.

Wasps indoors are not usually aggressive, but where many are buzzing about somebody is always liable to be stung. Wasp stings can be dangerous, so it is as well to keep vespas out of the house altogether. I have a friend who reacts violently to the mere presence of a wasp, and who is, during the season, permanently armed with an antidote on medical instructions. She is one of those who are hypersensitive to wasp stings.

My wife, who has never been stung, goes about in great fear of wasps, and orders come from her by the yard as soon as one enters the house. The invading wasp, for some reason, follows her as soon as she begins her run. The other day she rushed to the porch when a wasp buzzed past my ear, and slammed the door in apartheid gesture. Alas, the wasp was fast, and she succeeded only in shutting herself in with it! I had to go to the rescue.

Scotland's Tiger

Behold the Scottish tiger—*Felis Sylvestris Grampia*! See him padding high across a wild scree in the slanting morning light, big-fisted, wide-eared, long of limb and tusk, with ringed club-tail, and you've seen one of the lords of life: a cateran of fire and brimstone, implacably savage, reputedly untameable.

Cry Vermin! if you like. Throw up the gun or wave out the dogs. Or set the steel-toothed gin on the fallen birch that spans the burn, so that he'll be taken by a foot when he walks across. But see him at bay on a ledge of some rugged crag, after a run before terriers, flat-eared and bristling, with teeth bared to the gums, back arched and tail bottle-brushed, and you've seen explosive demonic cattiness in all its unconquerable fury.

You'll hear the treason whispered that there's no such animal. For shame! English red squirrels we have; even English ways. And Swedish capercaillies. And Japanese deer. And Norman, ex-Spanish rabbits. Our grey squirrel is an American. But the wildcat is our own—and real.

Hark to the wild pibroch of him, as he stalks through the gloom of the corrie when the moon is riding high. This is the devil's black laughter: hiss and crackle, scream and sob. The wildcat's skelloch is of the lonely places, of mountain and high forest. Beside him the caterwauling alley cat is a cartoon clown.

Yet there is some truth in the whispered treason. The tameless tiger of the hills will sometimes form a liason with the tame tabby of the fireside. So far as one can gather, the liaison is always the same way: the wild tom luring the domestic female to a rendezvous. The kits from such a mating are high explosive, especially if they aren't found until they are some weeks old. But they can be tamed. Some look like the real Grampia, but the tapered tail-tip gives them away.

Half breeds can make a big size. A tom I had scaled 10lb. at the age of seven months.

Regarding miscegenation of this kind, Professor L. Harrison Matthews has this to say (*British Mammals* 1952): 'In spite of this, and contrary to the belief of some writers, there is no indication of any general degeneration in the robustness of the wildcat of Scotland when compared with the remains left by its prehistoric ancestors, though individual animals may plainly show the effects of recent crossing by their small size and thinly furred tails.

'The effect, rather, has been to improve the stature and pelage of the domestic cats of Scotland, which are bigger, stronger and better furred than the house cats of southern Europe.'

Is the Scottish wildcat really untameable? Although the record supports the view, I have always had doubts about it. As far as I can discover, the untameable wildcats were all caught as kits with their eyes open. The late Harry Mortimer Batten believed they were untameable even if bottle fed from their blind days, although he did not say on what he based this belief. If this is so, it is a notable exception to the Lorenz principle of imprinting.

Batten had plenty of experience of captive wildcats, so knew what he was talking about. He knew of no cat, taken as an open-eyed kit, that became tame, even if it lived for twenty years. He tells of a pair that bred in captivity, producing kits that remained as untameable as their parents. Batten found that first crosses were also undependable.

This was not my experience with the only kitten of this type I have had. To begin with it was apparently untameable; but after seven weeks, a lot of work, and a few cuts, I was able to win it over. Then it became absurdly affectionate.

Tameable or not, the wildcat is a formidable animal, and a considerable predator, its prey range being about that of the mountain fox, up to and including roe deer fawns. I don't know if they kill many fawns—I know of only one myself—but they can kill them.

Then there are lambs. Like the mountain fox the wildcat will kill lambs, but the number they kill is as difficult to arrive at as the number killed by foxes. Refreshingly, one hears facts about this from time to time; more often one gets hearsay at two removes. But the fact is that lambs are killed. The question is: how many?

Keepers and shepherds have always tended to wage war against the cats of the hill, perhaps as much for what they might do as for what they did. At one time, fifty years ago, the wildcat almost went the way of the polecat. Now it is as plentiful as it has ever been in its recorded history.

Small mammals are, almost certainly, the beast's most usual prey: rabbits, hares, voles, mice and squirrels. Young rabbits and leverets are easy for it, and, in timber, it probably takes a lot of squirrels. I once saw a wildcat stalk and kill a squirrel; I've seen a house cat do it too. The style in both cases was the same. I should imagine that what a domestic cat can do a wildcat can do better.

The wildcat, like any cat, kills birds. It will take grouse and ptarmigan, maybe even a hen capercaillie if the chance is right. Against red deer calves I would put a big question mark. A weakly one would be within a big cat's powers; a strong one is another matter.

Yet red deer hinds react strongly to the presence of a wildcat. One day some friends and I watched a parcel of hinds trying to kep a wildcat on a rocky slope. The cat, flat-eared and snarling, was lost among the rocks. But for a long time afterwards the hinds milled around, stamping, with their ears up, obviously unsettled.

There is little evidence to suggest that the wildcat is fond of fish, or that it fishes to any extent. Here, the domestic cat appears to be an unsafe guide. Yet, a few years ago, a Ross-shire deer forest owner told me about one which had been caught and drowned in a fish trap. There were fish in the trap, and one can't imagine anything other than the fish attracting the cat to the trap in the first place.

The war against the wildcat goes on all the time. Like the Mustelids it is easy to trap, a favourite set being one on a log spanning a burn. Many are still taken in gins, although the gin is now illegal for this purpose. But, despite the heavy casualties, the cat holds its own, and more. It is far, far south of its one-time strongholds north and west of the Great Glen. The latest reports, completely reliable, place it on the perimeter of the industrial belt.

Wildcats measure from 2½ft. to 3½ft. from tip to tip. The first is small, the second very big; a good average size is three feet over all. Weight ranges from light-weights of seven pounds to the big ones of a stone or more. Colour varies, the striping being bold or shadowy; the fur is more or less frizzled; the ears out-pointing and set wide apart; the tail is clearly ringed and blunt at the end. Limbs are strong, and the teeth powerful.

A cat with a pointed tail is not a wildcat, however big or powerful. But it may well be a hybrid. And a hybrid is a pointer to the presence of the real thing. If the real thing isn't there, there can be no hybrids. Therein lies your problem.

The Scottish wildcat breeds twice a year; the European apparently only once. This double littering has been taken as a pointer to the admixture of domestic blood in our stock. But that seems to me to be a bit shaky. It is by no means certain that the European wildcat breeds only once. And what right have we to assume that the Scottish toms are the only philanderers?

Why shouldn't there be domestic blood in the veins of some north European wildcats too? The domestic cats are there; so why shouldn't the twain sometimes meet?

The trouble with wildcats is that you don't often meet one on foot and free in daylight; at night they're even harder to come up with. Most are seen dead in traps, or found dead on lonely Highland roads, victims of the modern motor car. I have found two in one winter on the road over Schiehallion.

Of live wildcats the best one usually hopes for is a fleeting glimpse; the close-ups are generally arranged by terriers when they run a cat to a dead stop on the face of some cliff or crag. That's when you really see and hear the fireworks of the bristling Scottish tiger. The long-range view through a telescope merely pulls the stalking tiger tantalisingly closer, but not close enough.

A cat run to ground by dogs may stay inside, paw-whipping any probing stick; or it may burst out like a charge of dynamite. The old Highland method with the stay-at-home was to tie the gralloching knife to the end of the stick, then try to dirk the sizzling cat.

A bolting cat can be dangerous if you happen to be peering in when it comes out. Men, usually by taking chances, have been clawed by wildcats; but there is a record of one man who caught a cat in his bare hands. There is also a record of a man who died after a close grapple with one.

Once out of the kit stage, it is doubtful if the wildcat has any active enemies other than man and his dogs. The relations between the mountain fox and the cat are obscure, but it would surprise me if they went out of their way to tangle with each other.

The golden eagle has been recorded as a predator on kittens, as it is on fox cubs. I've known an eagle clean out a litter of fox cubs; so far I haven't found a wildcat of any age in an eyrie.

In the days of the wildcat's scarcity—for example before the 1914-18 war—its last strongholds were in the remote mountain cairns of the north west. Nowadays it has many strongholds, and one of the great factors in its comeback has been the reafforestation of great areas of the country.

Whatever the wildcat's impact on sheep farming, there can be no doubt about its use to forestry as a predator on rabbits, squirrels and voles. The cat, like the fox, is a problem because of its different impact on two major forms of land-use.

Here, clearly, is a viable species in a period of increase and spread. Yet, despite this, not many more people are seeing many more wildcats alive and free. But more ears are hearing the eerie love-calls and the quavering war cries where the cats walk their secret trails in the dark.

Technique for the Small Fry

Blessed for once with a sunny Easter, the small boys from the town were gathered round the pond, with sticks for fishing rods, black thread for lines, and small worms as bait.

The quarry was the active, attractive, neat little stickleback, popularly known as the baggie.

When I was a very small boy, we had no high-falutin' distinctions between sticklebacks and minnows. Sticklebacks were baggies, and lived in ponds; the real minnows, which we called real minnows, lived in the burn.

And everybody knew because the information was handed down from one generation to the next, that you could keep baggies in a jar indefinitely whereas real minnows needed running water.

The catching of sticklebacks is a simple and ancient craft. Apart from trying to catch them by hand there were two approved methods, neither of which involved pricking the fish's mouth with a barb.

Firstly, there was the line and worm method. When the fish sucked at the end of the worm, like a man sucking in spaghetti, it was pulled from the water and dropped in a jam jar. Then the worm was recast to catch another sucker.

Then there was the ham-net technique. We used to haunt the grocer's shop at this time of year mooching ham nets. The net was stitched to a wire hoop, which was then fastened to the end of a stick, and we were ready to net baggies by scooping.

The entire net cost nothing more than the small amount of work involved. Nowadays most boys use ready-made nets of superior design and quality, but I haven't noticed that they catch more baggies.

The first tank I ever had was a half biscuit tin which smelt of perkins when I first filled it with pond water and pond weed. Believe it or not, I used to persuade baggies to breed in such a tin, and more than once succeeded in launching perfect dragonflies into the world from one.

Nowadays, with a tank half the width of the house, and a far greater knowledge of the ways of sticklebacks, I don't do any better and certainly don't get half the fun, because I know why I am doing this or that, and the end product is expected. The small stickleback of the ponds is a remarkable fish, just as remarkable as its big marine relative.

The small male fish who, in the breeding season, becomes red in the face (the goldie of my boyhood), has a very highly developed paternal instinct. He is, indeed, a better father than a great many mammals, and one of the problems of his life is to prevent any one of his several wives from eating their own eggs.

The male fish makes the breeding nest himself, using small pieces of pond weed and debris which he sticks together with mucus from his own body. The completed nest is a sort of cup, or barrel, not much bigger than a thimble.

Into this he drives any baggie he can corner, and when she has laid her eggs he drives her out again. He does this with several females until he has a nestful of eggs. Then he mounts guard over them, keeping all raiding baggies at a distance.

When the fry hatch, he watches over them until they are able to take some kind of care of themselves. But despite this, great numbers fall prey to such predators as dragonfly and water beetle larvae, not to mention birds and small boys.

The small fish, which we used to call pinheads, were never greatly in demand for the jam jar aquarium. The status symbol was the big baggie; better still, the big goldie. The pinhead was children's stuff. In fact, it was a pretty little fish which throve well with the minimum of care.

Small boys still go for the goldies. I suppose they always will. With a brightly coloured fish in the jar, it's hard to concern yourself with a nestful of fish eggs left without a policeman.

But, mind you, there are boys who, when they've heard the story, will put the red-faced fish back in the water. Until the following day anyway . . .

They Spell Danger for the Frog

Despite the biting east wind, the frogs move by hops and leaps to their traditional spawning grounds, and will not be turned aside by anyone, or baulked by landscape changes since they went into hibernation. I sat on the old hutch road, in the sun and out of the wind, and about fifty yards from the pond, watching two frogs moving in on their spawning ground. One was big, fat, goggle-eyed and yellow—a female; the other was smaller, darker, and a male.

There was no sort of relationship between them; they just happened to be going the same way.

The big frog made frightened, Olympic leaps when I approached. I turned her about, and stepped back. For some minutes she squatted quite still, in the middle of the hutch road.

Then she turned in three short leaps and was on her way again, in the right direction. I followed her, at a slow walk, and saw her reach the pond and enter the chill water.

Other frogs were already there—not the big battalions, but the advance guard—but there was little sign of activity: no croaking, no neb-showing, no amplexus—just man-shapes swimming with strong strokes of hindfeet in the depths.

Already, on the margin, were the bleached bodies of casualties, frogs which had died almost at the moment of arrival.

Casualties among frogs at this time are common, and dead frogs may be seen before, during and after the spawning period. This, however, has nothing to do with old frogs being killed by young frogs.

The so-called young frogs of this story are really the bull frogs, which are noticeably smaller than the females.

The powers of orientation in frogs are as remarkable as you will find in birds. They may travel only short distances, but their range of vision is strictly limited.

Many years ago I used to mark frogs at this pond—with little rings or daubs of paint—so that I could keep track of them. Frogs found within yards of the pond were taken 200yds. away and turned loose.

Two days later they were in the water. I have followed a frog for more than one hundred yards right to its destination.

At this time the enemies of the frog are legion, common ones being the hedgehog and the heron.

The hedgehog begins to move about at the time the frogs are coming on to spawning, and I have photographed one in the act of swimming after this kind of prey.

The hedgehog, of course, is usually a night hunter, or at least crepuscular, but when he is hungry after hibernation he will move early in the day and before dark, as he will do in autumn before going to sleep.

Some hedgehogs kill their frogs with a clean chop near the head. Others, especially the woolly-headed ones, still half awake, just catch hold anywhere and begin eating without bothering to kill the beast first.

I have watched this process several times, and am certain that a frog, so treated, can mew thinly like a small kitten.

On one occasion I came across a hedgehog in a ditch with a frog in his mouth. He was fast asleep, half coiled with the frog held tightly by the leg. The frog was wide awake.

Obviously the hedgehog had caught his frog when he was in a dwam and was too sleepy to eat it then.

The heron, when frogging at a pond, hunts in the early morning and at dusk, at least where I have watched these birds. Usually the heron stalks ashore when he makes a catch. Then he shakes his prey, perhaps batters it about a bit, before swallowing it. Other herons will swallow the frog where they stand.

Otters are extremely fond of frogs, and eat a great many of them. A bitch otter with a cub or cubs will, at this time, take them to a pond where frogs are spawning and introduce them to this prey. I have watched an otter catching frogs for her cubs. Some of the frogs were eaten whole; others were eaten down to the skin.

Tawny owls will take frogs from time to time, but in a long experience of this owl I have seen only a few such items brought to a nest. Foxes, on the other hand, will take frogs readily. I do not know if all foxes are fond of frogs. Some certainly are, and I have never known a cub which refused one.

Ravens and crows will take frogs when they can and individual birds of this group become adept at exploiting this food resource. It has been said that the raven can lure frogs to the surface by dapping on the water with his beak.

This would not surprise me at all, although I have no personal experience of the habit. But I have seen something of ravens and place them among the most resourceful of birds.

Young Frogs Face Many Dangers

The big heron was stalking about the pasture, high-stepping and slack-necked, looking this way and that, stopping every now and then to pick something from the grass. The pick-up was leisurely and casual; there was no coil of neck or lightning downthrust of javelin beak. Clearly the bird was picking up food, and equally clearly he was swallowing it as he strutted. But I could not see what he was eating.

There were puddles in the pasture, and the grass was wet. I thought at first the heron was finding worms which had surfaced during the recent rain.

The glasses told me nothing, so I left the car and climbed the fence into the pasture. The heron flapped up and away almost at once and I had the field to myself, except for three Ayrshire calves which decided to join me.

Presently I realised what the heron had been feeding on. There were young frogs all over the place—dozens of them—of finger-nail size, hopping this way and that. They were in the grass and in the puddles.

The calves crowding me close, saw them, and reached out with glazed muzzles to smell them. But when a frog leaped the calves jumped back, snorting; not afraid but startled. There was something Disneyish about it. I left the calves to the mystery of the frogs.

That was early in July, in a field not far from several spawning ponds, so there was nothing really surprising about the presence of young frogs at that time. A month later I met more young frogs, about thumb-nail size, in quite different circumstances.

Just back from holiday, I opened every door and window in the house to let the air through it. In a heavy shower of rain I took the dogs out, leaving the back door open.

A miniature whirlwind was sucking up leaves, twigs and debris in the drive and tossing them all over the place. When I came back with the dogs the back porch was littered with twigs, leaves, debris and three small frogs.

I called my wife to see, but instead of expressing surprise or interest she called me into the hall where two small frogs were hopping over the carpet. In the dining room the cream cat was playing with another.

Altogether I gathered, and put outside, seven small frogs. I cannot believe that seven small frogs decided to pay me a visit at the same time. I imagine they were all tossed into the back porch by the wind.

The young folks in the house were full of why, why, why? It has, I told them, been raining frogs. It rains frogs every year, several times a year in summer and early autumn. And it's all very simple.

The tiny frogs, half an inch or so in length, which have spent ten weeks or so growing from the tadpole stage, leave the water during a good shower of rain, after having spent some time using their new hindlegs in the shallows. At such times they may be seen swarming in fields and lanes, and on country roads.

Rain attracts them at other times, so that a number may be found in one place. My heron was taking toll of a swarm which had recently left the ponds; my visitors were frogs which had been some time afield.

When I used to live near a spawning pond I spent a great deal of time watching the young frogs when they were completing their metamorphosis. One day they would all be in the shallows, thousands of them; the next, after a night of rain, they would be ashore in swarms, so that one could scarcely walk without trampling them.

At this time, great numbers of young frogs are killed on the roads, and it would be interesting to know the casualties arising from modern traffic conditions. Nevertheless, the greatest number of casualties must still be caused by animals which prey on the swarms.

Hedgehogs kill young frogs, just as they kill old ones, but the youngsters, before they disperse, are extremely vulnerable, and at this time the hedgehog becomes something of a specialist predator.

But the hedgehog is not the only enemy. The following animals take young frogs: otter, stoat, weasel, badger, fox, rat, snake and polecat. Bird predators are hawks, owls, crows, magpies, jackdaws, gulls and ravens.

Gulls will gorge on young frogs when they find a swarm in a field. I once had a carrion crow who ate six at a time in summer, then hid as many more before he gave up catching them.

The young frog which leaves the water measures about half an inch in length. By the time it is due to hibernate it will still be under an inch long. A year later it will have doubled its size. At the age of three years it will be breeding.

But it will not have outgrown any of its enemies. The predators which levied toll on young frogs will still prey on the big ones. There is no period in the life of the frog when it is safe from attack, except during hibernation perhaps.

Yet frogs remain common, despite the predation. High mortality is balanced by high production rates. A female frog will lay between 1000 and 2000 eggs in spring. But only a fraction of the tadpoles hatched from these will grow into frogs, and only a fraction of the young frogs which leave the water will survive to breed.

The Homing Instinct

A big-bellied toad, in warted leopard skin, with shark mouth and a tongue like a flamethrower, was ambushed under a hanging buddleia spike in the garden, the sun glint in his eyes half closed under drooping lids. He cut off the flamethrower before I reached him, and sat there with his elbows out and his fingers crossed under him. When I tickled him under the chin he turned obesely away and waddled under the buddleia bush.

I always have a toad or two in the garden about this time, and again in the early part of the year when they come awake. A year or two back one lived for months in a flower pot in the cold greenhouse, disappearing before spawning time in the spring. The nearest spawning pond is a little over a mile away, so my present visitor must have crawled that distance at least, and spent the summer getting to me.

Toads are not great travellers, spending their lives perhaps within a mile and a bit of their home pond. But they have a great sense of direction, and when on their way to the spawning place in spring will not be turned off course, no matter how often you pick them up and face them the wrong way. They make a turnabout and go where they want to go.

I have taken toads from a quarry pond and released them, facing the wrong way, fifty yards along the old hutch road that was their line of travel, then followed them on their laborious crawl back to the water.

Before the First World War E.G. Boulanger was trying homing experiments with common toads. He collected pairs and singles from their spawning pond and removed them to a neighbouring pond that seemed equally suitable for spawning. Most of them quickly made their way back to the pond of their choice. You find this if you try to get toads to spawn under control. Unless you leave them with no way out they will be off the moment your back is turned.

Boulanger also took a number of pairs from their own pond and released them on a hillock midway between it and another pond where frogs were spawning at the time, but where no toad had ever been known to spawn. Every toad made its way straight back to its own pond.

In a homing experiment carried out on Carolina Toads, the toads were carried from the vicinity of their breeding pond and released at varying distances from it. Each toad was marked in such a way that it could be readily identified afterwards. Of the 444 toads released, 60 per cent of those put down at 300yds. from home were recovered there later. Out of forty-three toads released a mile away only eight were recovered, but they had probably been turned loose on strange ground. Two of the toads, released at distances of 300 and 850yds., were back in the pond within twenty-four hours. But how the toads found their way there is still a mystery. Toads travelling to a pond already populated might be guided by the croaking of those already there. But how does the first migrant find its way?

Nobody knows—yet.

Toad Sleeps One Hundred Days
—and Lives Twenty Years

He sits under the chicken coop, on a smooth seat of his own size and making, a big fat toad, warty and topaz-eyed; replete, relaxed, bulbous—a placid chiel waiting for the rousing alarm call of March, when he will head for his favourite water as unerringly as any swallow flying from South Africa to my barn.

He just sits. Barring accidents—and accidents are unlikely at this stage—he will sit there for a hundred-odd days, just ticking over, and asleep.

He may die, of course, but he may live for another twenty years. If he lives through next summer, and reaches the stage of hibernation again, he may return to the chicken coop, if I leave it there, which, for several reasons, is most unlikely. But he can have it this once.

Most people have a kind of horror of toads. The frog they can tolerate; the toad they find ugly and forbidding. Shakespeare didn't help them any, and the literature (the non-scientific literature in the main, but some of the old natural history books too) is full of dark innuendo about the toad's power to harm.

But this Bufo of the scientific naturalists is no daemon, and his near-evil reputation is a gross libel. He doesn't spit fire; he merely flicks out his tongue to catch his food. His warts, however, do secrete a poisonous fluid; but it is as well not to get too excited about this.

You can handle hundreds of toads, as I have done, without ever realising that the toad's warts secrete anything at all. The fluid is released only under great stress of excitement or fear. If you rescue a toad from a dog or a snake, or from the assaults of a crow, you will probably find that he handles wet.

This wetting agent is the toad's poisonous secretion, but wiping the hand is all that is necessary and is, in any case, almost a reflex action.

The poison of the toad has been well studied, and two types, bufotalin and bufogin, have been isolated from the granular glands. The action of these poisons slows the heart's beating until it stops altogether.

Taken into the stomach they cause nausea, sickness and a weakening of respiration. But these things only happen in the laboratory, when experiments are conducted with large doses.

Under natural conditions the beast which eats a toad will in all probability be sick, so that it will avoid toads afterwards.

I have never known a fox to do more than paw a toad around, but I have had a magpie which ate the toad's guts and left the rest, and I have seen a hedgehog do the same.

But just how much protection the toad's skin actually gives him is an open question. That he is tasty enough underneath you can demonstrate quite easily by offering him, skinned, to hedgehog, shrew or crow.

Many years ago I took a toad from a grass-snake which had begun to swallow it hindlegs first. I got a liberal wetting on that occasion. More recently I put a terrier off one.

The dog had bitten the toad, which was 'sweating' freely. The toad died, and the terrier began to drool, then put up froth, but it was perfectly all right a short time afterwards.

Toads shed their skin periodically, and moving them into new quarters seems to bring on a moult. I have seen it happen several times with toads brought from a pond and kept indoors. The skin splits along the back. The toad struggles out of it, and swallows it, and his new skin dries quickly.

The toad is far more phlegmatic than the frog, which is always exciteable, nervous and temperamental. It tames readily, which is well nigh out of the question with the frog.

It is—if the word intelligent can be used in this context at all—much more intelligent than the puddock which will jump wherever it is pointing. You can keep a toad in your garden and teach it to come to you for food.

It is as well that a toad is not as big as a dog; if it were, the man who kept it would have to go out to work for it, for its appetite is prodigious.

When it is in a gluttonous mood it will stuff itself with worms until they are literally coming out at the other end.

I have placed toads beside a wasp's bike without ever seeing one take a wasp with its tongue, but I know they will take bees, and the habit appears to be general if not significant. It is also widely known.

I remember reading in a novel of D.H. Lawrence's a long time ago (I can't remember which novel) of a gardener who killed a toad with the remark (as near as I can recall): 'Happen th'art good for slugs, but th'art not goin' to empty beehive into thy guts . . .'

Any toads I have had ate a wide variety of food, but earthworms were what I used most, mainly because they are easily and quickly available. Slugs, beetles and flies of several kinds were also offered and taken. Certain flies use living amphibians as hosts for their larvae, and the common toad is one which is victimised in this way by a species of greenbottle. Research on this fly and its relation to the toad was done by T. Spence at the Veterinary Laboratories, Weybridge, and it is an interesting story.

The female fly, when gravid, seeks out the toad in his retreat and deposits her eggs on his back or thighs. They are laid in a single cluster, and number between sixty and seventy, sometimes up to one hundred.

When the eggs hatch the larvae travel towards the head of the toad, entering the eye and travelling to the nostrils via the lachrymal duct. Some enter the nostrils direct. The toad's breathing becomes very distressed, so that he wanders about gulping air.

Development of the larvae is very rapid. They invade the soft tissues of the head, and in two or three days the toad dies. The larvae eat him, leaving skin and skeleton, then go to ground where they pupate. In a few days they emerge as flies to prey on other toads.

The eggs of this greenbottle have been found on the common frog in England (Smith 1951) but there are no records of frogs being killed in this way.

Solving The Great Eel Mystery

When I lived in a valley in the mountains of Castille, many years ago and before Spain's last monarch abdicated, one night I saw eels travelling overland like black snakes.

In those days, I saw and handled many eels, of all sizes, for the canal was full of them and many and varied were the traps set to catch them. There was an eel-catcher, in fact, who, when the canal was temporarily sealed off and drained, used to squelch through the muck catching eels. He bit them behind the head and stuffed them in a sack which hung from his shoulder.

Since then, I have seen eels ashore, but never the kind of exodus from canal to river that I witnessed that long-ago night in Castille. I was seeing eels on the first step of their journey to the sea.

Our European eel, the favourite grouse of anglers and the favourite food of otters, is only a British subject by residence for it is born beside its American relatives on the other side of the Atlantic. Its breeding cycle is the salmon's in reverse.

The common eel is found right round Europe, from Norway to the Mediterranean; but no spawn or eel containing spawn, or newly-born eel, has ever been found there. Eels were known to everybody—caught, studied and eaten—but the great query remained.

Where did they come from?

From horsehair? From pieces of straw? That was once believed, and Isaac Walton, writing three hundred years ago in a book that is a classic, believed eels were bred of a particular dew falling on the rivers in May and June, and 'out of the corruption of their own age, which exceeds not ten years.'

The horsehair legend was still current when I was a boy at school, though it had been completely exploded more than one hundred years before. It was known for a very long time that eels sometimes headed for the sea, and that millions of young eels, or elvers, were to be found at river mouths at certain times.

But nobody could get any further back in eel history than the elver, so there was plenty of scope for fanciful speculation. Anybody's guess was as good as the best marine biologist's—until the discovery of leptocephalus; a tiny, transparent ribbon of a creature that held part of the secret, though nobody guessed it at the time.

Leptocephalus was first discovered in 1763, but it was some time later before its real identity was established. Some specimens were kept alive long enough for them to grow into little eels. But the biggest part of the mystery was still unsolved.

Then, in 1904, Dr. Schmidt of the Danish Fisheries Board got busy after capturing a three-inch leptocephalus off the Faroes. The seas were scoured from there to the Mediterranean, with interesting results.

The wealth of data collected showed one important thing—leptocephalus increased in numbers and decreased in size with each move across the Atlantic. And smaller size, with greater numbers, suggested that the investigators were on the right road to the point of origin and dispersal.

So the advance went on, until the breeding ground of the eels was discovered near the Bermudas. And there all the eels of Europe were born, beside the allied

42

species of America, with which they mixed freely within a prescribed area.

The one has its eastern, the other its western limits; and in due course each goes its own way—the Americans to the west, the Europeans to the east.

During the long, dangerous journey to their European destination the little leptocephali change gradually in appearance and by the time they are ready to invade the rivers of Scotland and England they are recognisable elvers.

Presumably the spawned eels die. There is no evidence that they survive after breeding. The birthplace of the eel is also its graveyard.

It is a fascinating story, but so is the later part—from the arrival of the elvers to the departure of the adults, perhaps seven or ten years later. And the second part of the story has this advantage—it can be studied by anyone.

The eel has tremendous vitality and amazing adaptability. It can stand terrible mutilation. It is at home in streams, ponds, ditches and canals. Land barriers do not stop it; it conquers them by the simple expedient of crossing them.

Eels are very active after dark, and that is when you are most likely to see them on the move. Few fish are more savage or voracious, and they will move from river to canal and from canal to pond if the reward is food. A big eel is a powerful fish, with considerable power of constriction. They wreak havoc in rivers, and otters wreak havoc among eels. That is why I believe that otters are good for rivers; they are predators on the eel, and the big cannibal trout.

Barn Owl with a Cat-Like Style

The white owl launched from the gable pop-hole and pitched on a leet in the stackyard: a shadowy figure against the night sky. At first it sat tall, motionless, but presently it shrank down, swivelling its big head, cat-like. Then it faced back, staring out over its tail, looking in my direction.

I am sure it didn't see me; or, rather, it saw me without seeing me because I was sitting still and would have been recognisable only in movement. In any event, the bird turned to face front again, and resumed its head-swivelling as before.

After a few minutes, the bird hopped and pattered, with half-open wings, to the other end of the leet, and I could hear the crickle of claws on the frozen thatch. For a moment it was out of my sight. But it soon returned to its former stance, with wings held out, and claws scratching as before.

At this point a second owl flew from the pop-hole, threw up, and flapped, moth-like, round the head of the other before disappearing behind the leet. In a moment I had a shadowy glimpse of it as it banked and flew out across the field. I was left with one owl again, momentarily floodlit in the headlamps of a car coming round a bend in the road.

There would, I thought, be little stirring on the ground on such a night: frost with a powdering of snow. I found myself wondering how the owls had fared on the previous night, when the snow fell. Had they gone hungry to bed in the morning, or had they found prey in the building under cover?

Suddenly the owl on the leet did something I have not seen a barn owl do before. It flew down to the ground, and proceeded to walk round the bottom of the leet, stopping every now and again to peer and swivel its head, reminding me of a mousing cat. Cat-style was the only description I could think of for this behaviour.

The bird reached the corner of the leet and was soon out of my sight, and it seemed a very long time—actually five minutes by my watch—before it appeared again at its starting point. I had no way of knowing whether it had found anything on its foot-tour. I expected the bird to fly to its perch now; instead, it high-stepped round the leet again, pausing frequently. After the second tour, it flew to the top of another stack, where I could not see it so clearly.

The owl was reluctant to leave the stackyard, or appeared so, for it flapped from stack to stack, and once into a tree, before it finally disappeared in the direction taken by its mate. More than an hour had elapsed since it emerged from the pop-hole.

Both owls were inside long before daylight. When I knocked them up at 6.45a.m. both flew out. By 7.15a.m. they were back in again. Before I moved off to let them in, I made sure their roosting place had no pellets lying about.

That night there wasn't a pellet at the roost, so presumably they hadn't eaten for a spell. The next afternoon, however, I found two new pellets, both of them containing fur and a skull and bones of small rodents.

I have known this pair of owls for a long time, and in the nesting season have seen them bring a dozen prey items to their chicks between dusk and daylight. In snowy weather I have noticed how poorly they eat on certain nights, and two winters ago there were nights when they appeared to have nothing at all, if pellets at the living quarters are any guide. At no time have I seen either of this pair hunting inside the farm buildings, to which they had access at most times, and where there were always plenty of mice. This, of course, means no more than that they didn't hunt there when I was about to watch them; they may well have done so when I was not.

The behaviour of another pair, living on the edge of moorland a few miles away, is different in this respect. I have seen one of these birds, at different seasons, 'waiting on' on a rafter for prey to appear on the floor. Then the bird would swoop and clutch. When I was watching this pair from a hide during the nesting season I saw the male bird, after delivering prey, turn away and swoop to the floor to catch a mouse, so that he made two visits with prey inside two minutes.

During the terrible winter of 1946-47 I used to spend odd nights in the ruin with this pair, and was struck by the fact that they made a habit of hunting in the immediate neighbourhood, realising apparently the futility of hunting the drifts. I have seen both of them make a kill under cover in such conditions, but their hunting was not good enough to save one of them from death. I found the emaciated body before the winter was over, and when I opened it there was no sign of food in the whole digestive tract. The conclusion that the bird died of starvation is, of course, open to question.

Tawny owls and kestrels were, however, also dying at that time—I came across several—and since they were on the same ground, and all birds competing for the same kind of food, the cause seems inescapable.

Water Birds That Use a Slipway

The red-throated diver—which North Americans would call a loon—is a bird of the west, the north and the isles, breeding on small tarns and lochans, and not usually favouring the large lochs frequented by the black-throated species.

You will, of course, find the odd pair breeding beside a big loch—the kind of loch that requires a good hour to walk round; but this is not typical. Small areas of water, sometimes no more than pools, are more likely places; and, if they are lonely and remote, so much the better.

In summer, generally speaking, a diver on a small lochan will be a red-throat, and a diver on a big loch will be a black-throat. The two species appear to be the same size, although there is a slight difference.

The red-throated diver has a red throat in the breeding season, and the black-throated's is black, but this difference is not always noticeable, especially if you are some distance away or the light is poor, for red has a trick of appearing to be black. The red is, however, quite pronounced, and once spotted leaves you in no doubt.

The beak of the red-throated diver is slender and obviously up-tilted; the beak of the black-throat is thicker and straight. Apart from the throat colour the plumage of the back is a certain guide, for the black-throated diver is patterned in bold black and white, while the back of the red-throat is fulvous grey.

Finding the nest of a red-throated diver is as simple as finding the nest of a peewit. The bird nests at the water's edge, or on the water, so all one has to do is walk round each lochan with one's eyes open. If the nest is ashore, it will be within a few feet of the water's edge. Sometimes it may be left high and dry, many yards from the margin, following a fall in water level after the eggs have been laid.

The bird waddles and belly-flaps on land. It can, and does, walk after a fashion, but even then it has to flap after every other yard like a tired puppy. This is very noticeable if the bird's nest has been stranded by shrinking water and it has to travel some distance to its eggs.

Most of the nests I've seen, including three in the Hebrides, have been less than six feet from the water, with a slipway on the bank, up which the bird flapped from the water and down which she slid when alarmed or disturbed. This conservative habit makes the diver an easy prey for egg-collectors, a breed we still have with us, as ruthless as ever, although it may seem an anachronism in this allegedly enlightened age.

The loon of North America is noted for its so-called laughter. We don't go in for such levity of description in this country, and I cannot recollect, off-hand, any description of the vocal displays of the red-throated diver which accused the bird of laughter. Yet I have heard cries which, if I were searching for an easy simile, I would call laughter. But then, I have the most unselective ears in the world, and bird calls written phonetically are meaningless to me, outside the most obvious ones. I have to hear the sound, and commit the phonetic sin myself, and then it resembles no one else's description.

One of the diver sounds which I would unhesitatingly describe as laughter (because it sounds like that to me) is that usually described as quacking, finishing with a cackling like a hen. Well, I've heard hens laughing often enough.

The bird at the nest has a soft sighing and churring call, which has been rendered as *ooo* and *ur,* and that is as near the call I've heard as need be. I can also recognise the goose-calls which the red-throat can utter, and the wailing cries which have been described as resembling the cry of one in great pain.

Although it nests by tarns and lochans, the diver flies to the sea, or to bigger lochs, to fish. It flies high and fast, and, when pitching on the water, strikes with its breast and shoots over the surface with up-held wings. Sometimes it will come down from a height like a whiffling goose, with a great swoosh of wings.

Two eggs are the normal clutch, and they may be laid in a nest of moss and weeds or in no nest at all. The nest may be no more than a space trampled down by the bird, or a hollow on a hummock of the kind so often favoured by curlews. The female does the major share of incubating, the male relieving in short spells.

The main nesting period is June, although eggs may be found in May and well into July. The eggs hatch in something under four weeks, and the chicks are looked after by both adults for about eight weeks. Only one brood is reared in the season.

In Finland and Iceland the red-throated diver has been found breeding in colonies, with nests as close to each other as those in a gullery.

Return of the Pairtrick

The memorable moments so often arise fortuitously—a simple matter of turning left instead of right, or of changing one's mind unpredictably and going that way instead of this. I do it all the time.

I came in the other morning at people's breakfast time (a ceremony I hardly ever celebrate) and I said to this one, who had come to see me:

'What a great pleasure it is to see the partridges back after so long. I bumped into four adults with thirty-three chicks this morning.'

And he said: 'You'll get mair tae get worked up aboot yet. I'm mair concerned aboot gettin thirty-three men, or ten times thirty-three men, aff the broo. That's whit coonts.'

'Amen,' agreed I. 'I'm sorry I felt pleased aboot the pairtricks. Would you like me to do penance because I felt pleased at seein them? Or tell them they've nae bliddy richt tae be alive when there are men on the broo? Why don't you get lost?'

Later, when I was telling my wife about the morning, I found that all the cheer had gone out of it, although I had no sense of guilt whatever. Any more than I have when somebody says to me, which is regularly, that I shouldn't be feeding my dogs so well when there are so many people in the world underfed.

Anyway, there I was, out at the screich of light, figuring how to get the hay in during the Met Office period of guarantee, and looking over the breeding cows, and making up my mind to get six hours sleep one of these nights, and maybe claim overtime, when I saw the single partridge.

He came out on the trail, and I stopped at once and hunkered down. And what a horseshoe he had on his breist! What a handsome wee bird! And then came the other three, all adults, and I said to myself:

'Hell! Two pairs of pairtricks on foot at this time of year and nary a cheeper among them.'

And then the cheepers came out of the double hedge, in twos and sixes, and fives and singles, all thirty-three of them, and I said to myself:

'That's better.' I'm the great one for hyperbolic enthusiasms. They walked, as though on some parade, along the trail and I watched them through the glasses.

The adults stopped to dust-bathe—shuffle, shuffle, shuffle, turn on one side, then the other, and ruffle feathers with beaks. The chicks whisked around, like grounded bees, so small were they. And so fragilely beautiful. The first partridges bred on the ground since the end of the war.

A big brown hare, running long-legged and slow, came out of the double hedge, and passed through them. The old birds carried on with their dusting; the chicks went on with their whisking; and the hare long-legged on. And nobody was bothered. What a difference there would have been if I had poked my ugly neb in.

Bumble bees were buzzing about by now; stonechats and whinchats were flittering around with food in their beaks; larks were flickering overhead. Then the grass ahead of me, on the right, parted, and the black and white face poked out. A big brock, and a boar at that, homing late.

He came out—and how big he was—and the partridges stopped dust-bathing. They rose, and the chicks scattered. The old birds said bad words. The brock stood still, peering, and I'm sure he wasn't seeing very well in the strong light. Then the chicks came on to the trail again, fell in behind the adults, and the procession walked away, slowly.

Badger followed on, just as slowly. And just as I was saying to myself, he's going to get himself a partridge, he began to weasel on at speed, and the partridges drew to one side, and he passed on without one longing, lingering look behind. Don't ask me why; but that is what he did.

And the partridges kept walking on.

Festive Fare Sets No Problems
for The Wafer

The Wafer was sitting by the fire, pasting up pieces of stout brown paper into cones about four inches deep, like the pokes the wee shops used to put sweeties in.

His long fingers worked deftly. He was making his twelfth cone when his wife came in, carrying a heavy shopping basket.

'When are ye gaun tae kill that auld hen?' his wife greeted him. 'Three days tae Christmas an nothin for the pot yet. The weans were asking again this mornin.'

'We're haein phaisant,' The Wafer said, without looking up. 'You get a'thing else an A'll get the burd.'

'Phaisant!' she said. 'An you wi a gun that's broke? Are they gaun tae gie theirsels up or whit?'

'Jist aboot,' he answered her, and went on with his cone making.

His wife turned away in disgust, and began to empty her shopping basket. But the sight of The Wafer pasting paper exasperated her.

'Whit's this?' she asked. 'Pokey hats?'

'That's right.'

'Must be for wee heids thae.'

'They'll fit,' The Wafer said, and would say no more.

He was out long before first light. His wife heard him starting up his van, then saw the headlights swivelling as he turned into the lane. When the sound of the engine died away, she turned over and went back to sleep.

The Wafer drove for a mile and a half, then parked the van in a bay used for road grit. He locked it, and began to walk across the big grass park towards Hackamore Wood.

A few yards short of the drystane dyke that ran the length of the wood he squatted down and opened his haversack. He took out fifteen paper cones, a cellophane poke containing whole maize, another with a tin of bird lime in it, and a small piece of stick. He laid them all down on the frosted grass, and opened the tin of bird lime.

Into each cone he spilled a little of the maize. Then, with the stick, he smeared bird lime round the open top of each one. He replaced the lid on the tin, put the tin back in the cellophane poke, and stuffed it in his haversack. Then he began to spread out his cones.

He stood them, open end up, in the grass at intervals, all of them opposite the dyke at a point where many pheasants flew out in the mornings. Round each he scattered a few grains of maize, then he walked away 100yds. or so and sat down inside the wood, with his back to the dyke. He had now to wait for daylight.

It was still grey dark when the first cock pheasant crowed in the wood. Later The Wafer heard the beating of wings as the early ones flew down to the dyke. Soon he saw their dark shapes flapping down from the dyke into the open, and, with his eyes level with the top stones, he watched the field.

Suddenly one bird began to flap into the air, and fall. It went up and down, up and down, and the others long-necked to watch. The Wafer remained down and, presently, the bird began to stagger around in circles, with a pokey hat down over its eyes, held in position by the bird lime.

When a second bird began to leap and flap about, The Wafer considered it was time to intervene.

The Wafer caught them easily enough and carried them inside the dyke. There he broke their necks, covering his right hand with a piece of cellophane to protect them from the sticky bird lime. When the birds had stopped flapping, he cut off the heads and hid them inside the dyke, complete with their caps. He didn't want his haversack, or the bird's feathers, messed up with bird lime.

With the two pheasants stuffed in his haversack, The Wafer gathered the other cones and walked quickly back to his van. He could hear the pails rattling at the farm. A bus full of colliers passed by as he reached the road. Within fifteen minutes he was handing the pheasants to his wife.

'Whit happened tae the heids?' she asked.

'They were too sticky,' The Wafer said.

'They didny stick very weel then,' she answered him.

'They could hae stuck tae their heids if they hadny stuck hats on them, then you winny had them tae stick in the oven. Whit aboot no being a comic an gettin me some breakfast?'

Three Mice and a Kestrel

Gallacher was glad he had taken a notion to visit Mossrigg Strip that morning, when the young corn was green in the drills and the cuckoos were calling in the glen.

'It must've been something,' he would say. 'I kent the bird was ticht doon on six eggs and wisna even thinkin' aboot her. It was jist a notion. But if I hadna went, I'm share he wid hae whupped the lot.' And he would end with an earthy six-barrelled snarl that questioned the new man's personal cleanliness, his state of health, and the legality of his birth.

He had stopped at the end of the Strip that morning, with his hands in his go-to-hell pockets and the two collies at his feet. His grey eyes scanned his tidy acres, and rested with pride on the white well-kept farmhouse and buildings four fields away, beyond the road. The corn was green, the tatties were in, the hay was rushing up thick and lush and would be ready for cutting early. Suddenly, a harsh *kraaing* made him turn his eyes skywards, and he smiled at what he saw—a little kestrel tiercel harrying a big black carrion crow from the sacred boundary of his nesting territory. Gallacher watched while the falcon drove off the crow, then was minded to take a look at the nest while he was in the Strip. He knew it well enough, for one of the things he did each spring was to find out where the Mossrigg kestrels were nesting.

He dawdled along the south side of the Strip, poking with his stick in the hedgebottom tangles, for in the hard-bitten, half-century farmer the nesting schoolboy still lived. He was a hundred yards from the kestrels' tree, looking at a partridge nest which contained ten eggs carefully covered with leaves and grass, when the dogs growled with birses up, warning him of the approach of a stranger. Crossing the field, coming towards him, was a man he didn't know. And he was carrying a gun.

That got Gallacher on the raw for a start. Always a thrawn man, he was never uncivil without good reason; but when he was wound up he could shrivel the leaves with his tongue. Who was this who dared to go traipsing over his ground with a gun, in the month of May, without so much as a by-your-leave? The man came forward and spoke.

'Mr Gallacher? I saw you from the road. I wanted to speak to you. I'm the new keeper from ower by. I hope you'll excuse . . .'

Gallacher thawed out on the instant, and was glad he hadn't let fly with some hayfield snorter.

'Nell! Glen! Sit doon!' he said to the dogs. 'Was it something?' he turned to the keeper.

'Mr Gallacher,' the keeper leaned on his gun. 'I've been here only a fortnight. I don't want you tae think I'm butting in'— Gallacher was immediately on the defensive—'but I wondered if you kent there was—ah—certain—ah—vermin aboot . . .'

'Vermin?' Gallacher was still very civil, but cagey. He had been friendly with the old keeper for twenty years. He had respected him. But he wasn't very impressed with this one.

'Yes. Vermin. I just wondered. You know how it is. And I've the rearing field ower by . . .'

'What vermin?' Gallacher was away ahead of him now, with his feet on the ground again.

'Ah! I was sure you widna ken. It's hawks, Mr Gallacher. They're everywhere. So, when I saw you here, I . . . There's one now!' he said suddenly, pointing above the cornfield, where a kestrel was balancing in the wind, with depressed tail, scanning the trembling green below.

'That!' Gallacher laughed. And when Gallacher laughed you could hear him three fields away. 'Man, there's anither wan nae further awa' than ye could rin haudin' your braith, an' she's sittin' on six eggs this very meenit!'

'A nest? But these birds'll play the dickens with your partridges. Believe me, I ken . . .'

'D'you tell me that?'

'And my young phaisants; they'll be a sitting target. I've seen hawks joukin' about in the early morning . . .'

'That'll be right.' Gallacher was still civil.

The keeper was getting nettled. 'But you don't seem to understand, Mr Gallacher. Hawks . . .'

'Man, I unerstaun' fine,' said Gallacher banteringly. 'You don't like hawks! Funny thing is your predeceesor didna mind them a bit, and he raired a pickle phaisants in his time.'

'That's as may be, Mr Gallacher, but vermin's vermin, and it's my job to see . . .'

'Look,' Gallacher interrupted him, 'maybe I don't unerstaun' aboot vermin, but I understaun' this. This is ma ferm. I'll worry aboot the pairtridges and the hawks. There's two pairs nestin' inside ma mairch fences, an' God help the man who lays a finger on them. He'll need it!' Gallacher was warming up.

'Mr Gallacher, I didn't mean to be interferin'. But the farms around here are breeding places for vermin, and I get . . .'

'So that's it!' exclaimed Gallacher. 'You've nae vermin o' your ain, so your itchin' for somebody else's. But who feeds your rabbits? Me, an' the like o' me. An' rabbits tae me are the worst vermin.'

'But . . .'

'But nothing! I'll gie ye a word o' advice at nae cost. Leave thur kestrels alane. The next time ye come spierin' for me, leave your gun ahint; I'll cairry the guns aboot here. An' ye'd better get that poletrap doon I see on the edge o' the mair afore a wee bird cheeps tae the polis. Guid day!'

And Gallacher left abruptly.

'Noo, Wull, ye shouldna mak bad bluid,' said his wife, when he told her. 'You're too quick wi' your tongue, an' the man was only daein' his job as he sees it.'

'Bad bluid!' said Gallacher. 'A man seeks me oot, tries tae run the ferm, wants tae shoot the hawks, an' I'm supposed to say nothing?'

'There's a way o' daein' things.'

'I thocht we'd seen the last o' his kind,' said Gallacher. 'An' I'll bet the laird disna ken what he's up tae. But onyway, he can dae what the hell he likes on his ain grun, an' I'll attend tae mine. Look oot there noo!'

On a big leet in the stackyard a kestrel had just pitched and scissored his wings. It was Yellow Foot, the cock from the Mossrigg Strip nest.

Gallacher turned to his wife. 'Him an' his phaisants. I should've offered him a sovereign for every phaisant he could get in that nest, and ta'en a hauf-croon fae him for every moose. I could've made thirty poun' gin haytime! The man's an eejit, a pure bliddy eejit. That's the wee man doon, noo! He'll hae anither moose. An' I'm supposed tae scrub him an' let the mice mak meal o' the corn stacks!'

Yellow Foot, Gallacher's wee man, squeezed the life out of a vole on the straw-strewn ground, and flapped back to the leet with it in his claws, unaware that a farmer and an old-look keeper had been discussing his fate. The vole was his fifth from the Mossrigg stackyard that morning. Yellow Foot knew Gallacher as well as any wild falcon can ever know a man. He followed in his wake at ploughing time, superintended the tattie happing, hovered behind the binder at harvest, and hunted the stackyard all winter. Gallacher had erected a perch for him, a look-out post, which was simply a stack prop stuck vertically in the ground. And Yellow Foot used it daily. While Gallacher watched, he tore the head from the vole, swallowed it with much gulping, then grasped the body in a foot and launched from the leet.

He flew low over the road, low over the cornfield, low over the cows in the pasture, and pitched in his calling tree in the Strip in forty-five seconds exactly.

Wree-wree-wree-wree-wree! he called, and transferred the headless vole to his beak.

Wree-wree-wree-wree-wree! his mate replied, and flew out to meet him.

Yellow Foot stood tall when Kree pitched beside him, and watched gravely while she bowed, with spread tail quivering in greeting. They said *wree-wree* to each other, softly, then Kree took the vole from her mate's blood-splashed beak, and swallowed it with much blinking and neck-stretching. Yellow Foot whetted his beak, looked about him, then slipped sideways from the branch. Kree, with head tilted, watched him for some moments with one lustrous far-seeing eye. When he had drifted out of sight beyond the trees, she shook herself and fell to preening her breast feathers. This was the time of day when she took fifteen minutes off from her brooding.

She was raking her rump plumes, with one eye closed, when she was alerted suddenly by the piercing screams of Yellow Foot.

Kek-kek-kek-kek-kek-kek!

Kree launched away with a whicker of wings as Yellow Foot came pinwheeling down, flattened, and flashed towards the nest. He was above it, hovering with hanging feet, when Kree topped the pines, gold-tipped, and blue-green in the sun. And in a moment she was beside him, wing to wing, with talons clutching at the marauder on her nest.

The magpie had sneaked in when Kree was preening, having noted the vacant nest when he was winging home to his own. Now he was tilted forward on a branch at nest level, with rainbow tail speared and ebony beak stabbing at an egg. He had already pierced one, and yellow yolk and scarlet blood were glueing on his beak. But before he could crack the second egg the outraged falcons were upon him.

Now, a kestrel will not, under ordinary circumstances, assault a magpie; nor is it fitted to do so. In Mossrigg Strip, magpies and crows nested beside the falcons, and there was an unwritten law which laid it down that you didn't raid your neighbours. But such a truce is always uneasy, always liable to be broken under stress—or when the right kind of opportunity offers. So the magpie attacked when the opportunity offered; just as he was ready to get out without argument because he knew he was in the wrong.

But the kestrels had other ideas. They had no kind of knowledge of truces or laws. They understood only a pattern of behaviour that was routine. The pattern had been broken; a magpie was on the nest; so they went into the attack, angrily yet circumspectly, for they knew they were faced with a powerful and cunning adversary.

It was the magpie's willingness to retreat and call quits that gave the falcons their chance. As he turned to launch from the branch, Kree struck him on a wing, unbalancing him and making him lose his grip on the air. While he was trying to catch hold again, Yellow Foot took him on the face with his claws, blotting out one dark eye with a hind talon and clutching his neck in a foot. Such a grip the magpie could not shake off; but he had time to act before the claws choked out his life. And he acted.

He went right down to earth, with the falcon riding him down. Kree hovered behind Yellow Foot, and pitched on the grass when the locked antagonists grounded. Yellow Foot released his grip during the landing fankle, and flapped back beside his mate to consider the next move. That gave the magpie his opportunity. With legs braced apart, tail wedged down in support, and beak presented like a rapier, he awaited the attack. Kree flashed in, but was met by a lightning stab of the magpie's beak, which hurt her and removed two feathers from her breast. Yellow Foot struck, and was met by a similar riposte. Then, as if realising that one to one left the odds in favour of the magpie, they struck together. Kree recoiled as the magpie struck, but Yellow Foot was again on top, with the old grip, and another claw sunk in the magpie's remaining eye. The magpie called *chock-ock* when the darkness came, and in the same instant Kree was on his back clutching for a foothold.

They clawed at him, They screamed. They buffeted him with wings. But he was slow to crumple, and dragged them many yards before he stopped, with head bowed and beak pressed to earth. Together they flapped up, their fury spent, and flew to separate branches. The magpie moved, ground-flapping and crawling, and sought refuge under a grass-grown briar. And there he was found, and chopped, by a fox when the moon highlight was bright in Kree's eyes as she cuddled close on her four sound eggs against the chill of the night.

That incident changed the whole balance of power in Mossrigg Strip by changing the outlook of the kestrels. Any magpie in Mossrigg was, to them, the bird that had raided their nest. So a time of trial began for the hen magpie, now left to fend for her nestlings alone.

Each time she flew from the Strip on her foraging she was intercepted by Yellow Foot, unless he was hunting more than two fields away. He stooped at her, and harried her, and sometimes snatched feathers from her; but he was unable to injure her, or force her from the Strip. Doggedly, she carried on feeding her chicks, fighting back at each assault, and refusing to be intimidated.

The harrying continued while she was feeding her fledglings in the Strip, till at last she was forced to lead them further afield, to the shelter of a big wood over a quarter of a mile away. Yet, she had no cause to fear for her family, because not once during this war with the magpie had Yellow Foot attempted to molest her chicks.

One moonless night, when the peewits were calling as they swept on humming wings over the Strip, Kree felt movement under her, and knew her first chick had hatched. Yellow Foot came in twice in the morning with prey—a vole and a shrew—and by nightfall there were four chicks in the nest—white downy things that cheetered each time Kree changed position or lifted her wings.

Gallacher saw Yellow Foot more frequently in the stackyard, or hovering with uplifted wings above the cornfield, and guessed the bird was busy with a family. Then, for some days he did not see the falcon. The fact registered suddenly, and he remarked to his wife: 'Come to think o' it; the wee man hisna been in the stackyaird this twa or three days back. Must've changed his beat . . .'

But he wondered about it, and the following morning he visited the Strip to allay his vague fears. Yellow Foot at once flew out to circle overhead, screaming, and Gallacher was put off. There was no sign of Kree, but the fact did not strike him at the time, and he left without misgiving. He didn't know the story of the twa-three-day period when Yellow Foot was absent from the stackyard . . .

It happened at dusk, when the chicks were seven days old . . . Yellow Foot, roosting near the nest, heard stealthy footsteps on the ground, and lifted away when they passed beneath him. Presently, Kree heard them too, and peered over the side of the nest. Suddenly boots struck the tree and branches began to shake, and Kree flew to join Yellow Foot in the gloaming. A man was climbing the tree!

The falcons circled, screaming, flying high and fast, alarmed by the dark bulk of the figure at the nest. But the visit was brief. In less than a minute the man started climbing down, and the birds dropped lower to watch. They heard the thud of his feet on the ground, and watched him hurrying away through the trees. As soon as he was clear of the Strip, Kree flashed in to the nest, to find her four chicks squirming and cheeping. They were unharmed. Yellow Foot touched down for a brief moment, saw Kree straddle her nestlings, and flew back to his roost. And both birds settled down as though they had never been startled.

It is doubtful if Kree was capable of realising that she had three mice in the nest that had not been brought by her mate. In any case, she found them in the morning, but did not use them right away, for Yellow Foot came in with two fresh kills in fifteen minutes, and she fed them to her chicks. It was in the late afternoon before she used the first of the mysterious mice—feeding two to her chicks and swallowing the third herself. Then she settled to brood.

When Yellow Foot called to her later in the day there was no answer. For Kree was sprawled on the edge of the nest—dead. Her nestlings were huddled together in the centre—dead. The little falcon flew in to the nest, delivered his mouse, stared uncomprehendingly, then went to catch another. When he delivered the next one, he flew to his roost, and delivered no more. He knew . . .

And so it was that Gallacher found them when he climbed to the nest—a mere routine visit—two days after he had checked the Strip for Yellow Foot. Like Yellow Foot, he stared at the dead bodies uncomprehendingly. Yellow Foot circled, swearing, the pattern of his life restored by the presence of the man at the nest. And at last Gallacher tossed Kree's body to the ground, certain she had not been shot, and convinced she had died a natural death . . .

Till he found the button—a leather button—with a small piece of tweed attached—caught on a sharp sliver just below the nest. Gallacher looked at it long and wonderingly, for he had seen that button before, and knew where to find the jacket from which it had been torn . . .

Painted Ladies, Palely Loitering

This year's butterfly week in the garden of *chez nous* was from the 14th to the 21st of September, with Thursday the best day for numbers and variety. On the biggest of the three buddleias there were twenty-three Small Tortoiseshells, two Painted Ladies, and a single Peacock at 4p.m.

The other buddleias had Whites and Small Tortoiseshells, but the big one, twice my height and twice my armspread, had the weight and the brilliance.

By Friday the Peacock had disappeared, but the Ladies were still there, one on the high spikes palely loitering, the other restless in buskins of brighter colour. And presently the brighter one disappeared, leaving the pale one alone among the Tortoiseshells. It was there until the Saturday, always on the same bush, and active long after the Tortoiseshells had gone to rest.

Painted Ladies are migrant butterflies, which very rarely survive the British winter. If it were not for new immigrants in the spring they would die out altogether. The immigrants breed and produce the butterflies of August and September, and almost all of these die off in the winter.

In some years the spring immigration is on a large scale; some years it is very small. So the numbers vary from year to year. In good years the butterflies come into Scotland; more usually their spread does not go beyond the north of England. We have had them in the garden here only three times in ten years, and never more than a few. The Red Admiral is much more regular, and has a little higher winter survival rate.

The Painted Lady is one of the most widely distributed and widely travelled of all butterflies. It is found in most parts of the world outside South America. Europe's visitors come from North Africa, where the Painted Ladies breed in winter along the edge of the desert belt. In North America the Ladies winter in Western Mexico and fly north in spring, sometimes thousands of millions strong, many of them reaching Canada and even Newfoundland.

C.B. Williams has watched thousands of Painted Ladies flying through Egypt and on over the Mediterranean, headed presumably for Turkey and the Ukraine. Migration flights have also been recorded in Pakistan, the butterflies flying at a height of 17,000ft. towards the south-east.

In 1952 there was a great invasion of Britain by Painted Ladies and at the earliest date ever recorded. They were reaching the south of England in early February, and were latterly recorded from all over the country, including the Isle of Man and the Outer Hebrides.

One of the biggest migrations, mentioned by Williams, was in 1879. The butterflies reached England in June, and on 9 June large numbers were washed up by the tide at Bournemouth. Immigration continued through June, and in August and September a southward movement was noticed by the new generation of British-born Ladies.

The food plants of the Painted Lady caterpillars are mainly thistles, but they have also been found feeding on the common nettle and other plants. Eggs are laid from June onwards, and these produce the butterflies of August and September.

The Year of the Butterfly

This has turned out to be a year of the Red Admirals. On the last day of August there wasn't one to be seen in the neighbourhood; but in the first flush of September sunshine they were all over the garden—dancing on the warm waves of air, tiptoeing over roses and dahlias, then burrowing into the massed drooping spikes of the buddleias to syhpon the heady brew with fat drunken bumblebees and nervous hoverflies.

They were instantly recognisable from the other end of the garden, in black velvet robes with scarlet diagonals and splashes of white on the forewings, and scarlet trailing edge of the notched hind wings. They were unsettled that first day, rising in full sail at the first hint of human shadow or movement. But by the second day they were settled, drowsy with deep drinking, and paying hardly more attention to a probing finger than to the crowding bees.

Six gathered on one bush, rubbing shoulders with and dwarfing the Small Tortoiseshells. One was in the quiet scented depths, with wings closed; three were on the bush's heaving breast, with wings spread, as though suckling. And on the top was the sixth, balancing on a tossing spike, blue freckled at the wing tips like the sky seen through pinholes in the velvet.

Each day they were in the garden, about a dozen in all, from the first sun of the morning until the late coolth and long shadows. In the heat of the afternoon they would all fly to the garden wall, and rest, like coloured cut-outs pressed there. One, choosing the path as a resting-place, had the scarlet hind-edge snipped from a wing by a blackbird, and was immediately identifiable thereafter.

After a week they were all easily approachable, and in the end I could cut off a flower spike, and carry it away with a Red Admiral happily settled on it.

The Red Admiral is a migrant butterfly. There is a northward movement in spring, and a southward movement in autumn about the time my garden had its invasion. Some Admirals survive the winter in hibernation, but most that stay die off. The population of Scotland, and Britain, is kept up by immigration from the Continent. A great deal remains to be discovered about the movements of this butterfly.

Two small boys arrived when I was photographing the Red Admirals. They produced a small box and said they had a beast in it which the father of one had cautiously caught up and put in there. What kind of beast? I wanted to know. A big wasp they said. And at that stage I guessed, and opened the box, and sure enough the beast was big, black and tan, like a hypodermic with wings on, close to an inch and a half in length and with a tail (ovipositor) like a stiletto.

It was the giant wood wasp *Sirex gigas* which is a sawfly, not a wasp, and is sometimes called a hornet by those seeing one for the first time, which is a fair enough error considering its great size and its rather awe-inspiring appearance. But, as I say, it has a body like the barrel of a hypodermic, with not the slightest suggestion of the cut-in-half wasp-waist of the wasps and hornets.

Sirex is a wood borer, using its stiletto for drilling and laying eggs. It bores in pines and other conifers, and avoids hardwood completely. It will lay in felled or standing timber, and in some parts of the world can be a pest. In this country it is parasitised, in the larval stage, by an ichneumon called *Rhyssa persuasoria,* which has an even longer stiletto—longer, in fact, than the body of the Sirex. Rhyssa lays her eggs where she finds the larvae of Sirex, and the larvae of Rhyssa feed on the bodies of the Sirex larvae, consuming them as they grow. It is a nice example of biological control.

I told the boys this, and emphasised that Sirex was absolutely harmless to human beings. The lance was not a sting; it was for laying eggs. They accepted it, academically as it were, but still weren't handling the beast until I handled it first. Then they made free with it and will never be in any doubt again.

There was a funny story about a Sirex twenty-five years or more ago in my home district. One had flown into the office of the local Inspector of Taxes, and a picture of it appeared in a Sunday newspaper, with a caption which read (I think) 'Did it go there to sting the Income Tax Inspector?'

Not even Sirex could do that.

A Question of Identity

There were three of them, one of them carrying a pint milk bottle with a wriggler in it—a wee newt padding around like a puppy treading against its mother's breast.

'Haw! Mister,' the pint bottle said, 'see what we've got?'

'What?' I asked, being awkward.

'A lizard,' he said.

'It's a newt,' I told him.

'See?' one of the others told him.

'I said it was an eft,' the third said.

And at that I was puzzled, because never in my life, in my part of the world, had I heard the newt called an eft. Other places yes; hereabouts no.

'Where did you hear that name?' I asked the eft.

'At school,' he said.

The pint bottle stuck his nose against the glass and said: 'Well it looks like a lizard.'

'So it does,' I agreed, 'but if you come to the house I'll show you a picture of a lizard, and you'll see the difference.' They came with me and I let them see a picture of the common or viviparous lizard. And I talked to them.

The newt is an amphibian, I explained. It lays eggs in the water and the eggs hatch into tadpoles that grow up in the water and become newts. Like the frog, the newt has to go through a metamorphosis, and in the water. Water is a vital stage in the newt cycle.

The lizard, on the other hand, is a reptile. This one gives birth to fully-developed young, hence its name viviparous.

The wee lizards are replicas of the parents. All they do is grow. There is no metamorphosing.

Then I told them about autotony, the lizard's ability to part with its tail. Self-fracture we call it, and the lizard is designed to allow this.

A lizard roughly handled will part with its tail, but self-fracture can take place without any kind of contact, merely as the result of fright or stress. Some lizards will part with the tail without handling; others can be handled without self-fracture resulting. All one can do is be careful and gentle.

There are two lizards in Scotland, I told the boys: this one with four legs, and the slow-worm, with no legs. So, I said, you can always tell the one from the other at a glance, even when the lizards are big and the bigger slow-worms are wee.

Blowing in the Wind

The wind was gadding about in the beechwood, now meeting me in direct confrontation and finding the chinks in my armour, then stealing up behind me to send my hair over my eyes, or gusting withershins to take me by surprise when I stepped from behind a tree. It was playing tornadoes with old leaves and leaf flinders, and other detritus that a body couldn't have put a name to.

It was having a time, making waterfall noises with the tossing trees, and getting colder by the minute, as a man's back teeth can quickly tell. I sat down against a buirdly beech, with my back to where I thought it was in ambush at the moment, and it was then I saw the crow.

An ordinary carrion crow he was, with a braw purple mantle, but a little disarranged sartorially by the wind. I noticed he had a broken tail feather, because every gust set it twirling and it looked like a half propeller. Judging by the way he was capering around, he was enjoying the blow, and I got the impression that he had time on his hands and was in the mood for buffoonery.

If you've kept a wheen tame crows over the decades you learn how to read the signs. You also learn that there's more to a crow than the establishment credo of a blackguard, a robber of other birds' eggs, an executioner of nestlings, an unwanted midwife of sheep, a predator on pheasant chicks and ducklings and whatever—a black, evil, sinister, rotten, cruel, dirty brute of a bird. In my book a crow is a crow, and I expect a crow to behave like one, just as I expect a bee to sting me, or an up-ended nail to puncture the tyre of my car.

Moral judgements are for those who devise and evaluate moral codes. A crow is a crow and is stuck with it.

Anyway, I had guessed right with this one, because, presently, he picked up a worn, desiccated puff ball and tossed it into the air. The wind carried it a couple of feet, and he flapped after it, picked it up, and tossed it again. And again. And again. And again.

Then, being a crow, he became quickly bored and tried a new trick. He flew up with it for six feet or so and let it fall. Again the wind carried it—a little farther this time—and again he swooped to retrieve it. So there he was, playing a kind of yo-yo game, all by himself, a game with no profit in it, no food at the end of it. So he was playing.

Tiring of the puff ball he then picked up a strip of bark measuring over a foot in length, and about four inches broad. I know because I measured it later. This he hoisted and tried to walk with. But the wind caught it like a sail, and spun him round until it was over his face and bent along his back. He liked this, because he allowed himself to be blown all over the place.

All this time he was holding the bark by the end. Now he put it down and took it by the middle, then flapped up into the gale and let it go. It sailed for several yards. He flew after it, pounced, took it by the middle, and rose with it again. And again he released it to watch it sail away. This, too, he repeated again, and again, and again.

Next he flew into a beech close by me and sipped water from a crotch. (I looked afterwards and found there was a bowl of water there). This crow knew his wood in detail. What would he do next, I wondered? Well, he flapped into a pine on the wood edge and began to snap off dead twigs, which he tossed clear and watched being blown away by the wind.

Having had enough of this he flew to a dead birch tree and began to peel off strips of bark. He tossed the bits into the wind, then followed them down, trying to catch them before they reached the ground. Once or twice he succeeded, and, when he did, he flapped up with the piece and released it again.

All this took about twenty minutes, which is a long time to spend on a ploy, even for a crow. I was wondering what he would do next when another crow arrived, and called to him, and they flew into the beechwood together. I waited on for a bit but the buffoonery was over for one day.

Waiting for a Fox

'Are you waiting for a bus?' the man asked The Wafer who, as he put it, looked as less as it was possible like somebody waiting for a bus.

'Naw,' he answered, 'I'm waiting for a fox.'

'Waiting for a fox? You mean . . .?'

'I mean a fox, no a bus. Yon fox. The wan ayont the threshes oot there.'

'Where? I don't see any fox. You must be joking.'

'Never joke, me,' The Wafer said. 'I'm the seriousest man you'll meet.'

'I can't see any fox,' the man said.

'That's aboot maist folk's track record, so I widna worry aboot it.'

The Wafer was laughing as he said to me: 'The bliddy man still thocht I was at the kiddin. He was wan o yon educated eejits.'

Well, in the end, with The Wafer pointing, the man saw the fox, and became excited: 'Yes,' he said. 'Isn't it very close for a fox? It's sitting there looking at us.'

'That's whit I've been trying tae tell ye,' The Wafer said. I could imagine his abrasive friendliness when he said it.

'Why does it keep sitting there looking at us?' the man asked next.

'Bitch foxes always look at me,' The Wafer said. 'They like me. D'ye no think I'm worth takin' a gander at?'

'Bitch fox?' the man said. 'Come now; how can you possibly tell? Nobody can, I'm told.'

At this point I warn you, gentle reader, to wait for it. Even I could see it coming.

'Who told you?' The Wafer asked.

'Well,' the man said, 'I've read about it.'

'That'll be right,' The Wafer agreed. 'An I ken whae would write it. And its richt. Naebody can. But somebodies can. Me an' him, like. An' some ithers forbye.'

The Wafer grinned at me.

'That should mak your day. Somebody ither than me reads you.'

But the man wasn't so sure that The Wafer wasn't at it, as they say. 'How can you possibly tell?'

I'll tell you what The Wafer said to me in answer to that one.

'Man, she had a belly line like yon curtains you used tae see faulded in windaes, an' titted like ony dug that's sookin' pups.'

In the end they became very friendly. The Wafer is easy to like, unless you're the law. The man ended up wanting to buy him a dram, and was suitably amazed when The Wafer declared himself teetotal.

'I'll tell you whit, though,' he said to the man. 'I smoke, and widna mind wan o your ceegars.'

'Certainly,' the man said, and gave him one. 'You smoke a lot?'

'Like a diesel engine.'

'What was she doing anyway,' I asked, as he reminiscently smoked the cigar again.

'The way she kept stoppin, then drawin away, an her being milkin, I guessed she was coming in tae cubs, and was trying to draw me aff. An so she was. She had fower among some stanes no far frae the road.'

And he ended up giving me directions where to find them. Which I did.

Poacher Gets His Character

I was sitting reading, and still nursing a few persisting aches and pains, when he was announced, and came in, his silver hair well groomed, and himself lean, hawklike, and fit as ever. Indeed, except for the hair, which has bleached fast, he looked the same as he did when he was forking a motorbike in France for the British Army more than thirty years ago.

'You've a wheen phaisants aboot ye,' he greeted. 'A seen seeven cocks and three-fower hens feeding ayont the burn. A see you've got corn doon for them.'

It was the nature of the man, still is, to notice all such details without noticing.

'Whit's the book?' he asked, out of the blue. I'd never known him ask such a question afore, and it surprised me.

'You must be psychic,' I said. 'I'm reading about you.'

'Me?'

'Aye!'

I showed him the book, an oldie, worth its weight in gold for nonsense and insufferable condescension. It's called *The Keeper's Book*. Maybe you've heard of it. If you have, you'll ken the bit about poachers. Date 1907, before I was born. Perpetrator Sir Peter Jeffrey Mackie.

I read to my visitor as follows:

'The twentieth-century poacher is an ill-conditioned, drunken and slinking scoundrel, an enemy to law and order, without a particle of true sportsmanlike feeling in his veins. Taken as a class, poachers are a set of hardened criminals, careless of everything but their own besotted lives.

'The occasional poacher is a much rarer bird—a farmhand, a village loon, or maybe a medical student home for the vacation,' (poor medical student) 'but he is not so dangerous as his professional brother, who is a cast-off from honest trades—a grain in the sediment of society.'

The Wafer goggled, displaying the white anger of the fanatical teetotaller.

I went on: 'Drink has him, as a rule, in its grip. He has a shifty and congested eye and a tremulous tongue. He is, in the majority of cases, an arrant coward. Remarkably ignorant on most questions he is terribly acute on all matters affecting the poaching of game and, although a coward, is prepared at a pinch to get rid of another life to save his own.'

Shades of Lonavie.

I burst out laughing at The Wafer's congested eye; at this teetotaller who is as articulate as anyone I know, and who could have given this pontiff lessons on most things, including good manners.

'Does he say ocht aboot keepers, this bein a keeper's book?'

'He does,' I said, and I quoted the gentleman to one of the greatest poaching artists I have ever met.

'Knowledge, skill, perseverance in face of difficulty and failure, discrimination, firmness, order, courtesy and enthusiasm—these are the primary necessities for a good keeper. Knowledge of the technicalities of his craft, skill to carry them out, perseverance in face of difficulty and failure, discrimination in dealing with superiors, equals, neighbours and inferiors ... modified by a gracious civility, will all tend to his own, his master's and his servants' satisfaction ...'

And The Wafer said:

'Come on, you're making it up.'

'I'm not,' I said. 'So get out of here, you lazy, besotted, slinking scum with your congested eye and shifty look, and say Sir when speaking to your betters ...'

'Jesus,' he said.

'Amen,' said I.

Night is to the Owls—and Me

The runway through the wood looks like a footpath, but after it passes under a peeling windfall, then a low-spreading branch, it becomes clear that it could not have been trodden out by human feet. It is, in fact, a badger runway, padded hard and bare by the feet of many brocks. I follow the path through tall pines which glow blood-red in the sunset, then down the slope under the leafless hardwoods.

Eventually I reach the badger sett, which is old, with massive earthworks, and four entrance holes presently in use. At each entrance, the tell-tale five-fingered prints of badgers are clearly recorded in the soft earth.

Four other trackways radiate from the sett, all leading to the badgers' feeding grounds. Two trees show claw-scars where the brocks have scratched like cats. There are two dungpits in use, one beside a rhododendron, the other among withered bracken. The sett is a going concern.

I sit down on a windfall, with my back against the upturned roots. Twelve feet from me, in front and downhill, is a well-used entrance. The wind is puffing from the hole to me; it blows my tobacco smoke over my shoulder. I compose myself to wait.

The sun's rim dips, and I think of Coleridge. The afterglow pales, and I look at my watch. Fifty minutes from now I should be seeing my first badger.

Cobwebs of clouds shut out the sky, and twilight comes grey and brooding. The wind dies away and the wood is silent, but only for a few minutes.

Suddenly a cock pheasant calls *cock-up* Then I hear the frog-croak and chirp of a woodcock, and when I look up there he is flying withershins above the trees, slow-flapping in tight circuits, in roding display. He flies round and round and round, nine times in all, then appears no more.

Now a pair of blackbirds begin to scold, and their scolding is owl-talk which tells that an owl is stirring to wakefulness. And, presently, from the dark trees beyond the rhododendron thickets comes the quavering hoot of a tawny owl. For a moment the blackbirds become almost hysterical, then they fall silent. The owl is a-wing.

Darkness comes slowly, measured by the brightness of a match flame. In the field a hundred yards away a partridge calls *kaar-wit!* A waterhen *krook-krooks* by the burn. After that the night is to the owls, and me. And, I hope, the badgers . . .

I am now bedded down, as quiet as a slater under a stane. In my dark clothing I am invisible. My eyes are fixed on the hole twelve feet away. It is still a darker patch in the surrounding darkness.

At a quarter past nine a muted orchestra starts up in the hole, a medley of purrs, and grunts, and squeaks, and whimpers, accompanied by the thud of feet. The sow badger, I decide, is disentangling herself from demanding cubs, who are complaining loudly against any such disengagement.

A few minutes later, while the orchestra is still performing, a big badger heaves out on to the mound of earth, and turns about, facing the way he came.

It is himself—the boar! As big and brawny a brock as the low country ever did see.

With his back arched, and standing high on his legs, he does a kind of war dance, but the sounds he utters are not war cries. They are the purrs and throaty chuckles of a boar to a willing sow. The boar, indeed, is calling out his mate.

Still purring and chuckling he begins to back-track, and I keep my eyes open for the sow. The cub-whimper stops, and in a moment a black and white face peers out. The grey body follows. The sow purrs to the boar, who now turns away. She follows him to an open space under the trees and there they chase each other in figures of eight, making about as much noise as a pair of bullocks.

I am no more than a few yards from all this, but they suspect nothing. Presently they scratch themselves, then pad away nose to tail, and soon I lose them in the darkness.

Thus the honeymoon period of badgers is brief and seldom witnessed. At most times the brocks are silent beasts, little given to any kind of demonstration. But if a man visits them often enough at the right time he has to be lucky once in a while.

I sit on for a bit, smoking my pipe. Not long after the badgers have gone, the cock tawny owl flies over my head and pitches, crying *kee-wick, kee-wick, kee-wick!* He is answered by cat-calls, then there is a tushkarue of hoots and croodles as the hen takes the prey from him. The vocal exchanges stop abruptly, like a radio switched off, and the wood is silent again, with a silence that fairly hisses in the ears.

I am thinking of moving to my off-the-ground seat to await the return of the sow, but the purr and patter of rain makes me change my mind. The rain settles to a steady drizzle, and when I am half wet I wrap up and set off through the dark wood, and head for home.

Now-a-nights I don't accept soakings gladly.

Opportunist Extraordinary

When it comes to what they call brass neck the crow bears the gree. Nane blate, unless maybe at looking down the barrels of a twelve-bore, it is the opportunist extraordinary, with an eye like the Rontgen ray and an IQ as high as a Californian Redwood: full of ramgumption laced with low cunning; bold but eident for the unchancy: a hallanshaker, a gobbler of eggs, a picker of brains—a bloody nuisance as the man said.

And mair forbye.

The man said to me: 'By God that craw's no glaikit.' And I said: 'You're right, the craw that's hatched glaikit never grows up.'

He came to me this day and said: 'Somethin's et a the men's sandwiches and left nothin but the pokes. Could it hae been that big dug o yours, maybe? But it couldny hae been, for I've never seen it oot on its ain.'

And I said: 'Anyway my bliddy dug widna thank ye for your white pieces— she's got mair sense. It wid be some less weel educated dug.' And he agreed that was likely enough.

And that brings me to the craw, the carrion craw, or as they ca it in my pairt o the warld—the cairryin craw. I merely phoneticise: the cairryin has nothing to do with the bird's abilities as a humper.

Well, the man cam back this day unexpectedly, and there was the craw dragging a the pieces oot o the hut—jeely pieces, pieces wi bacon, pieces wi cheese, pieces wi fried eggs, pieces wi steak even. And the man unloaded all his waes on me, as though I, somehow, were the chief crow in charge of behaviour patterns and discipline.

'Well,' says I, 'it's like this, you'll hae to keep the hut door sneckit, or tak the key aff the bird.' And at that it was left for the time.

Next thing I sees this craw in the foxes' enclosure, nabbing bits of meat and whuppin them up intil a tree, and Glen, the dog fox, and Fiddich, the vixen, watching from their holes, measuring distances and chances, but careful not to rush out and not catch, thus making fools of themselves.

This caper continues but I predict the death of the crow by over-reaching, although, mind you, if it came to a game of tag between a fox and a crow I wouldn't like to predict who would be most often het. If you know what I mean.

62

And then this craw strikes up a relationship with the pine marten who is in a 14ft. high enclosure. The black boyo comes down and gives the marten the come-hither, and the marten climbs the tree and tries to get at the crow through the heavy mesh. But, of course, he can't. The crow, however, perched on the cross beam, can get at the marten's eyes, his beak being slender enough to go through said mesh. So you can figure that one out. I'm sure the crow has.

Not wanting a marten with an eyeless socket I blast a shot over the crow's head and scare him away. But for how long I don't know.

This crow made two mistakes, however, but they're the sort any crow might easily have made.

First he flew down to pinch some of the badgers' food. And he did. But he didn't know—how could he?—that these badgers are day-girls, so he almost lost his head when one of them shot out and grabbed at him. If those jaws had closed on any part of him he would have been short of that part. He thus learned a thing about badgers that few other crows can know. And he stays away.

Then he tried pinching the deer's rations one day when the buck was standing by. This time Bounce buckled down and tried to ram him against the fence. And almost succeeded. Then he chased the crow across the field. So the crow now knows not to threep with a roe-buck.

That's the way of things with the crow at the moment. He's in the way of learning other things, especially about ganders. So we'll see.

Hoodies Always on Alert

Thirty-four hoodie crows, all real greybacks, were spread evenly over the sunny side of the knowe, not hunting, not doing anything, indeed, except enjoying the warmth. Some had their wings half-open, some had them drooped; some had their backs to the sun; some had their heads cocked away from it, with neck feathers raised and eyes closed, in the fashion of birds basking in heat.

Now and again one would bow, with open beak almost touching the tawny grass, a sure sign that it was enjoying the invisible jets of warmth. In the shady places the frost was on the grass; down below gulls were walking on the frozen loch. Up here the crows had found a sun trap, and were sunbathing.

No gathering of hoodies ever looked more relaxed, more off guard; but, as the man said, greybacks have eyes in their what you and I sit down on. I eased the door of the car gently open a few inches, and thirty-four hoodies stood to; before the long, black lens of the camera could be pushed through the slit, all were in the air. And that was that.

My wife said: 'They looked half asleep, too!'

I said: 'Hoodies never sleep, even when they're sleeping.' Which, even if it isn't quite true, is always the way it seems.

We watched the flock straggle away along the hillside, then moved on uphill. A few minutes later I had the glass on a pair of hoodies about 60yds. from the roadside.

Both birds were stabbing and pecking at furry remnants of hare, which they held down with their feet against the pull of their beaks. On the roadside, not far from the car, was a shapeless lump of mountain hare, so we assumed the birds had lifted pieces from this carcase and carried them off.

This pair wasn't alarmed when I wound down the window and poked the glass out for a better look. For a few moments they stopped eating to look at the car, but when I drew in the glass and closed the window again they returned their attention to their prey.

It was a day for hoodies all right. Apart from a solitary kestrel, it was hoodies, hoodies all the way, and not a buzzard to be seen.

The next pair of hoodies which I stopped to watch flapped up from the roadside, just as the car appeared over a rise. I stopped as the birds swung away, and watched while they pitched on the brae face a hundred yards or so from me.

'I wonder what they were at?' I thought aloud.

My wife, who has long-distance eyes, said: 'A fish.' 'A fish?' I said. 'Away up here? Let's hope it's a nice big salmon.'

'It is a salmon,' she said, and I knew she had to be right. She was.

It was a salmon about the length of my arm, with empty eye sockets and a lower jaw like a butcher's hook. The fish was untouched as yet except for the missing eyes, which the crows had taken first, as is their way with sheep and hares.

The birds sat on the brae face watching us as we moved around the salmon. I didn't touch the fish. Instead I went back to the car. We were hardly seated before the crows flew down to the fish again.

'I'm going to get the same distance on the other side of them,' I said to my wife, 'and take a photograph of these two. The light will be right at the other end.'

I started the car. The crows rose and flew to the same spot on the brae face. When I had driven as far beyond the fish as I had been on the other side of it, I stopped. And waited. But the crows wouldn't come to the salmon. They flew over it, and round it, but they wouldn't come down.

So I drove the car ten yards farther away, and waited again. But they wouldn't come down. They were now thoroughly suspicious, and nothing short of my disappearance would satisfy them.

I toyed with the notion of taking the fish with me as a bait for use elsewhere, then decided against it. It came to me that it wouldn't look so good if somebody happened along as I was putting the fish in the car—the police for instance. I could hear myself saying I'd found the fish by the wayside.

However, on the way back, I made up my mind to have it. But someone else had beaten me to it.

Courage of a Cushat

I came along the woodside on a wet-eyed morning of now and again sunbursts, spitting out flies, and sending up cushats from the cornfield endrigg where they had been breasting down the stalks to get at the tap-pickle. Squeakers somewhere were going to have their gebbies crammed with green ickers by mid-morning . . .

A fox came out of the corn, beshacht and alaigert with weet, shook the water from his fur, scratched an ear, then saw me and sailed over the old drystane dyke as effortlessly as a swallow breasting the hedge.

His line was mine, so I climbed over in his tracks, and pushed through the pines to the big ridge that backboned the wood. At the big badger sett I sniffed holes, and got the heavy musk smell in one that told me there was a fox in it. I guessed it would be my reynard from the cornfield.

I walked along the ridge to a hiding place I had below a big tree, inserted myself, and unslung the binoculars. The seat gave me a bird's-eye view of the wood, for I was placed more than tree height above the ground. I swung the glasses and found my cushat on her nest in a pine. She had two squeakers and this was about their feeding time.

Although she was in a dark place, a jet of sunlight was picking her out, like a spotlight on stage, so she was set up for me like a colour slide on a dark screen.

The flies gave me a lot of attention, whining and rasping according to their breed, but I could safely swipe them for the cushat was forty yards away, and kenning nothing about my presence. A wren flew over, and pitched above me for a look, but the look told her nothing and she went about her business round my ears.

Even a carrion crow missed me when he landed on a branch in the next tree to bow on his perch and fan his tail, and utter harsh comment on the world in general. He went away without seeing me.

Still the cushat's mate hadn't appeared, and I was beginning to wonder if he'd already been and I was wasting my time. I had a roebuck visitor while I was wondering this. He came out of the pines, and up the ridge to my right where he was sure to get my wind. He got it and *boughed* dog-like and bounded away, but the cushat on the nest reacted not at all.

Then I saw the second cushat flying in. He pitched near the nest, big-bellied, in shadow, and the bird on the nest rose. She flew to him and he flew to the nest, where the squeakers were already on their feet awaiting him. He opened up, and they delved in, ramming their beaks down his throat. The other bird flew away, presumably to reload in the cornfield. I kept my glasses on the feeder.

He had finished, and was standing aside, his breast bedraggled with spittle, when I saw what I took to be a squirrel on a lower branch. But the beast was red and there are no red squirrels in the wood. The cushat saw the climber too and flew off with noisy slap of wings. But not far . . .

At least I didn't think so, for in a moment a bird came back and pitched close to the climber who turned out to be a stoat. The bird that came back could have been the one that had been sitting: I had no way of knowing. Anyway a cushat came to challenge.

The stoat climbed, almost as nimbly as a squirrel, and when he reached the nest the cushat flew to the edge and tried to clout him with a wing. She clouted him with a wing. But he grabbed at her and she swung away, down and round, and back, but this time only to the end of a branch.

The stoat grabbed a squeaker. He followed it down, and I lost sight of him long before he reached the ground.

I went crashing down the ridge and over to the nesting tree. The cushat crashed from the branch and disappeared. All I could find on the ground were a few bits of fluff and golden thread down.

Blackcock's Early-Morning Tournament

Half-past one in the morning, with the sky overcast and the darkness brooding on the hill. The air is almost windless, but chill. No bird calls. The month is May, and there is a threat of frost.

Leaping shapes cross the forest road in the car headlights. They are deer—hinds, calves, and knobbers, and a few staggies. They run uphill and are swallowed in the gloom.

I stop the car and cut the lights. This is the threshold of the deer forest—the home of the raven, the mountain fox, and the eagle.

Two of us leave the car and plod uphill, feet splashing in ditches and crackling in brittle brushwood. The brushwood, layered to the far top of the hill, is all that remains of the great forest felled in the war.

A quarter of a mile above the forest road is a green slope, dry and close-bitten, almost clear of brushwood. This is the lekking ground of the blackcock.

I go into my hide of sacking and brushwood. I pull in my gear and arrange my seat. My companion pins me in tightly, waits for my O.K., then leaves for the car. The time is 2.15 a.m. British Summer Time. It is now very cold. I pull a scarf round my ears and wait.

The first voice I hear that morning is the cuckoo's. Then I hear the throaty chuckling of a red grouse. At 2.45a.m. there is a flurry of wings and the blackcock arrive. They are barely visible in the darkness.

In a few moments they have sorted themselves out—each to his own stance—and the singing begins.

'Roo-roo-ruckoo-coo-roo-cucu-roo-cuck-oo-roo.'

The sound is loud, clear, and musical, like the 'ruckety-cooing' of dovecot pigeons, and can be heard that morning down at the car, which is a quarter of a mile away.

The birds soon become visible as moving blurs in the gloaming. They sing on their stances, with heads down and chests inflated like pouter pigeons, and they leap into the air, crowing. The crow sounds to me like *'co-whae'*. It may not sound like that to anybody else.

From time to time one bird struts from his own stance, with lyre tail spread, and advances into the territory of the next. The birds meet face to face. They advance, retreat, and parade sideways, keeping even distance apart. Then they may leap to the clinch, and knock the feathers out of each other.

During an actual clash the birds utter a variety of sounds—throaty, stuttering and cat-like. I would never attempt to put these cries into syllables. Sometimes three birds rush at each other, but usually one stands aside and allows the other two to clash.

On some mornings there is much challenging but little actual fighting, but real clashes are much commoner than the text-books lead one to believe.

Blackcock will stay on the Lek from anywhere around three in the morning till some time after sunrise, but activity tends to wane as the light improves. At the end of the session the birds walk off the Lek, feeding as they go.

The Lek is regularly visited in the evening, and there is often quite a lot of displaying, but my experience has been that clashes are rare at these sessions. Other observers have found the reverse to be true.

I think there is a great deal of variation in the behaviour of blackcock from place to place, according to weather, time of year, and the number of grey hens on the ground.

Greyhens come to the Lek in the morning, and are mated there. I have never known a greyhen come to an evening gathering. The arrival of a hen is often the signal for an increase in song and activity. The birds leap and crow, and take short flights all over the place.

The display of a cock to a hen (sexual display) is quite different from the cock birds' display towards each other (aggressive display).

The excitement following on the arrival of a hen lasts for only a few seconds. The males rise and touch down, rise and touch down and look like a lot of pinioned birds trying to escape.

An odd thing about some groups of blackcock is that they will fly from the Lek to another patch of ground a few hundred yards away and display there for perhaps fifteen minutes or more. Then they will return to the Lek and begin all over again.

Sometimes you will find a single blackcock 'coo-rooing' and crowing on his own, far from any Lek, and sometimes you will find one displaying in a tree.

The size of a Lek varies according to the number of birds using it, as does the size of individual territories. All the Leks I have seen have been on grassy ground, never on heather, though some have had heather in patches. There may quite well be Leks on heather, but I have never heard of them.

On his own stance, the blackcock sings and crows; he breaks into other territories to challenge and fight. After a clash he parades back to his stance and begins singing again or crowing. A bird, after parading from his own territory, may fight in two or three others before coming back.

By and large, however, and in spite of the variations, the pattern is set. The birds gather at a specific place and take up territories on it. They challenge and fight, and the greyhens are mated there in the mornings. The males leave after sunrise.

The evening performances are usually of shorter duration, but I know of cases where the birds have been on the Lek all night.

Off the Lek, it is possible to keep track of the company of birds right through the day. They feed together and show no indication of rivalry, but once in a while a bird will crow or croon for no apparent reason. If blackcock fight off the Lek I have never seen them do it.

Recent observations seem to indicate that much has yet to be discovered about the behaviour of blackcock off the Lek in their relations with greyhens.

The bird is 'officially' described as polygamous. I have always inclined to the view that he is, in fact, promiscuous.

This is a view shared by many other observers, and I have no kind of evidence of blackcocks leaving a Lek with a harem or associating with a harem anywhere. I have seen a single blackcock feeding with a single hen in the breeding season, but nothing beyond that.

Kenneth Richmond, on the other hand, who, like myself, has spent a lot of time this year studying blackcock, has seen blackcock (odd ones) behaving as if they were, in fact, monogamous, but I do not know the extent of this. In any case it is a completely new piece of evidence on the behaviour of blackcock.

I am wondering at this date if there may not be a connection between this apparent monogamy and a scarcity of greyhens in a neighbourhood. One almost inevitably thinks of the cock pheasant in this context, a polygamous bird who shares the nesting duties when scarcity of hens forces monogamy upon him. It is a thought, anyway.

My own records of observations on blackcock, going back more than eighteen years, contain nothing like this at all. Brian Vesey-Fitzgerald, the author of *British Game,* is definitely of the opinion that the blackcock has nothing whatever to do with hens off the Lek.

Yet one has to explain how it is that hen capercaillies, colonising new ground, sometimes interbreed with blackcock. Either the hen caper has to visit a blackcock Lek (and I know of no record of this) or the blackcock must mate with her off the Lek.

If he will mate with a hen caper off the Lek, why not with a greyhen?

The Grand Duc

I've looked all the British breeding owls in the face at a distance of a few feet, in daylight and darkness, at all seasons, and I must say that the long-eared is the one that makes the greatest impact with his stare, whether he's facing the way he's fronting, or looking back over his tail.

All owls stare, of course. You might even say they all look surprised. And they can't help it. But the stare of the long-eared owl is something special. His brilliant orange eyes, with their jet pupils, glitter like jewels, and when his ear tufts come up, erect, after he has pitched, he is an owl on his own.

Most years I get to grips with a pair of long-eared owls, and in most of the most I get up beside them, perched on an enclosed platform anything from five to twelve feet away. At one time an enclosed seat on a branch did me; nowadays I make me a platform with a proper seat. Ach: it's the ageing . . .

For a number of years I was with the long-eared owls every nesting season, and the familiarity made me unreceptive when an old friend of mine reported to me in the terrible winter of 1946-47 that he had seen a big owl, 'a byornar big owl, wi horns like Auld Nick,' in a tree near his cottage by the loch. And he said to me: 'Looks like wan of them eagle owls ye see pictures o.'

And I said to him, uncharitably: 'Ach it wid be wan o the long-eareds, lookin big in the minlicht.'

To which the old man replied: 'Weel, if it's a long-eared hoolet somebody must hae blawed it up wi a bicycle pump.' But it was no long-eared. It was an eagle owl, the Grand Duc of the French, as I very soon realised and established. And it was the second I had seen in Scotland. The last one I had seen was in Spain in 1929.

If the old man's choice of a bicycle pump as the alleged inflator of a long-eared owl strikes you as strange, let me explain that all he had ever owned was a bicycle, and he belonged to a generation that still called a bus a charabanc, which he did to his dying day.

I spent every night for a week out of bed looking for the Grand Duc, after the first night's sighting, when he was perched in a tall tree, backlit by the moon, looking like a beer keg, and indeed with horns like the auld black wan himself. It was a breathtaking sight, made all the more so by the freezing moon, and icy stars and iron-clad snow.

And we both saw more of him than we ever hoped, because one morning, in what the old man always referred to as the sma' hoors, the big owl fankled himself in the wire netting covering the henhouse window, and raised such a racket that he woke the house, and I had to come along and disentangle him. And the old man forgave the bird for trying to get at his geese and bantams.

'He's a right royal Mahoun,' I said in a poetic moment, and the old man, excited, agreed. He was a remarkable old man in many ways; and once, when some shooters killed five of his tame geese, and said they thought the birds were wild ones, he shrivelled them with his tongue, charged them for the bodies, and kept them all to eat himself. Well, with a lot of help from his son, and others . . .

Well syne the Grand Duc came along shouting the odds ae day near haun darkening, and swooped at, and struck, a Chinese goose, and damaged her somewhat. But the old man heard the cufuffle and tushkarue and gaed oot and rescued the goose, and carried it indoors for the first aid. And next day the Grand Duc wasn't to be seen. Nor the next. He had gone.

And the old boy said to me: 'He'd best be awa, afore I get ma wild up. I widna like to see His Royal Highness endin up wi his belly stuffed wi sawdust . . .'

And I said to him: 'Come on, you widna kill him.' And he said: 'Aye wid A. Weel, likely I widna. I like tae see things gaun aboot.'

He was a wee, auld and gey man.

The Old Place

My mind often harks back to the old place, which I had to leave in 1958, for there happened almost everything that could happen in any single household in the space of nine years and a bit.

There, many personalities in fur or feathers made their entrances and their exits (often happy exits I am glad to recall). There, at one time or another, for one reason or another, many of them contemporaneously, dwelt fox and roe and badger, stoat and wildcat, eagle and raven and hoodie and black crow and magpie. A polecat, too, there was who, for some undiscovered reason, we named Marmaduke; there was even a coypu. And the hawks and the owls were legion.

I can remember them all—the lordly ones and the lowly—and the manner of their coming. They came mostly young, but sometimes grown up: wounded, broken, maimed, deserted, or handed in by boys who had found them 'lost', which is just another way of saying that they had picked up an unattended youngster and lost it. Tiring of their charges, they would go to the police, who passed them on to me.

We nursed many such wildlings, my wife and I; and we learned much from them. The following story (one of many I have yet to write) is about two of those characters and the domestic animals with whom they lived.

The animal story is always a one-performance drama or comedy; there are no rehearsals and no repeats, unless the events are witnessed and recorded by man. What follows is given exactly as it happened—without plan, gimmick or rearrangement. Any interpolations are descriptive or explanatory, and nothing more: they appear in brackets . . .

Dramatis Personae:

Nip—a Border terrier

Fencer—a Labrador retriever

Tiger—a tom cat (entire)

Cream Puff—a tom cat (neuter and a teuchter)

Jock—a hoodie crow (also a teuchter)

Whisky—a magpie (who came to me in a box marked 'Whisky—magpie— dangerous', causing no end of puzzlement to the railway staff)

Scene:

My back yard on a blue and gold autumn day, after the corn has been cut but before the departure of the swallows.

Audience;

My wife, some friends, and myself.

Plot:

None. The characters made it up as they went along (the important thing being that *they* made it up).

There was I at my own back door, minding my own business and thinking what a fine day it was with the sky blue and the stubbles yellow in the sun—even although two jets were messing up the blue with their vapour trails—when who should fly on to my shoulder but Jock, my big boisterous hoodie crow.

Jock *kwarps* in my ear in that deep, harsh voice of his. He is hungry; or acquisitive; or both. He is also Highland (a real teuchter of a bird, and ass fly ass a fox, and man I would not be putting it past him to haf written the whole script himself).

I call to my wife that the teuchter is looking for something to eat, or maybe just something to hide. (Och, it iss the provident one he iss, that bird, knowing that the rainy day will come whateffer).

My wife brings out a piece of Swiss roll, and I give it to Jock. It coils out as he snatches it from me, and he chortles his thanks, or whatever it is he chortles when I give him something to eat (that gets round any charge of anthropomorphism).

(If I'd known what that piece of Swiss roll was going to lead to I'd have brought it out much earlier, and the cine camera too!)

Jock flies away with his roll, dropping a piece from the end of the snake on the way, and pitches on the ground at the corner of the meal store. There he finds a shallow hole in the ground. He lays his tit-bit carefully in the hole, tamps it down with a wicked, suspicious look in his eye, then looks for something to cover it over with.

He gathers short pieces of stick, some straws, and a small pebble, carrying them all to his cache in his beak. He criss-crosses the sticks and straws, and puts the pebble on top. He then finds a few old leaves and a small bit of slate, and lids the cache with them. He steps back and reviews his handiwork, chortling his satisfaction (or something).

(This hiding of food is a habit with the hoodie in the wild, and is inborn in Jock. Indeed, he remembers all his hiding places and can always find a meal when he wants one).

At this point enters Whisky, the magpie. (As I've said, Whisky was from out of town. He arrived nervy and suspicious, but soon responded to the treatment and settled in. His favourite ploys were stealing clothes pegs and pulling cats' tails).

Whisky sails down from the top of one of the big elms—a flash of blue and white—and snatches up the fragment of Swiss roll which Jock has dropped from the main body of the snake. With a flick of wings, he is up and away, taking the morsel with him. But Jock comes from that part of Gaeldom where men are men and hoodies one jump ahead of them; so he knows what Whisky is up to. He proves it by taking off after him and giving chase. He follows Whisky over the barn roof, but the magpie is nippy and cunning and escapes with the prize.

Jock returns at last and sits on the barn roof, swearing (profanely I should think) in his throat. He long-necks, looking for Whisky, or something (in the mood, I imagine, for venting his spleen on somebody or something), and sees Fencer.

(Fencer, my Labrador, is a delightful, sweet-tempered dog who is the friend of every living thing, even at meal time, and who has 'mothered' pheasant chicks, woodpigeons, roe deer, foxes, kittens, and Nip the terrier).

Fencer enters the arena, and plods across the yard. Jock sits up there watching him with an unfriendly eye. Fencer is almost blind but he has a nose like a foxhound. He smells Jock's hidden treasure and moves in to mooch—and gets a terrible shock!

Jock flies down from the roof, hop-skips across the yard, and pulls the dog's tail viciously, whereupon Fencer, good-natured always, makes a strategic withdrawal and lies down, with muzzle on forepaws, drooling. (He can drool like a leak in a roof on a wet night).

Jock stands beside his cache, but keeps his eye on Fencer. I am thinking he is thinking that Fencer is thinking about lying there innocently until the cache is unguarded. I become certain of this when the bird rushes suddenly at the dog and stabs him smartly right on the tip of his wet nose. The big dog yelps once, walks round the yard, then comes back and lies down in the same place. By then Jock is back standing over his cache again.

But the grey-backed teuchter isn't happy at all. (This is inferential; I have no way of proving it). He pulls away the coverings from his cache, retrieves his Swiss roll, swaggers round the corner of the meal store, and finds another hole near the old, fogyirdit drystane dyke. He puts his roll in this hole, comes back for all his coverings, and places them as before over the new cache. Then he mounts guard, still watching Fencer.

At this point Cream Puff enters. (Cream Puff is handsome and Highland, from the hill overlooking Killiecrankie. He is all-over cream-coloured, even the claws of him, and his eyes are hazel. The size of him and his teeth, and the set of his ears and the ringing of his tail, tell their own story of the night the wildcat walked the low ground to tryst with the hearth-bound cats of man). Cream Puff panthers superciliously past Fencer. Fencer sniffs, identifying with his nose. Cream Puff mews his greeting (he is a great pal of the old dog) and passes on. He isn't going anywhere that I know of.

71

Close on the heels of Cream Puff comes Tiger Tom, moon-faced and indolent, good-natured, a moocher. A commensal parasite on Cream Puff, he has never heard of symbiosis. He is grey, tiger-striped, fat as butter and full of bonhomie. He lies down near Fencer, rolls on his back, and yawns prodigiously.

Fencer raises his muzzle from his ebony paws and sniffs. This time the sniff is for Nip, the terrier, who has just come out of the byre to stretch her legs after mothering her growing puppies. (Nip is from the Glen of Weeping, from the shadow of Buachaille Etive. She is the colour of ripe wheat, short-haired and dark-muzzled, with a face like a vampire bat. She is the great rat killer, and once spent eleven days in a fox-hole, which is something else altogether).

The positions of the characters now are as follows—Jock, the hoodie, is standing by his cache, with the feathers of his head on end, watching. Cream Puff is lying up on the cornfield edge, atop the dyke, watching too (for mouse or vole which, apparently, he is able to catch to order). Nip is busy nuzzling Tiger, who is liking it and purring. Fencer is just lying there, keeping track of the lot of them with his nose. Whisky, the magpie, is sitting on a barrow, bright-eyed, also watching. Everybody is watching everybody else.

Cream Puff suddenly flicks a paw into the drystane dyke, and when he withdraws it there is a shrew sticking to the end of it. (He doesn't like shrews, except to play with, which doesn't do the shrews any good). He lies on his back, and juggles with the shrew, hooking it from one set of fish-hooks to the other. He is enjoying himself (I think).

But the shrew is far from dead, and Cream Puff is careless as only a well-fed, confident cat can be. The shrew breaks away . . . and runs . . . right into the face of Tiger Tom! And what d'you think? Tiger Tom being on his back at the time, is getting an upside down view of things. He leaps up startled, with his tail like a bottle brush, and whisks aside. The shrew scurries on.

But not away. Jock, who has been watching all the time (as we have been watching), swoops at it. He snatches it up, squeezes it in his pick-axe beak, then lays it down to look at it. (Maybe he iss not liking the taste of it, and iss for giving it back to the cat).

That line of argument has to be thrown out of the window when Cream Puff comes stalking up looking for his shrew, for Jock faces him, with hackles raised, ready to dispute with him. (You have to understand that this hoodie fears no cat that was ever born).

While the two of them are disputing—the hoodie chuckling in his throat and looking fierce—the cat trying to save face (anthropomorphic?) by pretending he is looking for anything but a shrew—Whisky flashes down, and up. And he's away with the shrew!

At this point we are all laughing outright, and the thought occurs to me that what we are looking upon as a comedy is a matter of deadly earnest and import for the characters concerned. It is, in fact, a serious drama.

With the magpie's theft Jock forgets about the whiskers he was going to be pulling, and takes off after the thief—right over the barn roof. (This has happened already, as you may remember). As Whisky tops the gable he collides with a swallow which is hawking flies. Swallow and magpie lose their grip on the air for a moment, catch themselves up before they crash on the tiles, and fly their separate ways. But the shrew has fallen on the roof, and is sliding down the tiles. It ends up in the guttering, but it lies there, for neither hoodie nor magpie has noticed. So it is, for all practical purposes, lost.

It stays lost. Jock looks all over the roof for it, but can't find it. The magpie is up in one of the elms again, out of reach of the crow's wrath. Jock flies down to the yard in a great rage. (We aren't guessing; we can tell by the struts, and the feathers, and the swears of him). He walks across and grabs Nip by the tail! (That's right, we say. Take your spite out on Nip).

In the past, Nip has taken a lot of that kind of thing from the big grey crow. She isn't taking any today. She chops at his face—a snap that would have left him owl-faced for life if it had succeeded in closing on the target—and chases him across the yard. But Jock, skipping with wings spread wide, always manages to keep that little bit ahead of her. As soon as Nip breaks off the chase, which means she has called truce, the crow is back straddling his cache again, chortling in his throat.

Presently, he collects more sticks and lays them carefully over his treasure; the covering now makes quite a noticeable heap. Then he notices that Cream Puff has hooked a mouse from the dyke (proper Apodemus this time). Right away he swaggers over to the cat, and stands by. He kens . . .

And, sure enough, Cream Puff begins to play with the Apodemus. And he is careless again, not being hungry. The mouse, inevitably, tries a break. Cream Puff squirms round to scoop in with a paw, but Jock is faster. He snatches the mouse in his beak, dances back, stops, and lays it down.

He stands over the prey, challenging. Cream Puff arches his back, mews, then stalks away wearing a look of utter boredom (anthropomorphic?) Jock kills the mouse, then flies with it to the porch roof, where he proceeds to stuff it under the tiles overlapping the guttering.

While he is there, Cream Puff (I told you he could catch mice to order) catches another mouse in the dyke. But this time he panthers into the stubble with it. He isn't subsidising any more crows. Suddenly, Whisky appears on the store roof. He sees the crow on the porch roof. He sweeps down to the hole in the ground, tosses aside the twigs and other coverings, and grabs the length of Swiss roll.

He is in the air with it before Jock notices. Having noticed, he gives chase, and they are away over the barn roof for a third time. This time there is no swallow to collide with, and they disappear, so I walk round to the stackyard to find out what is happening.

Jock and Whisky are sitting on fence posts, ten feet apart, each with a piece of roll held under a black-taloned foot. Did they share it; or did the crow steal half of it back; or did he retrieve a piece dropped by the magpie? (That we shall never know).

When I return to the yard, Fencer is sniffing at the hole where the roll used to be, with saliva dripping from his flews like a rainstorm. My wife pats him. (Vigil unrewarded). I feel sorry for him and go indoors and fetch him a piece of roll, which he bolts at once without ever tasting it.

While he is bolting the morsel, he has a crow and a magpie among his feet hoping for crumbs (they do that when he is eating at any time). Nip is watching all three, obviously feeling huffy because she has been overlooked, so my wife, whose dog she really is, goes in and brings her a piece of roll too. And that, you would think, would be that.

But as she turns to carry her morsel away (a habit of hers) the grey-back is in like a flash and has snipped off the bit that is hanging from the side of her mouth—clean and neat and quick. And he is away before the peppery little bitch can growl.

73

Such are the bare facts of what has become known, among those who saw it, as 'yon terrible cairry on.' The ultra-scientific critic, however unimaginative, has to accept it because it is a bare description of a piece of observed animal behaviour, with many witnesses. Yet I am sure that, if I had written this simply as a 'story', or part of a 'story', it would have given rise to much head-shaking and resigned talk about nature fictionists, the deprecatory term applied to writers of the animal story all over the world by certain 'scientists' to whom animals are nothing more than parcels of conditioned reflexes, and the story-tellers the fountainhead of anthropomorphism.

The great novelists of the past (those one was made to read at school) often finished their stories with a brief account of what happened to the main characters after the actual story had been told. I am in a position to do the same, or nearly the same, for the characters who were responsible for the terrible cairry on.

Nip, I am pleased to record, we still have with us. Cream Puff too. But Tiger Tom has gone where many country cats eventually go; he was killed by a fox in the running season—January. Big Fencer is dead; painlessly destroyed after he became totally blind and almost lost his power of scent, on which he depended.

Jock, the hoodie, returned to the wild, and may be back in the Highlands, or dead, for all I know now. Before he quit he developed the habit of eating the breast and brains of every crow and magpie shot by the local farmers. To the end he loved pulling cat's tails and I am wondering, if he is still to the fore, if he has ever tried the same thing with a true *Grampia*.

Whisky, the magpie, drifted off, too, We saw him every other day, then once in a while, then not at all. I have no reason to believe that anything happened to him at the time. He may be dead now. If he is dead, he can't have been picked up (unless he was shot, of course), for he had a ring on his leg, and I am sure the ring would have found its way to me, complete with bird.

In ending this account, I must record one other thing about the grey-backed crow. He was a potential killer, for he attacked all strange women on sight, with unbelievable savagery, and would have caused many a serious injury, or worse, if he had not been dragged off by my wife or myself.

Keeping Snakes Alive

He was curled below the boulder, somnolent in the gentle heat of the sun, slack-wound not spring-coiled, relaxing in the wide-eyed, unblinking day-dreaming of the snake, and I was sure he was not seeing me from behind his perspex guard because the head did not turn to bring a look at me when I moved my hand gently in front of him.

I came down close to him, until my face was less than a foot from his, and said: 'Hello.' Of course he didn't hear me, and he gave a slight quiver, like a person stirring in sleep. Then I lightly touched one curve of his body, and he came slowly to, and stared at me with those expresssionless snake eyes, and I kept my face where it was to find out what he would do.

He tightened his coils and raised his flat head a little, but he made no move to retreat and certainly none to strike. I spoke to him again, without thought for the waste of words.

'It's like this,' I said to him. 'I'm going to lift you and move you, and you're going somewhere else, because if you stay in this spot and move at the wrong moment, you're going to be clobbered for sure by people coming up behind me. Dost thou not ken, foul fiend, that everybody's heels are sore enough without the extra of thy strike?'

He was a placid soul, and did nothing except stare, and he did that because there was nothing else he could do. I inched in and ran a finger along the outer coil, forcing him into a gentle whirlpool of movement until his tail was clear. And, once it was, I took it between thumb and forefinger and gently hoisted him at arm's length, rising to my height as I did so.

I twirled him round and all he did was reach up a little with his head, his forked tongue flickering the while. Then I put him down and he glided away, but only for a few yards, and there he was lightly coiled again awaiting further speech with me, and I gave it to him straight, his deafness notwithstanding. 'Now look here, Mr. Adder,' I told him, 'you can't stay on this path. There's feet only 200yds. away and they'll soon be here. So unless you want your head flattened more than it is, and your curves broken into angles, you'd better go.'

But I couldn't persuade him to move. As I crept round him he followed with his head, turning on his own coils. So I picked him up again, and took out my binoculars, and put him in my binoculars case, and said 'Good afternoon' to the people who arrived, and watched them go; then I moved up the hill and spilled out my snake in a dry hollow, where he sat and looked at me. He was bucking for oblivion that one.

Three days later I was on the same hill with a mixed party of schoolchildren, and I took them by the route where I had met him. So I warned the youngsters to keep a look-out for adders, and to sing out if they saw one.

A small boy, the smallest of the party, found a snake almost at once, but it was not my friend of three days before. This one was a big female, dark in colour, and she hissed like a steam pipe when I picked her up. The whole party crowded round, interested, not afraid, although they had all been brought up to abhor adders.

The snake quietened down soon—adders always do despite their explosive nervousness at first handling—and before long she had been handled by every youngster in the party. They were full of their prowess when their parents came to visit later in the week, and the snake had grown a little in the interval (as fish do) but I had the feeling that no adder would ever be in danger from any one of them at any time. They knew that adders had more to fear from them than they had from adders.

One adder more or less is of no account. The reflex act of killing them on sight is something else. D.H. Lawrence realised that when he reacted to the snake that came to his water trough, and called it pettiness.

Stingless Stinger

There I was on the riverbank, overlooking a dark pool, with ferns at my back, bell heather at my side, and tormentil between my toes. The water in the pool was slack and deep, and in the dark depths I could see its false bottom of piled beer cans, the tourists' contribution to a beautiful river.

In midstream the water frothed and foamed round and over boulders, discoloured, like cafe-au-lait, but through the smother spouted fat jets of amber, like whisky. Whisky and coffee with milk: that was the mixture.

The play of the sun on the water dazzled the eye; the trees flapped limply in the heat. Nothing was moving on the hill, but behind me, on the road, cars hissed and purred by at the rate of one a second, an invasion of driver ants into the once remote haunts of the eagle and the wildcat, the peregrine and the raven.

I had been part of the reverse stream for several hours before opting out to sit by the river; in four hours I had covered about sixty miles, and I had been trying hard to make distance.

The dragonfly appeared suddenly and pitched on a lichened rock where its feet had a sure grip. It clung, and its quivering ceased, and it basked, and a thousand darts of light sklented from its gauzy, fine-veined wings.

It moved slightly, and there was a dry rustle from its wings, in which the sun kindled many-coloured flames. It was a beautifully fearsome harmless thing, a predator with jaws as relatively powerful as a hyena's, a powerful hunter in gem-like mail, born of the sun after long seasons in the water and with a life expectancy of a moon cycle. If it was lucky . . .

I sat there and watched it until it flew off on trembling wings, rustling as it passed close to my ear. Then it darted across the boil of foam, hovering and changing direction until it was lost on the far bank.

One doesn't see so many dragonflies nowadays, mainly because there are not so many habitats left that suit them. When I was a boy they seemed to be everywhere at this time of year, and it was a common thing to meet people with a dragonfly newly killed, pinned on their chest as a brooch.

Some folks called the dragonfly the horse stinger; my generation used the name stinging adder, although, of course, it could no more sting us than we could it. A common name, but not one I have ever heard used in my part of the world, is devil's darning needle. One has to admit that, despite its beautiful buskins, it hasn't got a pretty face.

When we were boys, trying to catch dragonflies near a pond (which is the place to seek them), we used to lay down a white handkerchief to attract them. It always seemed to work, but I have never been sure whether this was chance, or whether the white square really acted as a lure.

It takes a long time to make a big dragonfly—a couple of years more or less. The larva, like the final flying predator, is a predator—a fierce, insatiable one that eats tadpoles, small sticklebacks and other waterlife. It is an awesome creature, with hinged jaws that can shoot out like arms to grab the prey, reminding one of one of those modern earth diggers that bite great chunks of sod and swallow it for transporting.

After the larval stage comes the nymph, a tough shell with the fly inside. One day the nymph climbs up the stem of a water plant and sits there waiting, literally to burst. In the end the skin bursts, and the fly comes out, head first then bends backwards to draw out its long body. Presently the fly dries, and quivers, and comes truly alive, and soon it takes the air to start its life as flying predator.

Where Our Birds of Prey Have Found Sanctuary

The larch forest is hushed, like a cathedral without footfall; through its green vault the sun shafts from blue windows of sky, touching with colour the rash of lichens on the trees. The moss carpet is deep plush, where a man's foot can tread cat-like. It is a cool, sweaty place, roe-haunted, where the badger walks at night among dark, mossed stumps in shadowy hollows, and the war whoop of the tawny owl can be heard as it hunts for voles on the fringe.

'*Kek-kek-kek-kek-kek-kek!*'

The chattering cry incises the quiet—sharp as a razor, unalarmed, challenging. And it had to be the she-devil herself, sharp as her voice, the sparrowhen: razor-faced, sharp-shinned, slim-taloned, with yellow jewel eyes. She flickered round the slim larch tops, jinking like a swallow, negotiating with almost bat-like precision the fuzzy strata of tufted larch twigs. Then she pitched on a top twig that bent with her weight and machine-gunned us with contumely.

She was an old resident, with six old nests high in larch trees to prove her claim, and a new one with four downy chicks 26ft. up in another. So a man could understand her 'What the hell?' and vocal fusillade.

Presently she turned down the volume, and swore under her breath. Until she saw the nesting tree, her tree, being climbed. Then she lifted away and circled, swearing loudly and profanely, but she made no other demonstration. Bold she was, as her kind are bold—so bold, indeed, that I have had one swear in my face when I poked it from a hide in the next tree—but not for her or her kind the physical clash, the contact assault. Cuss words are the extent of a sparrowhawk's protest. She leaves surgery to the tawny owl.

In the early days she brooded her chicks for long spells, and sometimes she slept over them, while I sat in the next tree and watched. The cock bird was the hunter, and usually she flew off to take the prey from him when he arrived near the nest. But on a few occasions, when she was off somewhere, he flew in to drop it in the nest.

A bantam compared with his mate, he was jimp and braw buskint, slate-grey of crown and wings, with breast barring of bright chestnut. Wee he was, but weel graithed as they say, with a glitter in his fierce eye and wearing a permanent frown.

Two thirds, or thereabouts, of the sparrowhawk's prey are birds, with the other third made up of insects and small mammals like voles, mice, young rats and baby rabbits. But I have seen a small weasel brought to a nest. The cock bird plucks the

77

prey before it is passed to the hen, who carries it to the nest in her feet.

A sparrowhawk's plucking place, or plucking stool as we used to call it, may be on an old stump or a rock or mound, but it is just as likely to be in an old nest, which I suspect was the case with the present pair, for there was scant evidence of feathers elsewhere, except for a few beside a mossed stump. The nest of the year is used as a plucking place for some time after the young birds have flown.

The sparrowhawk has had a remarkable history. Like most other birds of prey it was (and still is by the die-hards) classed as 'vermin'—that discredited, inadmissable word. When the change came with the Protection of Birds Act 1954 nothing changed for the sparrowhawk. It was outlawed as a delinquent and thrown to the wolves.

Then, when its numbers began to decline seriously (arguably influenced by the use of organochlorine pesticides) it was given special protection. So now the former 'vermin' hawk is classed with the elite—the eagles, harriers and the peregrine falcon. It has the legal status of the osprey.

But not everywhere. In too many places it is still given the salute of the twelve-bore, on or off the nest, and its eggs and young destroyed. In the State Forest it is safe, like all the other birds of prey, for the Forestry Commission's instructions to its officers at all levels are printed bold and clear.

In the forest there are also hen-harriers. Hen-harrier is a rather confusing name because it means speaking of the cock hen-harrier and the hen hen-harrier. Change is in order, and heath-harrier or moor-harrier would be better, though neither would exactly describe the birds in this forest, which nest in knee-high, waist-high, and even man-high plantings as a matter of course, and sometimes in timber as high as a house. Six pairs, widely scattered, nest in the forest, on the ground as all hen-harriers do, sometimes well hidden by ground cover, sometimes not at all.

Here is one nest in a blanket-size clearing, concealed by a routh of ferns under a man-high spruce. There are six young in the nest, two of the same size and age at the jay-winged stage, in helmets of down; the next three days younger and smaller; the others progressively smaller as you would expect from chicks hatched at three-day intervals.

The result is that the youngest chick, not much bigger than the egg it came out of, is a midget compared with the big two, more like their offspring than their brood brother or sister. There is no food in the nest; there has been none since the day before when it was also scant. Two inches of rain have fallen in the past two days. The chicks are hungry, with flat crops.

At long last the grey cock bird arrives over the open ground below the trees and the hen flies to him wailing her greeting. She sweeps up under him, flies with him closing, and then, foot to foot, he passed the prey to her in the air. At other times he drops it for her to snatch.

Within minutes the hen drops on to the nest with a woodcock chick in a foot. She stands on the edge with the chicks in a half circle in front of her, and begins to tear the prey for them. Five chicks receive a mouthful; the smallest is missed, crowded out. The big two receive a second portion. And that is the way of it. There is no 'fair' sharing. When food is short it is not spread around. The following morning the smallest chick was dead. The hen tore it up and fed it to the others.

That was last year. Voles were low on the ground, and the wet weather made for bad hunting. The harriers killed mainly birds but not enough of them. In all nests there was famine, and in all of them chicks were dying and being eaten. Only the intervention of the forester, supplying rabbits, saved the broods.

This season was different. Six pairs of hen-harriers hatched full clutches and reared full broods. Voles were up so much that a man could see one at two minute intervals on a walk over the hunting grounds.

Hen-harriers feed mainly on small mammals and birds, but they also take slow-worms and frogs, and, like the buzzard, will sometimes pillage the nests of ground-nesting birds, taking nestlings, or even the eggs. But this seems to me to be exceptional, as I believe it to be in the buzzard's case, and forced on the harriers by scarcity of more usual prey.

Much has been written about the hen-harrier's aggressive display towards intruders on its nesting ground, but this is far from being a habit with all of them, or even most of them. In the past two seasons I have met only one bird (out of eleven pairs) that made a real demonstration, hurtling in low and sweeping just over the intruder's head. Even this bird never pressed home an attack, as the nesting tawny owl often does.

The hen-harrier has spread its range and hefted its numbers remarkably in recent years, despite persecution on many grouse moors, where it is still shot, protection laws notwithstanding. In State Forests it has sanctuary.

Kestrels are as common as harriers in the forest (a statement that might raise eyebrows in view of the past status of the two birds) but not any commoner. This is the real mousing falcon, about two thirds of its food being voles and mice. Some kestrels kill a higher percentage than that, others a lower; it depends on where they live.

A sixth of the bird's food can be insects injurious to man's interests. It also takes frogs, young rats and baby rabbits, and can handle small weasels. Like the short-eared owl it will build up its numbers in forest areas during a period of vole abundance, and then its food is mainly voles.

Unlike the sparrowhawk and the hen-harrier the kestrel does not build a nest. On crags it will lay its eggs in a simple scrape on a ledge, usually with some shelter from blaeberry, fern, bluebells or other vegetation. In trees it lays in the old nests of crows, less often in those of raven, sparrowhawk or buzzard. Six eggs are quite usual in some places, but clutches may number four or five. Young kestrels, like young sparrowhawks or harriers, defend themselves by lying on their backs and presenting their talons to the enemy. But you'll get far worse pricks pruning roses.

The buzzard is a big predator, well known for its mewing cry and its habit of gliding and soaring. Although looked upon as something of a duffer and a lazybones, the buzzard should not be underestimated. It can take birds like grouse or pheasant when it has to, or when it wants to, but does so only exceptionally.

More usually it sticks to rabbits, voles, mice, and mutton and deer carrion. Like other birds of prey it takes what is easiest to find and capture, and that will include the sick and the dying. Pressures prompt. On the island of Eigg I knew a pair that habitually raided the nests of small birds, but that was following myxomatosis.

Buzzards are common in the forest, where they nest in old oak trees or on crags. Voles stand high on the list of prey, equalled or exceeded only by carrion, of which there is an abundance on the open hill in the form of dead sheep. It seems that, in sheep areas, the buzzard, like the eagle, depends largely on carrion for most of the year, but especially in winter. I have seen a nest so piled with sheep wool that one might have thought a beast had crashed to its death on the ledge.

Big brother, the golden eagle, has not done so well, despite the strictest secrecy on the part of the forestry staff. The bird hatched out two young but these were spirited away when they were a week or so old. The eyrie was on the perimeter of the forest area, far beyond planting and ditching operations, and not easy to keep and eye on.

The typical owl of the open moor and the young plantings is the short-eared, and there are several pairs. This year one bird laid nine eggs and hatched seven young; her hunting range held a big population of voles. Voles were her main prey, in fact the only prey we found on the nest.

The short-eared owl has long been noted for its habit of invading an area in strength during a vole plague, when these mammals may increase to several hundreds per acre. The owls lay big clutches and rear big broods. When the vole crash comes, that is, when their numbers drop back to normal, as they eventually do, the short-eared owls move out again. Their young, crawling on the ground, are eaten by foxes and other mammal predators, which then suffer a decline in numbers themselves. So the short-eared owl provides a good example of predator-prey relationships, both ways.

This bird is also notable for its distraction display when it has young, a habit more usually associated with game birds. The bird flutters across the ground to lure the intruder from her young. Some short-eared owls will even go through the ritual when flushed from eggs.

The other forest owl is the tawny, which is also thirled to voles. The tawny likes hollow trees to nest in, but will use the old nests of crow or sparrowhawk. It will even nest on the ground, or in disused rabbit burrows, but this is exceptional. This is the owl that hoots, and gives the ringing war-whoop: *'Kee-wick! Kee-wick! Kee-wick!'* It is also the owl that will make a strong, silent demonstration against an intruder near its nest at dusk or during darkness, striking hard with its talons.

It can hit a man's ear as hard as a fist, and a few are experts at ophthalmic and peripheral surgery. Not all of them are like that, but many are, so it is wise to be careful. Sometimes there is a warning of impending assault, when the bird clicks its beak loudly and sharply, but it will do this without assaulting and it will assault without doing it.

These are the birds of prey in one forest of the west. In other forests there are others. Some have the merlin, or the peregrine, and somewhere, sooner or later, one will get the osprey back. It has been seen in some parts. It is a safe prediction that all of them will be given living room; the Forestry Commission plays fair with its birds of prey, as it now does with most of its wildlife.

The Cleverest—and Laziest—of Dogs

I was looking at a photograph of him on Christmas Eve, taken just after the war, which showed him at his favourite occupation of doing nothing—unless you can call lying on the garden path, with one of my tame pigeons picking grain from under his nose, doing something.

I suppose if he'd been a man, we'd have called him a layabout. Certainly there was nothing he liked better than being unemployed.

Yet he was the most teachable puppy I ever had, and one of the brainiest gundogs anyone ever fired a shot over. He was also a good looker.

The only time he would walk a yard more than he had to was if there was a puddle of water in front of him. He would walk round it. Confronted by a burn six feet wide he would have walked round the parish to avoid crossing it.

He was one puppy you could depend on to sit, and stay seated, until called in. He would have sat for the afternoon at one end of a field while you sat at the other, and when he did come in to the recall he came as though he had fifty-sixers tied to his feet. He was as slow as treacle on a frosty day.

But he had been at the head of the queue when the brains were being handed out; no question about that. People used to say to me he was too brainy to see any future in working like a dog for a man.

Once he had shown he could do a thing, he had made his point, and saw no reason why he should go on doing it. He was the intellectual type—but he was the laziest dog I ever met.

If he had been just another handsome eejit, I would have forgiven him. But he had everything except the will to do.

He could find anything—dummy, dead rabbit, pheasant, or even my bunch of keys. He would range 200yds. downwind, never letting his feet run away with his nose, and find what he was looking for with an almost deadly certainty. Then, having found it, he could never see any sufficient reason why he should have to hump it back to me.

Mind you, he always brought it. You could depend on him to find and bring back, so long as you didn't mind his arriving like a slow goods train.

I tried to put some pace into him by sending him out with a bouncing black dog. A little competition, I thought, would do the trick.

But he went at his own pace, letting the wild black dog run all over the place while he worked down slowly on to whatever was lost and picked it up. Then he looked for somebody to carry it home for him.

He was as certain as tomorrow, as slow as the mills of God, as likeable as could be, and too lazy for words. If I had been one hundred years old, he would have been the perfect dog. And yet . . .

Many a time I put him on a line after two other dogs had given up, and sure as the coming of night he would find what they couldn't find and bring it back as though he hated the very taste of it.

For a dog who hated getting his feet wet, he swam tirelessly when there was an end product. And then would avoid every puddle after he had delivered the goods.

He just didn't like work, or exertion, or discomfort. Usually a clever, eager dog has to be slowed down. This one you couldn't speed up.

He never jumped the gun, or ran in, or chased after exciting scents. All he wanted was not to be asked to go. He knew it all, and he was never so content as when watching other dogs doing the work.

I was very fond of him.

The Mighty Atom

The terrier almost went off her head the day the whittret spat in her eye from a hole in the drystane dyke. Spat here doesn't mean expectorated; the spit was a cat's spit only cattier.

The whittret is a mighty atom—a bantam sable, a mink in miniature, a 2½ to 4oz. mustelid delivering a thousand volts, a firecracker in brown topcoat and white waistcoat who can put the fear of death in many a grown man.

This one was a fingerling, not much heavier than the tobacco in my pouch. A tiny tot full of tantrums and terrible tricks, who had to be a her because she was so small, who could have retreated but didn't, who stayed instead to spit in the terrier's eye and crackle cuss-words in her ear.

I could, as they say, have died laughing. My wife, at the kitchen window, refused to come outside, saying she could see it all from where she was. Maybe she could, but she couldn't stand the scent of battle.

The terrier sniff-sniff-sniffed along the dyke, gathering the whiffs of whittret. She poked in hard where whittret and smell joined company. Then: *yelp!* She drew back, shaking her head, swearing in her throat. Little Beelzebub had bitten her on the snout!

It was then I sat down to watch, pleased to do so without interfering unless the whittret broke cover and was in danger of being chopped. That I couldn't have. I was interested in whittret behaviour so long as the whittret lived to entertain again.

By now the terrier was boiling and grumbling like a volcano. She ran from this chink to that, while the whittret—white-breasted, with boot-button eyes and snout like a ripe blackberry—poked her face from holes above or to one side of her, snake-darting into view then snake-withdrawing again.

Once, indeed, the midget mustelid came out, head and shoulders, and peered down at the dog, who was up to the ears in a hole. When she looked up the whittret withdrew like a cut in a movie film. She was an electric spark, quicksilver, seemingly worked by elastic. The dog could hardly follow her. The dog was in a dither.

After a while I thought, well, I would see what I could do. I crawled on hands and knees, peering into the dyke cavities. I didn't see the whittret. But I became aware of her when she tissed and nickered in my face, and there she was, just inside, swearing at me and making threatening gestures.

I take off my hat to the clan mustela. Here was two ounces and a bit of whipcord and sinew spitting the odds at a thirteen stone man, and not a syllable of it empty talk. She meant every word of it.

Boot-high terriers, or horse-high men, were all the same to her. Blown up to wildcat size she would have faced a Bengal tiger. I have a liking for Davids who take on Goliaths.

After a bit I called the terrier off, and she came reluctantly. She had a lot of spite she wanted rid of, but I said coorse words to her, and she realised whittrets weren't in the book.

And then out on top of the dyke came Tom Thumb, doing everything but beat her chest. She sparked from one end to the other, upended, cart-wheeled, whirligigged, stood on her head, climbed down backways and up frontways, while the terrier boiled over into my ear, and swore terrier swear words which had never been sworn before. She was deaf to my whisper-snarls to hold her wheesht.

There are, of course, those who will say that the weasel's behaviour was a displacement activity. After her fright she had to act in some way, even a foolish or meaningless way. But I preferred to think, and still do, that she was playing 'Damn your dog, and I can be in another hole before she leaves your oxter anyway.'

I know hard-boiled, sophisticated zoologists who would agree with me, although maybe some of them wouldn't say such a say in the dreary decorum of a scientific paper. But who cares about scientific papers when there's a whittret dancing a war dance for an audience of a girning terrier and a laughing man?

Mind you, it isn't so funny when a whittret has a swing on the end of your finger, with its teeth through your nail. On the other hand, if your finger isn't where it oughtn't to be, no whittret will seek its acquaintance.

By now you will have gathered that I am a partisan of weasels. I am. I know all about their annoying, irritating, sometimes destructive ways.

But I can stand the annoyance, and the irritation, and the occasional destruction, considering them a reasonable trade for the sight of a quarter-gill featherweight who is always prepared to cut homo sapiens down to size, when the occasion arises.

Lore and Legend of the Hare

Every time I look a hare in the face I get the impression of a sly, crafty animal with a superior smile—a split-lipped, buck-toothed grin—and when you add the goat eyes and the big whiskers you get what the late Cheyney would have called an old-fashioned look. She leers. You wouldn't be surprised if she winked at you. Or giggled in your face. Or even if the hare lips made hare talk.

You'll see that I keep using the feminine gender. So do a lot of people. So did a lot of people. The ancients had a lot of trouble with the hare. They thought it a magical beast, a witch, a beast of ill omen. They seem to have had a lot of trouble making up their minds whether it was he, or she, or both. Or a changeling. Or even a hare.

The notion that hares could be of several sexes—and there's one to stop the traffic!—or of interchangeable sex, or hermaphrodite is very old. Sir Thomas Browne thought that buck hares could give birth to young. In Wales, where it is still called St. Monacella's lamb, it used to be thought that it changed sex by the month, being of one sex one month, and the other sex the next.

'The hare,' Pliny wrote 1900 years ago, 'is a hermaphrodite, and reproduces equally well without a male . . .'

Archelaus stated that a hare is as many years old as it has folds to the bowel.

The association of hares with witchcraft is one of great antiquity. Witches, it seems, were very fond of masquerading as hares. There's an old Scots legend about a hare being wounded, and lost, then a witch being found the next day dying of the exact wound. Away back in the late Twenties I heard, in the Cantabrian Mountains of Spain, that a witch, being wounded when masquerading as a hare, would have the exact wound when she was restored to witchness again.

Then there's the matter of superfoetation, which means becoming pregnant while pregnant. Pliny, in his classic *Natural History* Book V111, had something to say about superfoetation in the bawd or maukin (hare to you): 'The hare,' he wrote, 'is the only animal beside the shaggy-footed rabbit that practiceth superfoetation, rearing one leveret while at the same time carrying in the womb another clothed with hair, and another bald, and another still in embryo . . .'

Pliny's great work contained mountains of fancy, but he was right about superfoetation in the hare, even if he did let himself get carried away on the conveyor belt. Only in the past 100 years has research proved him right . . .

The hare has a far longer history as an animal of the chase than the fox. Xenophon, the Greek general and historian, was hunting hares on Mount Pholoe three centuries before Christ, and writing about them.

Edward Duke of York put the hare in a class by itself. In his book *The Master of Game* (which was a rehash of an earlier work by the Norman Count Gaston de Foix called *La Chasse*) he wrote: 'It is to be known that the hare is the king of all venery, for all blowing and the fair terms of hunting cometh of the seeking and the finding of the hare. For certain it is the most marvellous beast that is'.

It is doubtful if anyone would rate the hare as highly as that today. Beaglers would probably be inclined to agree. Greyhound owners probably wouldn't argue about it. But, though many people still like to eat hare, she isn't eaten to the extent she used to be.

There must be some Manx folklore about eating hares. My friend the late Brian Vesey-Fitzgerald told me that no true Manxman would eat hare, although he probably wouldn't admit why. One wonders if the taboo still lurks in the Manx psyche.

The rabbit, like the hare a lagomorph, is unlike the hare in that it practiceth not superfoetation. Instead it has a built-in birth control system known as re-absorption, which means that the young die in the uterus and are reabsorbed into the mother's tissues. Brambell discovered that this was common where rabbit numbers were high, when up to 60 per cent of pregnant does might reabsorb. Embryonic losses also occur in the hare.

The brown hare does not burrow. It favours rolling, wooded country, so is common on farmlands up to the limit of cultivation. It makes its sleeping nest (form or seat) in a grass tussock, among threshes, or in woodland. Its young are placed in such forms or nests. The leverets are born fully furred, and with their eyes open, unlike young rabbits, which are born blind and naked. Young hares are able to run about almost from birth.

In March, traditionally, the brown hare goes 'mad'. Then you will see hares running about on their trysting ground, chasing and boxing. Presently a buck goes off with a doe, following her all over the place. Once he has mated her he goes back to the tryst. The doe is on her own from the time she is mated, and rears the family herself. She will produce two or three litters a year; even a fourth in some seasons.

Because of its long hind legs the hare is at its best running uphill, and that's the way it usually goes. Downhill it will run a diagonal. It has its favourite exits from and entrances to the fields, so is easily poached if you know how. But that is another story.

Rabbits in the Rhubarb Patch

If you never heard of a rhubarb hide before you're about to hear of it now. It's a hide built of rhubarb, and a hide is a place one hides in so that sharp-eyed birds and mammals won't see one.

There was this doe rabbit I wanted to film with her young, at the entrance to her nesting burrow (or stop), which was in a big wild garden close by a great stand of shot rhubarb. The rhubarb was as tall as I, and in places taller, with wine-red and green stems as thick as my wrist and seedheads like a riot of cream astilbes. Viewed from rabbit level the stand was like a wee rain forest.

The rabbit was seeing the wee rain forest every day, hopping through it and hiding in it, and kenning every smell of it: so what better material could there be for a hide to film her from? Asks he of the prepositional ending . . .

But rabbits—like foxes and hares and badgers and deer and other siccan ground dwellers—have good noses for interpreting any message the wind sends along. So I had to get myself off the ground to make sure that the scent of me would pass over the doe rabbit's head. Table height would be fine at the distance from which I would be working. But where was a table?

Woman of the house said there was an old rickety table in her husband's workshop ahint the house, and an auld rickety chair forbye that I could have for a seat. And that was it. My son draped me in rhubarb, and from my rhubarb hide I worked on the two hours of dry weather that ensued before the young rabbits quit the nest.

Nowadays, or nowanights, one doesn't go to the length the late pioneering Kearton brothers went to when they were photographing a nesting bird. They thought nothing of getting inside a cow skin to photograph a peewit, which can be done as easily from a barrel, a tractor, a cloth hide or a tattie bogle.

I once photographed an oyster catcher from a tattie bogle, and why not? The bird was sitting on her eggs only a few yards away from it, and wasn't in the least surprised when one day it spoke to her.

As for the barrel. I spent many fruitful hours in one of the big, fat-bellied beer ones they used to have, watching water voles at the burn near where I had my rhubarb hide.

At the old mill near which I once lived I photographed wren and swallow from inside a 2cwt. hessian grain sack which the miller pulled over my head. It must have been the only time it held a man instead of oats or barley.

Maybe you can remember the old horse-drawn wagons they had for carrying in the corn stooks? Well there was this family of stoats that used to play back and forrit on it. I had the farmer build some stooks round me and syne the stoats were playing over me. And you should have heard the croons and croodles of them.

A man can now buy prefabricated hides of camouflaged cloth, with jointed poles, bracing wires, press-stud peepholes, and all the rest, but I don't use them. Except for a few basic items I like to live off the land. It's surprising how little flat ground there is around for such hides to sit on. And in trees, they're no use at all.

For settled work over a period, green cloth pinned to wire netting is the basic material. The netting prevents wind flap. For the rest, one takes what is lying to hand, the stuff the subject is familiar with; like flags or threshes for a waterhen, or

85

heather for a merlin nesting in heather.

I used to have my friends working for a couple of hours making kite tails of heather and woodrush when I was working on golden eagles. The dressing was put in position and the hide built inside. I've done the same thing with peregrine falcon and buzzard.

With some subjects you don't have to go to any such trouble. A box big enough to hold one is all that is needed for rooks; the trick is building it maybe fifty feet from the ground. Any sort of cover that can stand up to wind and weather is good enough for birds like the kestrel, long-eared owls and tawny owls, jackdaws and woodpeckers, cushats or sparrowhawks.

I've never been what one would call an enthusiastic photographer: in fact camera talk bores me to sleep. But when I'm taking photographs I always keep one thing in mind. The subject always comes first. In most cases it should never know one is there.

The Robin

Everybody knows the robin, red-breasted and buxom, like a potato on matchsticks—the garden Lazarus, the kitchen intruder, figurine on the unleaned-on garden spade, picker of worms from claggit boots, dancer in sugar basins, nester in coat pockets, old kettles and tin cans, the commonest fauna on Christmas cards, the national bird of Britain.

British robins are noted for their tameness and friendliness, characteristics not seen in their Continental kin. English literature, since Chaucer's day, abounds with references to the robin and its tameness (it was Chaucer's 'tame ruddock'), and it is this tameness towards people that has made it everyone's friend, just as people's friendliness has made it tame. In the relationship between people and robins the Pax Britannica is virtually unique.

Despite Chaucer's tame ruddock, the earliest reference to the robin's tameness is a Scottish one. At Culross, in the sixth century, the holy Serf was feeding and gentling the robin that was one day to become immortal on Glasgow's coat-of-arms, the robin that 'was wont to receive its daily food from the hand of Serf, the servant of God, and in consequence of this had become familiar and at home with him. Sometimes it was even wont to rest upon his head, or face, or shoulder, or in his bosom, or to sit by his side as he prayed or read.'

The account is from Metcalfe's *Ancient Lives of the Scottish Saints,* a translation from *Vita Sancti Kentegerni,* written by the monk Jocelin of Furness in 1125. Serf was in the field as a robin tamer 1400 years before that other famous robin tamer Viscount Grey of Falloden, master of ellipsis, Foreign Secretary at the time of the Kaiser's war, whose memorial is the Edward Grey Institute of Field Ornithology at Oxford.

Serf was the teacher of St Kentigern, better known as St Mungo, the founder of Glasgow Cathedral. The Saint and Serf's robin are still part of Glasgow's coat-of-arms. The robin motif was introduced by Robert Wyschard, who became Bishop of Glasgow in 1272. The bird, Serf's, was represented on Wyschard's episcopal seal, and is the earliest known illustration of a robin.

But why should Serf's robin have become associated with St Kentigern and therefore Glasgow? Kentigern, as a boy, was the favourite pupil of Serf, and was hated by the other boys as a result. Plotting to discomfit him, they hit on the idea of killing Serf's pet robin and blaming it on Kentigern. In the event they decapitated it. Kentigern 'stood aloof' and had no part in the affair. Jocelin goes on:

'The old man took the destruction of the bird very ill, and threatened to avenge its death on its destroyer very severely. The boys, therefore, rejoiced, thinking that they had escaped, and that they had turned on Kentigern the punishment due to themselves, and that they had lessened the grace of friendship which Serf had hitherto entertained towards him . . .

'When Kentigern, the most pure child, learned this, he took the bird into his hands, put the head to the body, and impressed upon it the sign of the cross, and raising his pure hands in prayer to the Lord, said: "O Lord Jesus Christ, in whose hands is the breath of all Thy creatures, rational and irrational, give back to this little bird the breath of life, that Thy blessed name may be glorified for ever." These words spake the Saint in prayer, and immediately the bird was restored to life; and not only rose safely with untrammelled flight in the air, but flew forth in its usual way to meet the church.'

Scots poetry seems to have been much later than English in discovering the robin as a subject, and references in Scots appear like grains of sand on the beach of English literature. James Melville (1556-1614) was clearly a robin tamer, and tameness was his theme in his 'Robin at my Window':

Puir progne, sueitlie I have hard ye sing
Thair at my window on the simmer day;
And now sen winter hidder dois ye bring
I pray ye enter in my hous and stay
Till it be fair, and then thous go thy way,
For trowlie thous be treated courteouslie
And nothing thralled in thy libertie.

Come in, sueit robein, welcum verrilie,
Said I, and down I satt me by the fire:
Then in comes robein reidbreist mirrelie,
And suppis and lodgis at my harris desire:
But on ye morn I him perceived to tyre,
For Phoebus schyning sueitlie him allurd,
I gave him leif, and furth guid robein furd.

It will be seen that Melville uses 'progne' for robin in these lines, although the name is more usually applied to the swallow. It will also be noted that he uses 'robein' without the redbreast, which is the modern style, but for more than a century before Melville's day Robin Redbreast was standard English usage. Redbreast, not Robin, is the vernacular name for the bird throughout Europe. But Melville, although early, was not the first to use the nickname without the descriptive. According to David Lack, distinguished scientist, robin expert, and prodigious researcher into robin lore, the first use of Robeen (Robein) was in 1549, by the author of the *Complaynt of Scotland* —'Robeen and the litil vran var hamely in vyntir.'

Here we have an association between the robin and the wren, one that we are all familiar with in other contexts, and which is a part of folklore—the supposed marriage of the two, God's cock and hen. There is some pithy Scots rhyme on the subject and, as in the English, the wren is made out more than a little bitchy.

Which of us doesn't know:

'The robin cam to the wren's nest,
An keekit in, an keekit in:
"O weel's me on your auld pow,
Wad ye be in, wad ye be in?
Ye'se neer get leave to lie without
And I within, and I within.
As lang's I hae an auld clout,
To row ye in, to row ye in".'

Then there's the tail piece from 'Robin Redbreast's Testament' (David Herd's *Scottish Songs* 1776):

'Now in there cam my Lady Wren,
Wi mony a sigh and groan:
"O what care I for a the lads,
If my wee lad be gone?"
Then Robin turned him round about,
E'en like a little king:
"Gae pack ye out at my chamber-door,
Ye little cutty-quean".'

The same relationship crops up in 'The Wren She Lyes in Care's Bed,' even to the use of cutty-quean for the wren (David Herd's *Scottish Songs* 1776):

The wren she lyes in care's bed
In care's bed, in care's bed;
The wren she lyes in care's bed,
In meikle dule and pyne—O.
When in came Robin Red-breast,
Red-breast, Red-breast,
When in came Robin Red-breast,
Wi succar-saps and wyne—O.

Now, maiden will ye taste o this,
Taste o this, taste o this;
Now, maiden, will ye taste o this,
It's succar-saps and wyne—O.
Na, ne'er a drap, Robin,
Robin, Robin;
Na, ne'er a drap, Robin,
Gin it was ne'er sae fine—O.

And where's the ring that I gied ye,
That I gied ye, that I gied ye;
And where's the ring that I gied ye,
Ye little cutty-quean—O.
I gied it till a soger,
A soger, a soger,
I gied it till a soger,
A kind sweet-heart o mine—O.

Burns used the robin as no more than a fill-in name. Scott did more with it in
'Proud Maisie'. James Grahame (1765-1811) writes of the bird at length in *The
Birds of Scotland,* and again there is the reference to its tameness:

Ye lovers of his song, the greenwood path
Each morn duly bestrew with a few crumbs:
His friendship thus ye'll gain; till, by degrees,
Alert, even from your hand, the offered boon
He'll pick, half trustingly . . .

In another part of the same poem Grahame writes of the robin's association
with the Babes in the Wood—covering their bodies with dead leaves—a tale of
English origin, and not very ancient. Grahame, writing of children at the school
mid-day break, says:

In midst of them poor redbreast hops unharmed,
For they have read, or heard, and wept to hear,
The story of the Children in the Wood;
And many a crumb to robin they will throw.

The red breast of the robin—functional in robin display—has been variously
associated in legend with fire, blood, foreboding and death. The story that the bird
got its red breast from the blood of Christ on the cross is not part of the folklore of
these islands, and was adopted in modern times. I heard it many years ago in
Castille. Lack says it is a Breton legend. It could have reached Castille from there,
as it reached England in the nineteenth century. Fiona MacLeod tied it to St
Columba, who was kind to birds, although there seems no reason to suppose that
Columba knew of any such legend or that he had any particular relationship with
the robin, in the way that Serf and Kentigern had.

The robin is Britain's national bird. Will it be Scotland's? When the great day
comes shall we choose the bird of Serf, and Kentigern, and Glasgow, and
(apocryphally) of Columba, for whom the bird (according to Fiona MacLeod)
sang:

Holy, Holy, Holy,
'Come near, O wee brown bird!'
Christ spake: and lo I lighted
Upon the Living Word.

Holy, Holy, Holy,
I heard the mocking scorn!
But, Holy, Holy, Holy,
I sang against a thorn!

Holy, Holy, Holy,
Ah, his brow was bloody,
Holy, Holy, Holy,
All my breast was ruddy.

Whatever we decide, the robin on Christmas cards is there to stay. Political changes will not change the relationship between men and robins. Christmas is the time of goodwill, and there is a fund of goodwill between these two. One might call it a symbiotic relationship. The robin cheers by its presence at the back door in the darkest days of the year, and people reward its presence by the present of food.

I had always thought that this was enough to explain the popularity of the robin as a Christmas card subject, but Lack the indefatigable gives another, and highly persuasive, explanation.

At Christmas one of the most welcome visitors is the postman. Until the mid-nineteenth century the postman's nickname in England was Robin, and he wore a red uniform until 1861. Christmas cards of the day often showed a robin redbreast with a letter in its beak or lifting the knocker of a door. The robin redbreast remained after the colour of the postman's uniform was changed to blue in 1861. The conclusion, according to Lack, is inescapable: the bird is a symbol.

In our slavish way we have copied the motif, overlooking the fact that we have a far, far more significant reason for adopting the robin at this time.

P.S.—If you want to attract the robin lay out meal worms. The robin will become your slave and you'll become a Serf.

Those Red Deer Legends Are Dying

Thrice the age of a dog the age of a horse;
Thrice the age of a horse the age of a man;
Thrice the age of a man the age of a stag;
Thrice the age of a stag the age of an eagle;
Thrice the age of an eagle the age of an oak tree.

Thus the old rhyme of the Gael. And in the Highlands today the ear can still hearken to rhyme, ignoring research, and hear the great tales of olden times: when the proud high-antlered harts were the quarry of kings: when King James V, with 12,000 men killed eighteen score harts in Teviotdale and thirty score in Athole, along with roe and roebuck, wolf, and fox, and wildcat.

They tell of the time when 2000 Athole men, gathering the red deer in Mar and Badenoch, Athole and Murray, could drive a herd, numbering a beast for every man of them, to delight the eye of that Mary Queen of Scots who died at the hand of Elizabeth of England.

They tell yet, in the forests, of the milk-white hind of Loch Treig, that MacDonald of Tulloch knew: she who was never fired at and who lived for more than 160 years in the wilds of Lochaber.

In Badenoch roamed the Great Stag—the Damh Mor of the legends—who lived for two centuries, a great and magic beast.

Always there are the stories of the great deer of the past: who roamed the ancient forests through many reigns; who outlived the Chiefs and their sons; who were still young when men had grown old.

The kings and the chiefs have gone; only the deer remain. And the legends.

But the legends—'thrice the age of a horse the age of a man; thrice the age of a man the age of a stag;'—are dying under the remorseless scrutiny of modern research. The life of a stag can be measured by the rise and decline of antlers, as a salmon's is measured by its scales, or a tree's by the rings of growth.

At twelve the Highland stag is in his prime—a Royal if ever he is going to be one. Thereafter he begins to fail in prowess and antlers, and is old before men have attained to manhood. A score of years he may have, or more, but he will not be seeing the hills as he did.

In the wild, there is no place for the aged and the ailing. In the days of the Great Forest, when Scotland was a kingdom, the wolf harried the deer herds—as the wolves of today harry the caribou of the tundra—keeping them fleet and strong, so that death by time and misery were alike rare tragedies.

Today, man kills the deer selectively, for the wolf has gone; the stalker, treading the remote places with glass and rifle, grasses with a bullet the ailing and the old, the barren and the unsightly.

But the red deer has fallen on evil days. Like the great falcons, once the pride of kings, he has been reduced to the status of vermin—without protection or close season; butchered by gangsters; execrated by many; disowned by Governments; a pawn in the hands of political jugglers—honoured only by those who stalk him, and by the few who would still preserve something of our heritage of beauty in the wild places so long raped and disfigured by thoughtless men.

The great Caledonian Forest that the Romans knew has gone. The deer forests of the twentieth century are treeless barrens, home of the mountain fox and marten, wildcat and eagle, into which the new woods of spruce and larch, planted by a new generation of men creep slowly.

The wild glens, the naked slopes, and the high ridges are the territory of the red deer; true descendents of the great beasts of the Pleistocene—the largest and noblest land mammal surviving in Britain today.

At birth he is an awkward, sprawling bundle of legs, unable to lift his sleek head, unable to stand; wet and squirming; snorting and choking with the gleet in his mouth.

His mother stands over him in that first hour, baa-ing softly to him; then her gentle tongue, expert and soothing, licks him till his brown, spotted coat is curled and dried.

Blue and blood-red and purple are the mountains in the sunrise, with light and shadow creeping, and ravens croaking in dark ravines where wild waters rush.

In those days he lies still, chin to ground, unmoving except for a tremor over ribs following a deeper breath. Without twitch of ear he listens, hearing the song of the mountain blackbird, the *'chakking'* of wheatears, the *'prukking'* of ravens, the faint mewing of buzzards.

He is beautiful, the calf of the red deer; and his soft eyes glow without highlight under sweeping lashes.

He has enemies—the prowling mountain fox, most cunning of all four-footed things; the Highland wildcat, giant of frame and terrible of tusks; and perhaps the lordly eagle, watching from his crag above stupendous screes.

But the days of weakness are brief, and when the little calf is running strongly with his mother he has little to fear except man himself: man who is at once his only enemy and his sole protection.

Yet, of all our wild animals, the red deer has the longest childhood; he grows quickly but matures slowly. As a yearling calf he will still follow his mother, even when she has her new offspring at foot.

At three, he may still be running with the hinds—immature, a knobber unstirred yet by the fire of October, but experiencing a strange urge to join the stags on their ranges.

Among deer, the knobber is still a growing boy. In the forests, where there is an ethic in deer-stalking, he is not considered shootable. So, since man himself is the only enemy he has to fear, he is safe, and will live and grow unless cut short by accident or famine.

The knobber rising three becomes a brocket (or a staggie) when his new antlers grow, and he is a stag by name at six.

But at six his antlers will still be light—a promise of things to come. He may grow to be a switch; or a good-headed stag; or just a rag; or perhaps a Royal with brow, bay, tray, and three-point top.

He will, if he never grows antlers, be a hummel, and as such will grow into a heavy, powerful beast unloved by men: an uncrowned king of bone and muscle fit to topple the proudest head in the forest.

When first he takes the rut it will be as a flanker, a skirmisher, a raider of harems, a cutter-out of straggling hinds. Only when the days of roaring are far gone will he try to play the master and rut the last hinds of the season.

But his day comes at last, when the fire kindles in him, and he is in his prime.

Clear, vitreous blue are the day skies, with the wind snell from the north; at night, the peaks are etched, indigo and glowing purple, against fiery sunsets.

The peat hags, brimmed, are pools of crimson fire on the shadowy levels. And when the moon rises, frosty-brilliant, the hills are a spectral wonderland, mirrored dark on silvered waters.

These are the days of roaring. Then he strides down from the high ground—maned, swollen of neck, burning with the fever of the rut—to wallow in the peat hags and emerge like some primeval monster while in the south sky Orion glitters. He roars and gasps and grunts, and the hills echo with his thunder.

But there is much sound and fury, and little blood. The terrible fights of the legends are rare.

Mostly they are formalised displays of great punctilio and bluster. Sometimes they are long-drawn-out; sometimes a stag is injured; sometimes the injury is serious.

But most of those who take the rut survive to take it again; unless cut short by the rigours of winter, or by a bullet in the next session of stalking.

The Grouse Harvest

In the days before the war, when Americans in dambrod tweeds and smoking cigars like atomic submarines came to kill the grouse and fleg the hoodies to death, the Twelfth was something. The newspapers devoted some yardage to it. It was always good for a leader, a polemic, or a eulogy. Even the urban unemployed knew about it, and would discuss the prospects (that is still the word) or argue about the rights and wrongs of sitting in a butt and shooting driven birds.

The Press, and now television, give it a whirl, but this is mostly out of habit. Nobody bothers much about the Twelfth nowadays, unless maybe the twelfth of the month before which has much more significance for everybody. It puzzles me why we still bother. After all they shoot deer, and pheasants, and partridges, and blackcock, and capercaillie, and ptarmigan, and hares and rabbits, and ducks and geese, and nobody hears very much about them.

I was looking at a short piece on *News at Ten* showing the grouse shooters on a Yorkshire moor, and apart from a meaningless discussion on costs, the thing that struck me was the poor quality of the retrievers. Labradors have either gone back a lot, or they picked the worst for the show. Getting the bird from the dog was a tug of war, with a man using both hands to grab and pull. I wonder how he would have got on had he been carrying a gun?

It used to be axiomatic that you received game with one hand. The Labradors I knew came up smartly and almost threw the game at you.

The old days of dogging a moor have long gone, except here and there where you'll still find a lone ranger out with spaniel or retriever for a whole day and maybe collecting a brace, or a brace and a half, or two. This is strictly man and dog work, with hard work and a small bag, but it isn't an efficient way of harvesting grouse.

Driving grouse to ambushed non-walkers strikes many people as somehow unfair, unsporting. In fact, it is the efficient way of harvesting grouse. Good moors produce a surplus of birds and the most efficient way to kill the surplus is by driving. It is a method of cropping, and sport hardly enters into it, although those who shoot look upon it as sport and pay plenty to enjoy it.

Surplus grouse are expendable. They die from one cause or another. In the breeding season the birds share out the moor into territories from which all surplus birds are excluded. It would be foolish not to harvest this surplus. Of course, a grouse moor is a man-made climax, and it is valid to argue that it could be made into something else. But so long as it is grouse moor there is a harvest to be reaped, and every justification for reaping it.

Man takes the lion's share of the grouse surplus, but there are other predators who take a cut. Often man doesn't crop the whole surplus, and this is used up in other ways, predation or disease.

Eagles and peregrine falcons kill grouse, big grouse. Merlins and harriers take cheepers. The buzzard will take a grouse. The hoodie sucks grouse eggs. The fox will kill grouse. And there are others, opportunists mainly, whose presence or absence makes little difference one way or the other.

The peregrine is used in falconry, flown at grouse from the fist, which is a more ancient sport than shooting, and with much more appeal to those for whom the bag isn't everything. And it is more difficult to train a falcon than to learn how to use a gun.

Away back in the fifties, when I was doing a stint on peregrine falcons in Argyll, I watched a tiercel miss a grouse with two strikes, then follow it down to where it was hiding in the heather. I stalked in close to the bird and watched him walking around, like a man on snowshoes, high-stepping, and peering under the tall heather, trying to find the bird he had lost. He was persistent, and from time to time disappeared under the tall heather screen. But the grouse lay close and in the end the tiercel gave up and flew away. I had seen nothing like it before and have seen nothing like it since.

The Ethics of Shooting

Back in 1937, in a brilliantly evocative but warped book, *Sporting Adventure,* J. Wentworth Day wrote of Tom de Grey, the sixth Lord Walsingham:

'I doubt if we shall ever see Lord Walsingham's like again. His fame today rests on that amazing bag made on August 30, 1888, when, using a pair of hammer guns, he killed 1070 grouse with 1510 cartridges in a day ... The physical strain of firing 1510 cartridges in a day is far beyond the normal endurance of the average shot. Most men get a gun headache after the five-hundredth cartridge.'

Such was Day's yardstick of greatness. Well, I don't know about the gun headache bit, because I doubt if I've fired 1510 cartridges in my life; but I would think his Lordship's feat qualifies him for the ranks of the great butchers. One wonders how many people, back in 1888, could have afforded to buy 1510 cartridges, let alone blast them off in 449 minutes of one day.

Sighing for those days, Wentworth Day tells us: 'Men had more leisure and more money.' Which men? 'Or rather,' he goes on, 'the money was in the hands of those who appreciated good shooting, and were brought up to it.' So there!

That was still pretty much the way of it when Day was writing and getting steamed up about stockbrokers and beer barons usurping the place of the squire and the yeoman, and wishing Lloyd George had never been born. How he longed for the old days, when shooting was for the elect, and the elect were killing anything from skylarks to Norfolk ospreys, or bustards or night herons. Or trying to kill 1070 grouse in a day, with or without a gun headache.

Sport has become democratised since then, especially at the shooting end of the huntin', shootin', fishin' spectrum. But has that made it any better? One doubts it. Anyway, if a thing is wrong *per se,* democratising it doesn't make it right. I'd be opposed to otter hunting, for example, even although it became the exclusive pastime of the proletariat.

My own view about shooting is that as long as shooting is legal, and certain things can be legally shot, there is nothing wrong in someone shooting what can be legally shot, in places where he or she has a legal right to be doing it.

It bothers me not at all that some people will rear 1000 pheasants in the summer so that they can shoot them in October. They could rear them, then thraw their necks, which is what I would do, supposing I ever wanted 1000 pheasants, which I very much doubt.

Ah! some people will say: but shooting hand-reared pheasants isn't sport. What the hell has sport got to do with it? Personally I'm all for the unsporting shot: for shooting the sitting bird; because that way you're more likely to kill it outright.

Those who plead the unsporting bit are really supporting the idea of taking the difficult bird, when you're more likely to wound than to kill. The so-called sporting shot is a demonstration of personal prowess with a gun; not a way of ensuring that you kill your quarry cleanly. Does it make any difference whether the pheasants are hand-reared or wild-bred?

One may question the mentality of the person who rears 1000 pheasants so that he can shoot them when they grow up; but for God's sake spare us the sporting shot bit.

The Highland deerstalker waits until his beast is standing still, and broadside if possible, before he shoots it, because he wants to make sure of killing it. He doesn't take it on the lam at a maximum range. And it is a matter of protocol that he follows up a wounded beast to kill it. There is an ethic in deerstalking; none at all in modern style deer poaching.

The method of culling the deer herds of the Highlands compares more than favourably with methods in any part of the world. Deer farming is no substitute. Some people get sport out of deerstalking, and others object to some people getting sport out of it; but I can't see that this changes anything. The deer have to be killed, and stalking with the rifle is the best way I know of. If we all became vegetarians tomorrow the deer would still have to be killed.

Grouse shooting comes in for a lot of stick, and I must say that sitting in a grouse butt waiting for driven birds has no appeal for me. But it's the way to kill more grouse and people who do it say the birds are more difficult to shoot (the sporting bit again).

The fact is that grouse are territorial, and regulate their population density. The surplus are going to die of one thing or another: disease, starvation, predation. So I can see no objection to the hominids taking their share. The real question is whether it is right or wrong to have so much ground managed for grouse production; and there's an arguable case even for that.

It doesn't bother me that people shoot grouse; but it puzzles me why somebody should want to kill 1070 of them in 449 minutes. It does bother me that people who shoot grouse, as they're legally entitled to do, often shoot other things, like merlins, short-eared owls, sparrowhawks, peregrines and eagles, which are protected by law. They also kill mammal carnivores as a routine, which is a waste of time. And this is where I part company with the shooting people. And this is where things haven't improved much. And this is where things haven't changed much.

It's getting that you can hardly turn a corner without looking down the barrel of some kind of weapon, often, all too often, in the hands of kids who are a menace wherever they turn.

In the old days the birds of prey got it; in these days the birds of prey still get it; but so do the swallows, robins, blackbirds, willow-warblers, swans, windows and people. Shooting is getting absolutely out of hand.

One has to mention the Wildfowlers' Association of Great Britain and Ireland in this context. This organisation has a strict code, and disciplines its members; it also contributes greatly to conservation measures. And there are a lot of people

95

outside WAGBI who are responsible and ethical. But there's a lumpen bunch over whom there is no kind of control.

I would like to see all weapons under the control of the police, and registered. I would like to see all shooting parties with at least one capable dog along to find the runners. I would like to see the shooters of protected birds clobbered with penalties as severe as those for poisoning a salmon pool.

The law is still too elastic in favour of the sportsman, and it ought to be the other way round. And something really ought to be done about kids running around firing pellets in all directions.

Clearing Station

There's a clearing in the big wood, with old pine trees gathered around and a tall birch tickling one of them in the armpits, where a man can sit and enjoy the dusk, and sniff the pine sweat, and become one with the old stumps and blaeberry cushions and mossed hummocks, and wait . . . And wait . . .

The clearing is a crossroads, a hub, a junction of many trails, with its own routine and timetable: a timetable no more unreliable than most timetables nowadays. One is hardly ever surprised. Mostly one expects the unpredictable.

Like the tawny owl on his pine roost, who materialises from a curl of bark and yawns from ear to ear; then whets his sickle beak; then clicks it like castanets when he realises that there's something down there. The click is for anger; tawnies at nesting time are angry birds, and this is the very heart of his territory. So he goes clickety-click for a bit until he decides I can be ignored.

That's when he balloons up where his Adam's apple would be if he had an Adam's apple, and then he lets it go—the wild, quavering, bubbling hoot that is song and signal. He's signing on for the night's work, and about ready to go. In a moment he goes, down and through the trees. Soon I hear his war whoop, and I know he's away until he makes a kill for his hen on the nest.

Before or after the owl there's the woodcock. His roding circuit is round the clearing. Round he goes, withershins, with frog-croak and chirrup, oaring along above the trees, with slow flap at deceptive speed, in clear silhouette. He goes out of view, but I watch the sky space and soon he's round again: *croak-croak: chissick*. His wings are in slow motion; his pace is not.

Now a blackbird begins yelping, and the yelps are for the owl. This is owl-speak. A man can depend on the blackbirds kindling up when the owl flies.

A pair of carrion crows begins talking. It's amazing how much crow talk goes on at darkening. Their nest isn't far away. I am listening to the crows when the woodcock suddenly flashes past, barely seen, and when he tops the trees there is another one with him. The hen has taken off to have a burst with him. A sitting hen will often do this late in the day.

The woodcocks nest every year in the big wood, but their early nests are often robbed by the crows. They come on to lay again, and these nests are more successful because by then there's more cover and the crows are busy anyway.

It was in this wood, years ago, that a friend and I put four woodcock chicks in a hole with smooth, sheer sides, then sat back and watched through field glasses while the woodcock flew down four times and out four times, and when we returned to the spot the four chicks were out of the hole and squatting on the leafage again. They could never have made it out unaided. The bird got them out all right, but I am not at all sure how she did it.

There used to be many more pheasants in and about the wood. There aren't so many now, but a man can still hear the drumming of a cock pheasant at dusk in spring. It's a distinctive sound, like the owl's hoot or the woodcock's croak or the blackbird's swear words.

For years I had a small platform, about ten feet high, on the edge of the clearing. It got me off the ground and out of the noses of badger or fox, and even roe deer. You could get to grips with the birds whether you were aloft or grounded, but with the mammals it was, and is, safest to get off the ground. I used to put titbits on the ground near the platform and watch fox or badger coming to eat. They can be quickly taught to come looking, and a regular food supply, with no obvious strings attached, quickly overcomes even the fox's suspicions.

From the platform I often watched the badgers homing in the morning. The fox would come for his pick-up in the evening but never reappeared in the morning. In spring I watched the woodcock's morning circuit from it, and sometimes the bird would fly past me on his way into cover. Once a woodcock sits down it is hard to see. The plumage is excellently camouflaged.

The roe were sometimes a problem. If a buck came in and lay down in the clearing about sunrise one felt obliged to stay aloft until he went away. A man materialising above his head would not have reassured him about the clearing as a lying up place.

Crake of the Wild

The Sunday name for the corncrake is the landrail, and, of course, it's a rail, and it lives and nests on land, so the name is apt, succint and accurate. But I much prefer corncrake because, although the first part is only partly accurate, the second is onomatopoeic. The corncrake *crakes* and I've heard a parent tell a wean to stop *craking* when neither has heard, heard of, or seen a corncrake.

The call of the corncrake is rasping and vibrant, and carries some distance on a calm evening. In the days when I had them all round my ears, and they were rasping and running about at dusk, and until darkening, hardly anybody thought of commenting on them, unless just after they had arrived, when someone might say: 'I hear the corncrakes are back.' I can't remember anyone saying: 'I see the corncrakes are back.' Probably because few people ever saw one, and most wouldn't have recognised one if they had.

But there were plenty who would lay down the law about them. One was that the birds couldn't fly, overlooking the fact that they managed to get to the Mediterranean and North Africa. And back. Another was that the birds were ventriloquists, able to throw their voices. This was a reaction to the fact that a bird would call from six different parts of a hayfield in as many minutes. It wasn't, and isn't a ventriloquist, but it is a smart runner, and took its voice with it . . .

I can't recall anyone miscalling the bird because of its harsh cry, nor complaining about being kept awake by it. What surprises me today is that when a bird turns up in some place from which it has been absent for years somebody is on the phone to me at once, either directly or through the police, to ask: 'What can be done about it?' I usually reply: 'Send it to me—I'd love to have it.'

What is it about us? In an age when one can't escape from noise; when the ears are beaten constantly by what is euphemistically called 'background music,' in pubs, hotels, restaurants, streets and houses; when, to paraphrase Mark Twain, absolute silence is worth ten dollars a minute—somebody will go off the head when a rare bird turns up and does its natural, innate, inescapable thing.

How different it was in Blair Atholl some years ago, when Fergie Ferguson was head stalker there. He came to me to say the estate boys had come across several—not one, but several—corncrake nests in a tiny planting that was smothered in nettles. The men were hand-cutting the nettles (no herbicide syndrome there) when they came on one nest, then another, and another. They left a clump of nettles round each nest, and stopped working there for the time being. And I got the word from the stalker.

All these nests hatched off successfully; the birds had perfect peace to rear their young; but they didn't come back the next year. Probably one of them landed in Ayrshire, in a field beside a certain house, whose occupant was on the phone to me complaining bitterly about the noise, and what was I going to do about it! I?

How different from the farmer near Balfron, who phoned me to say he had corncrakes in his big hayfield. He had indeed. And the man was grand pleased about it. He held off cutting the hay until the last possible moment, then dogged the field before cutting, to give the birds the maximum chance to get clear. He was a big, tough, realistic farmer, as shrewd as they come, but in the diastole and systole of him there were tolerance and concern. He once put a tractor on short time because there was a pied wagtail with young on the cylinder block! She got off . . .

If you think of the way the collared dove has spread around since the war, that is the story of the corncrake in reverse. Once so common that nobody could have conceived of a countryside without it, the corncrake has retreated and retreated until most people in the countryside know it not. Its strongholds today are in the North-west, the Inner, Outer and Northern Isles. Always there are birds turning up here and there, to nest in places that haven't known them for thirty years or more. One year's craking, then silence again.

In 1974 a pair nested at Cumbernauld—if there were others I didn't know about them, but I think I would have heard—but there hasn't been a crake since then.

The bird is small, a handful in fact, and when seen in slow, laboured flight, isn't usually recognised for what it is. It will lay up to twelve eggs (the Blair Atholl birds laid ten). The hen can be bold towards a human intruder at hatching time, and I had one that dabbed at my feet while yakking away at me with feathers on end. A less than bantam trying to fleg an outsize hominid! When I mentioned this to someone, it was relayed as: 'The bird attacked him.' How fond people are of being 'attacked'.

Out for the Count

Counting gannets isn't everybody's business; certainly it isn't mine. I'd rather sit back and look at them sitting, or carrying in nesting material (the gannet is a compulsive carrier of nesting material), or crossing beaks like swords, or feeding young, or flying around, or diving, or whatever. The business of counting them I leave to men like my friend Jack Gibson, who practises medicine when he isn't counting gannets, which means he practises medicine for about 364/365ths. of the year.

But the counting day is a day and Jack Gibson has been counting the gannets of Ailsa Craig since his student days, which is a very long time. Every other year I make the trip with him from Girvan, a trip that is usually postponed at least once if conditions are not right for landing. But in the end the trip takes place and the gannets are counted.

This year the breeding population of gannets on Ailsa Craig was close on 14,000 pairs. Close, I say, because I have forgotten the few the number was short by, but Gibson has it in his book, a book that will give you the population right back to the war and before.

Nowadays Jack Gibson counts the birds from a boat lying offshore. He is able to do it this way because long ago he divided the cliffs into natural sections, like pictures on a wall. All he has to do—it sounds easy—is to count the number of birds in each frame. He does this by lying on his back in the boat, aiming the binoculars at each frame in turn, and counting. He checks each count and is hardly ever out between the first and the second.

When he has finished with one frame the boatman goes along to put him in position for the next. And so it goes on. Well within the hour he has made his count, and you can take it his margin of error is small, so small that it can be overlooked. The count is, of course, of breeding pairs.

After the count the boat completes the circuit of the Craig and we go ashore for a spell up among the birds on their cliffs. This year we spent three days among them, in a mixed grill of sunshine, gales and rain. We walked once right round the shore (there is a sort of compulsion to do this on any small island, although I don't always see the point of it) but I spent the rest of the time mainly sitting among the birds, looking at them, and taking some photographs.

For the past several years the gannet population of Ailsa Craig has been rising steadily, after having gone down from a previous high. So there is the appearance of a cycle.

Besides the gannets there are razorbills, kittiwakes, guillemots, gulls and some puffins on the Craig. The puffin population has suffered a steady decline.

I got the impression that slow-worms were going down, too. In previous years they were easily obtained; this year we couldn't find one. Of course, this could mean no more than that we were unlucky. But we were unlucky with the tiny shrews, too. Our Longworth traps didn't catch one, although at one time they were literally everywhere. There was, however, no scarcity of rabbits.

Not a single Soay sheep remains. They were quite heavily shot down, apparently quite unselectively, and the remaining animals failed to build up again. This is a great pity because the Soay is a distinctive ancient type, and Ailsa Craig is the kind of place where the breed can be easily maintained without bothering anybody.

Gander Guards His Nest Eggs

The big Canada gander came back from his grazing session and swam at speed, trailing a wide V wake to the islet where his goose was nesting. She was hard down on her eggs, with her neck laid along her back and her bill almost touching her tail. The gander took up his sentinel post at the tip of the islet, a spread-arms' distance from the sitting goose, with his feet planted firmly on a big stone just under the surface, so that he appeared to be standing on the water.

I was sitting on the shore with the dogs—the big German Shepherd bellied down at my feet, and the terrier poking her nose into vole runs in the bank. The gander knew them as well as he knew his mate on the nest, so he paid little attention to them, but when the terrier started to dig he decided to come over for a look.

He swam in tight against the bank, heaved himself out of the water between the terrier and the other dog, and began to pluck grass. The terrier looked at him for a moment, distance about 15in., then went on digging; the big dog, with the bird almost walking on her tail, yawned and put her chin on her forepaws.

I spoke to the gander thus: 'Give you a week or so, my boy, and you won't be so palsy-walsy with this pair, you'll be beating them up and driving them from the pool.' Thus spoke this one. I was recalling how, last year, he beat me up with great gusto just before hatching time. And he knows me as well as he knows the dogs.

Syne he swam out to the islet again—not to resume his stance on the hidden stone, but to float alongside. He was floating there when I saw the big water vole swimming from my right, just below the bank. It was its wake that drew my attention before I saw it. I clucked the terrier to me and she stayed in my lap until the vole swam past.

Then the vole dived, and I watched to see where it would surface. It surfaced like a submarine about two feet from the floating gander, whose neck went up like a giraffe's in surprise. He paddled aside and honked. And the vole dived again. I watched it surface beside the other islet and scurry for its waterline burrow a few yards away.

A while later I was sitting on the other side of the pool, with the dogs curled beside me, and the gander walking around doing nothing in particular. He would peck at a green blade here, and another there, but he was merely passing the time. Then the shrieking and croodling started in a patch of threshes.

Up went the gander's neck. Up to my face went my 8 x 40 Zeiss binoculars. And up out of the threshes leaped a pair of weasels in close embrace. Down they went again, and the threshes shook. The gander, long-necking, watched with what I can only call lazy interest. He wasn't in the least put out by the furry fankle leaping around like a flea. The terrier looked around, then looked at me, and I said: 'No!'

I watched the whittret pair for the few minutes of their grappling game. Then everything was quiet again, and the gander stalked back to the water. The vole (a vole) surfaced in front of him, and this time he scooshed at it with his chin on the water, and forced it to dive again. Then he flapped his wings, and *kronked.*

Then he went ashore, stalked over to the other nesting Canada goose, and beat her from her eggs. Then he went back to his standing stone. And the other goose came back to her eggs and settled down. The gander was now a bit steamed up, and I thought he might vent his displeasure on the terrier who was again at the water's edge sniffing vole smell. But he merely sailed past her with a perfunctory *kronk* and a look that seemed to say: 'Oh it's you again.' Anthromorphic; but that's the way it looked.

I called up the dogs and walked away along the bank, with himself sailing alongside. But he wasn't in the least angry. He was merely seeing us on our way.

Eerie Crooning of the Shearwaters

At nine-thirty in the evening, with the sun warm and the clegs launching a Nivelle-style offensive (with far greater success), Dr Lockie and I set out, heavily kitted with clothing and cameras, for the high face of Hallival. We were going to spend the night with the shearwaters.

The first lap was a long, gentle, sweat-sticky climb through a waving sea of moor grass—the much-talked-about, kenspeckle molinia. Heather, other than patches of bell heather, was rare, probably because of much burning; in hollows and damp places grew bog asphodel, Scottish asphodel, cotton grass, orchids purple and flagrant, lesser butterfly and twayblade.

Over the first ridge, topped with bare, fissured rocks we went, and into a dark, boulder-strewn hollow of dried water channels and cracked peaty flats; then it was a steep ascent through great, jumbled rocks and scree on the face of Hallival. It was a scene of desolation, like a moon landscape imagined by Blake.

Among the rocks on the steep face were the burrows of the shearwaters. We found one right away, in which a bird sat brooding a solitary egg, so we rigged up the cameras and sat down behind a rock. It was growing cold.

The light persisted; obviously it would never become completely dark. And darkness was almost essential for the shearwaters. Though they gather offshore long before sunset, they are nocturnal at their breeding stations. Even on nights of bright moonlight they are not keen to fly in from the sea.

The afterglow paled, but did not fade out. It spread like a stain behind the hills, slowly, and I said to Lockie: 'The sun's going to meet itself rising.' We were prepared for disappointment. A kind of twilight did, however, descend on our part of the hill, though we could see everything all too plainly still. The cold became so intense that we had to put on every stitch of clothing we had with us, even unto oilskin jackets. The wind came up like liquid glass and there was no escaping it.

Suddenly there was a crooning and crowing, throaty, dove-like and resembling nothing you ever heard, and round our ears came the shearwaters, flickering like great bats, but swift and erratic as swallows. Birds flying to burrows were greeted with croons and crows from birds in burrows. We photographed a bird at a burrow containing a downy chick. But the light was coming back quickly and all too soon the mountain face was deserted again.

Discoveries of shearwaters ringed in Britain show that they will fly a hundred miles or more from their breeding station. The birds take turns at incubating, sitting from one to four days or more. The chick is fed once in 24hrs. It is deserted after sixty days, and starves for nearly a fortnight before it flies down to the sea.

The shearwaters on Rum have been ringed in the past, and Dr Lockie has ringed birds on Eigg. I don't think there have been any recoveries of marked Rum birds, so it isn't quite certain where they feed.

We made our way downhill at 3 o'clock in the morning, reached our bothy after 4a.m. and crawled into sleeping bags. We had our heads down for four hours, then, fed and kitted, we took a car and set out for Harris on the other side of the island.

On the long, winding climb over a road like the bed of a burn, we made seven miles per hour jolting and lurching all the way, and saw hoodie crows as tame as rooks in a garden. It was astonishing to see those wild, wary birds rising from the road a car length in front then flying to a boulder less than a stone's throw away. Obviously the hoodies on Rum haven't smelt gunpowder for a long time.

Again the vegetation was molinia, molinia all the way, with almost no sign of heather, and ridges and knowe-tops bared to the naked rock. Deer, single and in small parcels, watched us all along the road. Stonechats, wheatears, meadow pipits and cuckoos flew and pitched close by.

On our left was the imposing mass formed by Trallval, Ainshval, Sgurr nan Gillean and Leac a Chaisteil—rugged, bleak, skinned and treeless, with a tremendous corrie and wild screes. Then we could see the blue, sparkling waters, the Dutchman's Cap and the flat, faint skyline of the distant Hebrides.

We wound down to Harris, with its extensive, well-marked, grass-grown lazy-beds and its wide raised beach on which cattle beasts were moving. Sheep dotted the slopes which showed tremendous gullying, and the low ground was ripped and gashed by rushing waters.

Down on the flat, near the Bullough mausoleum, is a solitary house, in which live the shepherd and his housekeeper, the only inhabitants of this lonely spot. Their contact with Kinloch is a Land Rover; their neighbours are the sheep, the deer and the cattle.

Here the low ground is green, and each morning the shepherd, who is from Islay, drives the encroaching blackface sheep back to the ridges. He doesn't like to see them crowding the low greensward in June.

Harris is a lonely, remote place, with a wild beauty of its own. The house receives great punishment from every winter gale, but that day the green place was serene and sun-kissed, with the bright lights glancing on blue waters.

Here a man can sit, soaking in the sun among the sea pinks, and imagine a once considerable community, planting their lazy-beds and cutting winter fodder. Now there are sheep, and deer, and cattle, and a droll, quizzical, hospitable shepherd from Islay who despises the tall red deer.

Since we were going eagling on the morrow, we asked the shepherd about the golden eagles of Rum. 'What about your lambs with the eagles?' we asked, avoiding a leading question. The best thing I can do is give his exact reply.

'Eagles don't take lambs,' he said, 'unless maybe a dead one.'

Chit-Chat to a Black Cat

I met a cat in the forest—a black cat with two white splotches on its face, and leopard eyes glowing in the day's ninety seconds ration of sun. It was a fox-tailed cat, with short legs, long whiskers and a sinful leer.

I had just been reading in my *Scotsman* about the First World War veterans revisiting the Somme battlefields on the fiftieth anniversary of that terrible massacre, and how they were limping along with their first-war soldier's French. And I had been thinking of the delicious hybrid French my father spoke, as he later spoke hybrid Spanish. And now here was this chat black avec the eyes d'or . . .

'Good morrow chat!' quoth I, remembering Shakespeare, although I knew right away that this chat was no fool. 'What seekest vous in this forest on ane rotten day like this mon dear feline?'

The cat didn't say. He shone his golden lamps on me, and glared, not so much at me as through me, arrogant and superior in his cattishness.

I tried to wheedle him out of it (masculine gender he was).

'Come, come, mon grand cheety cat,' I remonstrated in what I considered a fetching voice. 'Seeky vous perhaps the petit Microtus agrestis, field vole to you, the wan avec the wee tail? Or are you just having a shower?'

He turned the wick up in his great golden lamps, and sat there unanswering, aloof, remote, even supercilious. He was as sleek as a blackcock, and the white patches on his visage reminded me of the white eyes on a blackcock's shoulders.

It irks me when a chat goes sullen or superior on me, so I got out of the car, ignoring the fine drizzle of rain, intending to put the come-hither on him, to show him how with it I was when it came to chats.

But when I approached him, slowly and making speech, he lit the danger wick in his golden yeux, and slowly but certainly his black ears flattened and folded down the sides of his face, so that he reminded me of a Nubian goat. His eyes had the jewelled glitter of anger, and presently the wail came from him, warning me that he had four fistfuls of fish-hooks to argue with.

I hunkered down on the forest road, watching the glaring eyes and the twitching tail. Now he was crouching low, with his lips curled, and in a moment he was growling and hissing. What do I do now, alors, I asked myself?

'Come off it, puss,' I remonstrated. 'Je suis votre ami; or are you too thick in the head to see it? Dammit all, cat, I've got three cats of my own and none of them speaks to me like this.'

But he wasn't impressed, and I knew by the way he gathered back on himself, showing his teeth, that he was prepared to be rough. So I drew back.

I went back to the car and watched him from the window. He was half standing now, and his ears were coming up again.

'Well, mon dear cat,' I said to him, 'now that you've put the homme doon the road, what now?'

After some minutes he rose to his feet, turned about, reached into the grass, snatched up a dead bird, and made quick time across my front, with his tail bushed and his rump down. So I had come on him when he had just caught himself a prey. That explained his rejection of the Peace Conference, being the kind of cat he was. I have one a bit like him at home, but much older.

It was cat day in the forest, for not long afterwards I met another one, much nearer the village. This was a big smoky grey beast, heavily striped and flecked. He had a moon face, and a ruff, and his tail, too, was short. He was the wildcat type, but not at all wild. I addressed him in his native tongue.

'What's wi ye, baudrans?' I greeted him, getting out of the car.

He stopped on the road and looked over his shoulder. I carried on speaking to him, and when I saw the tail rise to the vertical I knew I was there. He lifted a paw, spread his fore-claws, and mewed. So he was friendly. I walked over to him and stroked his arched back. We had a long conversation.

I decided, rightly or wrongly, that twin-spot was running wild, a renegade from man's hearth, but that this one had a home to go to. Cats being individuals, however, I realised that it could just as easily have been the other way round. Or both could have been on the feral fling.

The Cat That Liked Birds' Nests

Three years and four nesting seasons ago, a pair of magpies built a great nest of sticks in the top of a pine tree just over my garden wall, and before long the hen was sitting on five eggs, not stirring even when one of us was rattling around below her feeding the poultry.

For about ten days there was hardly a cheep from the pine tree, nothing more, in fact, than the occasional *chakking* between cock and hen. Then one day there was the most terrible racket, and when I went up to see what was wrong both birds were flying around, pitching in trees, yattering and shuggying in obvious alarm. Yet there was nothing to be seen.

Had I disturbed a prowler at the nest? But what? I walked away, expecting the birds to quieten down, but the racket went on as before. So I called on my son, and had him climb to the nest to see if the eggs were all right. I could have climbed up myself, but there is a saying about keeping dogs and doing one's own barking.

When my son got above the roofed-over nest he looked down at me and called out: 'It's a cat's nest!'

And so it was. Our white cat was curled up in it, fast asleep on five eggs, and reluctant to move out when rudely awakened. The boy managed to get her out without a scrimmage, and we discovered the eggs were quite undamaged. The magpie returned to them later and went on with her job of incubating.

I subsidised the magpies heavily that year. My poultry feeding troughs were near the nesting tree, and the birds helped themselves to pellets throughout the day. They would sit by in the early morning, waiting for the poultry to wander off, then come down and fill up with food for their nestlings.

The following year the pair (or a pair) nested fifty yards away, in the spinney, but this nest was shot up after hatching, and the young were all destroyed. The next season the birds nested a tree nearer my garden, but this nest was robbed of eggs, and no second clutch was laid. This year the birds are back in the old 'cat's nest', and have hatched off five chicks.

They had to do very little patching up of this old nest to make it serviceable.

They carried a few sticks to strengthen the roof, and put in some lining; otherwise the nest was as good as it was four seasons ago. Although I have no poultry at the moment, I see the magpies down at the empty trough early in the morning dabbling about for food that isn't there.

When the young were a few days old I propped a sixteen-foot ladder against the tree and went up to look at the possibility of photographing the birds. I decided where to site an eighteen-foot pylon for a hide, and came down to telephone the scaffolding contractors who help me in such matters. Ten minutes later I went back to bring in my ladder, to find the white cat on top of it and the magpies flying around swearing in her ears.

I added a few newly-minted ones myself, at which she proceeded to come down, slowly, demurely, front end first, without putting a foot wrong. As soon as she grounded she came over with her tail up and her back arched, and rubbed against my legs. What can you do with a cat like that?

Now I have a problem (a familiar one), because as soon as I put up a pylon the white cat will try to spend all my working time on top of it. This means she will have to be shut up while I am upstairs, and watched when I am down. Wherever I work, there will she be also. And it was ever thus.

One day, when I was photographing woodpigeons in the garden, I was puzzled when the birds sat back in another tree, refusing to come in. Then I discovered the white cat was on top of my hide, almost at my ear. She had slipped out when the door was opened and hadn't been missed.

If there is a chimney sweep on the roof, or a slater, the white cat is sure to appear on the chimney or on the ridging. She can get up where any man goes, and come down without any help from anybody. When I was building my garden wall I had her on the tip of my trowel at every course. She is what you might call a man-type cat.

When she was a kitten, a long time ago, there was a man in the house putting a fireplace in the front room. After the job was finished we heard a mewing sound from behind it, and of course it was the white kitten, playing at Edgar Allan Poe. We had to make a hole in the wall to get her out.

And if you think that couldn't happen twice you don't know our Snowy. A year or two later the joiner built her in behind a cupboard, and part of it had to be stripped to get her out.

The Small Fry

If you like tracking down the origin of names, try 'terrier'. 'Terra' is Latin, and means ground. A terrier should be a terror who goes to ground.

Nowadays, too many prefer going to town in motor cars. Today's toughest terriers would hardly get a civil greeting in fashionable dogdom.

Scots of old bred some of the nippiest and nebbiest of terriers: thrawn and thrang, venomous and vaunty, varminty spitfires who could, and did, go where any fox or badger went.

Now we breed too many faded replicas who have little in common with their forbears except their genealogy.

Scotland's terriers—some of ancient lineage—are the Skye, the Dandie Dinmont, the Scottish (Aberdeen), the West Highland White, the Cairn and the Clydesdale, the last almost extinct.

Of the others, only the Cairn is still widely used as a foxhunter, and the number of hard-hitting, hard-snapping, hard-swearing, hard-working Cairns is dwindling.

The Cairn still has it; the West Highland hasn't lost it. The others are on the retired list, suited to what they do, which is probably what my present Border terrier does, and that is nothing. She bears an honoured name, but is a sad memorial to a great breed. She is also a delightful wee dog.

Nowadays you are more likely to find the average Highland foxhunter with Jack Russell or Border terriers, or their crosses, than with Cairns.

This is not because such foxhunters are anti-Scots, but because the best terriers today are Sasunnachs!

Of course any breed produces its softies, even when they come from the ranks of the toilers.

My jaunty little Border bitch, Cumberland bred, bears as much resemblance to her volcanic, gurrying forbears as mice to men.

When she kills a mouse it's by accident during a peace conference. She's had rat teeth in her hide, but has never holed a rat with her own.

We don't all want terriers who can work.

Still, it's nice when your wee dog can go into your meal store and reduce the rats without breaking a whisker—better than having a cartoon caperer who is always being bested by the local mice.

Working terriers are used on all the sheep runs of Scotland. Their main job is to go down foxholes, goad the vixen into bolting, and kill the cubs. And well they do it.

Judged purely as a method of killing foxes this is, in my view, the most humane and expeditious.

The foxing small-dog is cut from a special mould. He has to be wee enough to get down a foxhole, but not too wee either.

He has to be fast and fearless, a girner and a sticker, prepared to fly to the clinch with the fiercest hill vixens, and take a pasting in a teeth-to-teeth battle.

He has to fight foxes, think foxes and dream foxes, and go in cold to a snarling brawl below ground when ordered.

The hardest cases seldom wait for orders. Restraint is more often necessary.

This is when the men are separated from the boys, the toffs from the toilers.

106

Many a half-pint terror comes from the battle with its face red with blood.

Many a sewing job has to be done on the hearth after a hard day on the hill. But the small dogs are always ready to go back. They are Oliver Twists always looking for more.

I know one battle-scarred veteran, honourably retired, with the best chair in the house, whose speed today would hardly match a tortoise's.

In him the fox motif is complete. His very name is Sionnach.

He lost his front teeth by a form of dentistry unique in the annals of dogdom. A big vixen, breaking from his grip, went off with them in her hide before they blasted her with gunshot.

He once took a battering, and a near-drowning, from a big stag when he fastened on to its nose in a pool.

He is white-faced now, and creaks at the joints, but you still have to grab him by the tail when he smells a fox in a hole.

He is a cross Border/Lakeland: English by English again. And how I hate to admit it!

This Sionnach hands you a paw, then growls his thanks for a tit-bit. It is the only language he knows.

From the same kennel came the late Tarf the terrible, a dour, curmurring kind of fox terrier, tireless and fearless, who would have taken on a wolf in foxskin, but who couldn't be trusted to kill very small cubs below ground.

He had a soft spot for the weans.

I've seen some varminty Cairns at work—boot-high and big-tusked, hard-headed and hard-coated, tough as pig iron, and with jaws like steel traps.

You'd swear they'd be just a mouthful for a big hill vixen, yet a tot of the kind will clash ivory to ivory with a bristling bitch fox and win.

For sheer fire and frazzling brimstone you've to see a real Border at work. Here is one of England's best—all flint and fight and dauntless courage, that hasn't yet been coiffured and cossetted into a clown.

An old Border bitch I had, now dead, killed rabbits by the sackful, and a garron's weight in rats.

Not being a foxhunter, indeed having reservations about all present methods of fox control, I kept her from the foxes.

But she got to them sometimes when my back was turned, and once she was buried for eleven days in a foxhole. She would have gone ratting when she was dying on her feet.

Then there are the parsons, the Jack Russells, named after the English parson who bred them.

Stump-legged, trap-jawed, layered in muscle, they are among the most redoubtable terriers working today, and are in great demand.

I know a hard-boiled, cheery foursome gainfully employed today: two youngsters, their mother, and their grandmother— a family unit with all the ingredients needed in a well-baked terrier.

They are the right size, have the right jaws, the right teeth, the right outlook, the right mixture of brawn and brain, and the right kind of indestructible courage.

What of the fox terriers, as originally named? There are a few of the old stamp working yet, short-legged, low set caterans still up with the best.

But you don't often see them now, and they'd be kicked out as fakes from any gathering of modern fox terriers.

Almost all of these working terriers would face anything, including a badger, but only a fool would knowingly let his small dog go down to a badger.

Unfortunately, there are such people.

The main job of the terriers is foxing, an activity sanctified by authority and considered necessary in the interests of sheep husbandry.

I question the efficacy of this form of control, but it is approved by majority opinion, and is more humane than poisoning, trapping or shooting.

Of course, dogs get hurt, and sometimes more than slightly. Foxhunting like this is a messy business.

It is often a murderous business.

But it is a job of work, and they call it predator control, although to date it has controlled nothing because other factors redress the balance.

Still, given that foxes have to be killed, this is, in my view, the best way to do it. Only the dogs get any fun out of it.

Undeniably, there are individual cruelties, as when a captive fox cub is offered to young terriers for a worry.

But such men are few in this business, and you meet the type in all professions, and certainly in all forms of hunting.

Have we the right to submit small dogs to this kind of punishment, even when it is difficult to keep them away from it?

If man is defending his flocks and herds against predators, I think the answer must be 'yes'.

But if routine killing of foxes is largely a mistake, as I think it is, the answer surely has to be 'no'.

Behold the Pig—a Product of Man

Behold the pig! Any pig. Let it be the Tamworth in autumn tan, or the Large White, or the Essex or Wessex, or crosses pied or polka-dotted, or the modern sophisticated Landrace, or the tailor-made American: all have one thing in common.

They are the products of man's ingenuity, descendants of the shaggy Wild Boar, massive forward, shilpit aft, irresistible as a locomotive, and with terrible tusks, like curved dirks—the only animal, they say, that would drink at a waterhole flanked by tigers.

Yes—behold the pig! Man has evolved it, then subjected it to every conceivable indignity, of cramped quarters, foulness, and contumely.

He has speeded its maturity, shortened its life span, and extended its breeding season. It has responded, and adapted, to them all; yet man, most men, can do no better than insult it, use it as a synonym for clart and gluttony.

The pig, the healthy pig, is a good trencherman, but to accuse it of gluttony is to confuse it with man. Gluttony is one of the seven deadly sins—of man, not of pigs—and if men were not capable of gluttony it wouldn't have been listed.

To accuse the pig of being a clart is a form of self-justification by man, for it is man who makes a pig dirty. The pig is a cleanly beast, that will keep itself as clean as a badger or a ferret if given the chance.

Piglets will remain as satin sleek as new washed babies if they have elbow room and a clean bed. They will keep it clean.

When I bred pigs it used to give me the pins and needles on my spine to hear some superior sophisticate use the epithet dirty.

'Why do pigs like to roll in filth?' one such asked me when he saw my sows having a mudbath in the open. When I explained that the relatively hairless pig suffers from sun scald on hot days and insulates itself with mud, and that I provided such baths as a policy, I was greeted with a knowing smile that meant— of course you'll always make excuses even for dirty pigs.

When I showed the same person a litter of pink, near aseptic piglets, with a sow who looked as though she had been dry-cleaned, I was asked if I washed my pigs.

Many mammals are inherently cleaner, and cleaner in practice, than a great many human beings. Badger, ferret, stoat and weasel are all meticulous about hygiene. A rat will groom itself many times a day, but the idea of bathing all over is a comparatively new one in human society. The easiest thing we ever house-trained was a little pedigree piglet my wife brought up on the bottle.

It has been said that a man takes after the things he has most to do with; it is truer still that a pig takes after the man who has most to do with it. A clarty man will never make a clean pig, because the beast can only be as clean as it is allowed to be.

From this you will have gathered that I like pigs, and I do. I am very fond of pigs, and I have been on terms of considerable camaraderie with a few. One of my best friends was a pig.

Her name was Susan, and I had her from the time she was a piglet until she died at home. I kept her on because she was a friend of mine, refusing to sell her down the river.

Susan, who always appeared as well groomed as those females who advertise soap, had a number of families, either eight or nine at a time, tailored to her own cut. All of them turned out well. Susan, in fact, was always in the blue ink, which is one reason I was a friend of hers. But there were other reasons, the main one being that I liked her.

When she was out on grass, between litters, she would follow me about, lie down to have her ears scratched, and hold confabs with me. I never understood half of what she was saying, but she certainly had plenty to say. She would rush to greet me, squealing, and almost poke in my ribs in seeking the apple or whatever she thought I had hidden on me.

I always sat up with her when she was having a family, and she would give me a friendly grunt now and again as she had a spare minute, while my terrier sat by midwifing the piglets. Old Susie and the terrier were great pals, too.

Return to Good Husbandry

I was breeding Large White pigs, and running a flock of 200 Light Sussex poultry, when the antibiotic craze began: the inclusion of powerful anti-microbial drugs in feeding stuffs for pigs, poultry and other livestock to promote growth and prevent disease.

That was the song the experts sang, but I didn't join in the chorus. Instead the wee alarm bells in my head kept me awake at night.

It was the old mystic mucker in me, and it seemed to me that the implications of this revolution were staring us all in the face. My suppliers said something like: 'Come on man, everybody's using the new mixtures, and all the experts are agreed . . .' etc. But I declined and said I'd still have my stuff the old way.

I was being optimistic, because the old way gave place to the new way that I distrusted, indeed rejected; so I had to do what we used to do—make up my own feedingstuffs mixtures from scratch, without the penicillins and the tetracyclines. I was looked upon as something of a nut case, although I discovered, when lecturing all over the country, that a great many people felt the same disquiet, and for the same reasons.

Well, it's all old hat now. The waking-up process began ten years or so ago when the committee under Professor Michael Swann advised the Government to curb the general indiscriminate use of antibiotics in animal husbandry.

The Government acted on the advice and brought in controls. Penicillins and tetracyclines were banned altogether in foodstuffs, and were to be prescribed only by veterinary surgeons where necessary.

It was all so predictable. The indiscriminate use of antibiotics ensured that the non-resistant bugs were knocked off, and selected out the resistant strains to replace them. The result has been a massive increase in drug resistance. The salmonellae have become almost totally invulnerable.

So where do we go from here? Will more experts give us more and more potent drugs so that we can achieve the same thing over again? Or are we going to get down to doing a little sane, ecological thinking.

A couple of months back *New Scientist* highlighted this problem and made a telling quote from the *Veterinary Record* (vol. 104, pp 511 and 513). Maybe I liked the quote so much because it's the form of words I've been using like a gramophone record since the beginning. Here it is:

'Events of the last couple of years indicate that it may not be purely alarmist science fiction to suggest that we could be considerably nearer to a return to the therapeutic status of the pre-M&B 693 days of the early 1930s than we realise.' I'll have a dollar on that one.

In my view it was all a big con anyway. I couldn't see the logic of drugging a healthy animal to make it healthier. Did those drugs promote growth?

Perhaps when husbandry was bad they did. But not with a healthy animal in the hands of a good husbandman. My neighbours used to argue that the drugged foodstuffs promoted higher egg yields in their poultry. Yet none of them, over the years, came near my henhouse average with Frank Snowden's Light Sussex. So you can choose your pick, as they say.

And what about the food chain? This was one of my worries. If you ate animals fed on antibiotics where did that leave you later on if you needed them? I think that one has been well answered in the past few years.

In my view the use of antibiotics in feedingstuffs has created serious problems for animals and people, and achieved nothing that could not have been achieved by good old-fashioned, top-drawer husbandry.

Master of the Surprise Attack

I was motoring along the country road, dawdling on the lookout, as I often do, and suddenly there she was ahead of me, on my right, flying the hedgerow bottom with flap and flicker of wings, about three feet above the frosted grass. She was a sparrowhen, at the old trick of the hidden approach.

But I could see nothing ahead, or rising from the other side of the hedge. I accelerated slightly and got on to her tail—almost. And there I stayed for about another 150yds., saying to myself that she had to be on to something.

Before the bend, where two earthmoving machines were parked, she lifted suddenly, and corkscrewed over the hedge, and she was in among a group of starlings that burst up round her ears. And, dexterously, she had one in a foot, and side-slipped away with a leg down, and pitched in the field with her catch, while the other birds formed circus and drifted away.

It was neat, precise, and well-timed. She must have seen the starlings feeding, and made her long approach on my side of the hedge, as a matter of tactics. It was the classical sparrowhawk attack—the strike from ambush, or hiding, or from the hidden approach.

I put the car on the verge where there was a gap in the hedge, and there she was with both feet on the prey, already plucking feathers from it, and every now and again lifting her head to look around with her bright, hard, yellow eyes. I watched her pluck and swallow, pluck and swallow, until she rose and flew into a nearby tree. And then I climbed into the field to look at the little ring of feathers, and the starling with breast eaten almost clean.

It is an everyday occurrence, of course; But one doesn't see it happen every day. Even when one is working with a pair of nesting sparrowhawks, one doesn't see it all that often.

The sparrowhawk is notable for the strike from cover, from the other side of a hedge, over a corn stack, round the corner of a building. And it is something to see one pursuing a small bird in woodland, jinking round trees with the most perfect agility and timing.

One has to assume that the hawk looks at the situation before making an approach, although there are times when the dash from hiding is made speculatively. At least one sees a hawk doing this from time to time when there is nothing on the other side. But mostly there is.

The cock bird, who is more brightly coloured than his mate, is also very much smaller, so small indeed that he looks like a little boy beside her. Being bigger she is also more powerful, so she can attack bigger prey. Even a woodpigeon is not safe from attack by a sparrowhen. In fact some sparrowhens kill quite a lot of woodpigeons.

111

When the hawk kills a woodpigeon, it strikes and rides it down, the body being too heavy to be carried. It is unlikely that she kills the bird in the air. Death takes place about the time it hits the ground.

If the hawk misses with her strike she will pursue a small bird headlong, flying it down in straightforward, open fashion. As a result, sparrowhawks often end up in strange places—cut up, stunned, or even dead.

I watched one following a starling into a building one day. The starling knew where the bit of glass was missing from the window pane, and flew through. The pursuing hawk hit the glass and fell cut and stunned. We had to revive it and dress its sore face.

Another flew into the stable, through the open door, and hit the wall, breaking its neck, while the bird it had been chasing sat on a rafter and watched.

Another flew into a byre after a sparrow, and finished up in a grip, stunned. When it came to, it flapped up the grip soaking itself with urine and clarting its feathers with dung. My wife and I had to bathe it, scrape it and generally coiffeur it, until it looked like a hawk again. It was a week before I let it go.

Then there was the one that pitched into a hedgebottom after a bird. It missed the bird but entangled itself in a rusting bundle of barbed wire and wire netting. In its struggles it got itself inextricably fankled, and I had to use wire cutters and a lot of patience to free it. In return it struck at me; and I was left with three bleeding knuckles.

Goat of the Air

Habitat destruction is like when you drain and plough a hundred acres or more of threshy pasture that haven't known a furrow since the First World War. Creatively, it is the reclamation of a habitat loaned out to marginality for half a century. You pull down and rebuild, destroy and recreate, but not the same thing. Either way you upset something.

When the threshes go, and the wet and boggy become dry, there is no need to ask: where have all the snipe gone? Or the coal-heads (reed buntings) for that matter. Or the whaups.

Month after month, in field after touzy field, the plough has been turning dour furrows, followed by a storm of gulls, lifting and falling away in relays, like bubbles blown from the tractor's exhaust. They are harvesting the rich store of soil animals turned out by the plough.

Come spring, come real spring, a man will be able to dog the whole hundred acres, back and forrit, without flushing a snipe, because by then there won't be a square yard to suit a snipe's feet, or the kind of place a snipe likes to nest in.

There will be something else for one to miss, albeit on a small canvas. Maybe few people will notice the difference. The farmers will, I am sure. And a few others. But in the main we are not given to noticing the disappearance of something until our attention is drawn to it.

There's going to be another hundred acres of sky at least in which no snipe will be heard bleating as it sweeps high in dipping circles on its display flight. Goat of the Air is the snipe in Gaelic; the Gaelic of it is *Gabhar-athair*. In the old language it was also Goat of the Frost, and Kid of the Air. Goat and Kid; these two. The metaphors are felicitous, apt and accurate.

The bleating or drumming of the snipe; so like the bleating of a goat or a kid, was once a far commoner sound than it is today. It was the sound of the wild boggy places, the marshes and the low-lying water-logged pastures, and many and many a time, and many a time again, I have lain on my back and watched this tiny instrumentalist on its glad, high-flying circuit, a mere speck, lark-size, in the haze of blue.

Instrumentalist it is. The bleating sound isn't vocal. It is produced by the wind playing on the two outer tail feathers, which are held out, like lesser wings, during the snipe's display flight. Drumming we always call the display, although bleating better describes the actual sound.

The snipe's nest is often well hidden in thick growth, sometimes almost as open to the sky as a peewit's, always on damp ground, often in actual contact with water, occasionally in a clasp of marsh marigolds. The bird will lay up to five eggs, olive-brown in colour and boldly blotched, which are very large for the size of the bird.

By any standard, the snipe is a shy, wary bird, active mostly at or after dusk and hiding away mostly by day. A bird put on the wing by day will fly far and fast, and of course the display flight can be seen at almost any time.

The flushed snipe rises with a *craik* of protest before zig-zagging swiftly in going-away flight. Some snipe will crouch well down until almost trodden upon, although this is usually more characteristic of a sitting bird. A snipe hearing the approach of man or dog usually springs with time to spare.

A snipe that has been sitting for some time treads out a track, six feet or more long when walking off to feed and walking back to her eggs. She pitches at the end of the track when coming back, and walks to the end when leaving her eggs. This track is distinct, but not obvious to eyes that don't know what they are looking at.

A snipe with newly hatched young will drop right on to the nest as often as not, just as she will drop beside her chicks after they are away from the nest. They leave it as soon as their down is dry, or as soon after that as the hen decides to go. At hatching time the cock bird will often crouch a few feet away, for half an hour at a stretch, and he helps to brood the chicks after they are on their feet.

Territorial Disputes

The wind, having rocked the sea to sleep, was breathing gently cool. Not a white mane tossed on the blue waters. The Dutchman was wearing a cloud feather, fleecy but unruffled. In a tidal pool a heron stood, statuesque and motionless, like a figure on a Japanese screen. A buzzard, tawny in the sun, was playing kestrel above the roadside, holding station with swept-up wings and fanned tail.

A dead Blackface lamb was lying awash in the smooth channel that wound across the silver sand to the bay. There was a hole like a cup in its belly. It was a heavy lamb, big-hoofed. Beside it stood a great black-backed gull, pulling violin strings of gut from the hole with its powerful beak and enlarging it by expert surgery towards the ribs.

Then it was suddenly dispossessed by two of the bhoys, a pair of hoodies, as genuine greybacks, in bonnets of shot ebony, as you could ever wish to see. They swooped, with suitable commentary, and the great gull, himself of the 16in. gun class, yielded the body without the hint of an argument. He flew to the water's edge and waited.

The hoodies worked on the eyes and the hole, one of them standing up to the thighs in water and bracing himself with a foot. A man could see the gobs of gut being excised and swallowed. They ate on and on and I found myself saying to my companion they must have brought their tapeworms with them.

But eventually they flew away into the trees to whet beaks and preen, and it was then the big black-back sneaked back to take over. It got no further than two nicks with its scalpel. The grey and black bhoys sailed in on a converging attack and swept it clear back to the tideline. They flew back to the dead lamb. And began to eat again. There was no doubt now: they hadn't left their tapeworms at home.

Three times in the course of an hour the hoodies flew off, and the black-back came, and the hoodies returned to drive him away. When we left, the hoodies were in possession, still eating.

It turned out to be a day for bodies driving other bodies away from something for some reason.

A peewit, for instance, standing in amicable association with a drowsing Blackface lamb, took wing at once to drive a gull from her home acre. Then she pitched beside the lamb again. Then a hoodie appeared, trespassing on her air -space, and she was up at once to drive it away too.

When seven starlings pitched close to her she flew at them, and they scattered but she pursued them in the air, outflanking the outflyers, and herded them like a collie gathering sheep, and when she had them in a tight bunch she harried them clear off her ground. It was as perfect a job of herding as one could expect from any bird.

Next it was a heron, a big one with a knee-length fringe. He came down beside a pool, and at once the gulls came along, five of them, to drive him away. One at a time they swooped at him, and each time he pulled his head down, coiling his neck, ducking. When they had all made their first pass they wheeled to come in again. The heron took three steps then opened his wings like blankets and retreated down the estuary.

Heron was even driving away heron. Bird one left a pool and flew 100yds. Bird two came along and took his place. Bird one returned and drove bird two away. Then bird one flew off again, and bird two came back. But bird one returned and drove it off again. And so it went on until, after three such double moves, bird two didn't come back to the prohibited spot.

Late in the day I returned to the dead lamb in the channel, hoping I might find the black-back in possession. But the hoodie pair were there, still swallowing. Viscera are not very sustaining.

Blue Jim

When Pate the poacher discovered that his young Bedlington bitch had been mated with a greyhound, he immediately accused his wife of letting her run loose when his back was turned.

'Noo she's no' only wasted,' he shouted at her, 'she's in pup tae a groo an' the pups'll likely kill her.'

Betsy the Bedlington crawled under the hole-in-the-wall bed when Pate began shouting, for harsh words hurt her like a physical blow. She lay trembling while Pate booted the coal bucket up and down the kitchen floor, and swept dishes and cutlery from the table into the hearth. Then, when he threw the door wide open in dramatic gesture, asking God to strike him dead if he didn't have the stupidest wife in the parish, the Bedlington thrust past him and bolted into the night.

Through the wall, in the but-and-ben next door, Pate's mother-in-law listened to the crashes and the shouts, then hurried to the village on her bicycle to fetch the police.

Pate had quieted down by the time the policeman arrived. His wife had her apron over her face, drying her eyes. The constable stood with his back to the door, surveying the coal scattered over the floor and the broken delft in the hearth.

'Who's been assaultin' who?' he asked, a bit awkwardly.

Pate grinned. He said: 'Naebody's been assaultin' naebody, Sam. Ask the wife. Ye ken me better'n that. I've scattered a wheen coals aboot an' broke a few dishes, but it's ma ain flair-head an' ma ain dishes.'

'That's richt, Sam,' said his wife. 'There's been nae assault, an' nae breach o' the peace, if that's whit's in your mind.'

The constable sat down. 'An' whit was the donnybrook aboot?'

'Betsy,' said Pate. 'She's wasted. A groo got at her.'

His wife said: 'Pate thinks I'm tae blame because the dug's in pup, when the truth is it was his ain brither took her oot an' let her get away. She's a guid dug, ye ken, Sam.'

The constable laughed. He was a fair man, popular enough, and not hell-bent on tapes.

'Now, look here, Pate!' he said as sternly as he could. 'Ony mair o' this an' you'll be in trouble. Real trouble. This is the first time, so we'll let you go. But, damn it, man, you're wife's mair important than a dug!'

'That bitch cost me ten poun',' Pate defended. 'An' noo she's in pup tae a groo She's wasted!'

The constable shook his head. 'Where is this wonderful dug?' he asked.

'She's no' in,' Pate told him.

'When will she be in?' The constable had the words out before he could stop himself.

'Hoo the hell should I ken?' growled Pate. 'She didna leave ony word.'

'That's enough o' the bliddy lip,' the constable warned.

'Well, don't ask bliddy stupid questions,' said Pate.

At whelping time Betsy didn't die. She gave birth to six puppies under the hole-in-the-wall bed, on a dreich January night, when the constable was at the police dance and Pate was in the laird's coverts with ferret and purse nets. Two of the

115

puppies were smoke-coloured, two were brindles, and two lemon-and-white.

When Pate saw them he said: 'There's a queer cleckin if ye like. A real Heenz's mixture. Ah, weel, we'll see . . .'

'Ye've saw a' ye need tae see,' his wife said. 'So get a pail o' watter, an' get them droont, or syne we'll hae tae bed ootside tae make room for the dugs in the hoose.'

'The morn,' Pate compromised. 'I'll dae it the morn.'

In the morning he filled a bucket with water, put a lid on top, and thrust four puppies underneath. While they were drowning he sat looking at the two he had left with Betsy. Betsy licked his hands when he touched them. They were the two blues. Pate was thinking ahead, getting ideas about them.

'I'm keepin' two,' he told his wife.

Pate called the smoky pups Blue Jim and Blue. Both were dogs, whip-tailed and smooth-coated. Blue Jim was bigger than Blue, more active, and Pate thought he'd turn out more fierce.

Betsy was given plenty of porridge and fresh milk, and all the rabbits she could eat. Pate didn't buy milk; he had an arrangement with the milkman, who gave him all he needed in exchange for two pairs of rabbits a week. In return for the occasional pheasant, the miller gave him oatmeal. The pheasants, like the rabbits, were poached.

The pups throve amazingly. Blue Jim was a growler and a bully from the day his eyes opened, always pushing Blue away from his teat-hold although Betsy had more than enough milk for both of them. The more Pate looked at Blue Jim, the better he liked him.

As soon as the pups could lap, Pate gave them cow's milk in saucers. Then he fed them chopped raw rabbit. While Blue chewed his pieces of rabbit stolidly and quietly, Blue Jim growled all the time he was eating, and bolted more then he chewed. Pate's demands on the milkman increased every week, until they were met at last by a request for extra rabbits. Told to take it or leave it, the milkman accepted, grumbling; he knew this was the first time he had been on the wrong side in the bargain, and realised it wouldn't be for long.

Until they were eight weeks old the pups looked pure greyhound—long-legged, long-tailed and long-muzzled. There was nothing of Betsy to be seen in them at all. Then their hair began to bristle and twist, betraying their breeding.

Pate began to walk them for short distances. On the road he kept them coupled, but when he was nearing home he turned them free to run in the big field behind the house. In the field, Blue Jim attacked Blue, snapping at his pads, bowling him over, and reaching for his throat, until Blue became so afraid of his litter-brother that he would lie on his back, in submissive posture, as soon as he was thrown.

Pate was coming back with his pups one day when he met the village policeman at the road-end.

'Whit pups are thae?' the policeman asked, interested.

'Betsy's.'

'The wans bred fae the groo?'

'That's them.'

The constable stroked Blue, who wagged his whip-tail. Blue Jim growled as the hand came down, and the policeman drew it back.

'This is a girner, eh?' he said. 'In a wee while he'll take the haun' aff ye.'

'He's a' right,' Pate said. 'I'm keepin' him.'

The constable patted Blue again. The pup reached up, trying to lick his face. 'Whit's this yin's name?' he asked.

'Blue,' Pate said.

'I could dae wi' a pup like that. He'll be big. Handy for takin' aboot at night.' The constable kept stroking Blue.

'It'll cost ye,' Pate said.

'Come on, Pate,' the policeman laughed. 'Ye'd gie a pup tae a freen, sharely.'

Pate grinned. 'Me a freen o' the polis! That'll be right. A poun' an' he's yours.'

'Ten bob,' said the policeman.

'Are ye share that widna leave ye bare?'

'Hell,' said the policeman, 'I'm no' a man o' means like you. I canny live aff rabbits like you.'

Pate fished a length of string from his poacher's pocket, uncoupled Blue, looped one end of the string round the pup's neck, and handed the other end to the policeman.

'Here! Take the bliddy pup,' he said. 'So long!'

He turned into the lane with Blue Jim, dragging the protesting pup savagely to heel. The policeman held tightly on to Blue, who was crying, eager to follow.

At harvest time, Blue Jim, who was now more like a deerhound than anything else, killed the cat from next door. He snatched it from the doorstep, chopped it and tossed it in the gushet under the spiggot. The cat kicked, exhaled a thin mew, and died.

Pate's mother-in-law saw it all from where she was putting out clothes on a line strung between two trees. Screaming, 'My cat, my God, my cat!' she rushed to the doorway. When she saw her cat was dead, she picked up a slate and threw it at Blue Jim, who bristled, and faced her, growling in his throat.

The screams of her brought Pate round from the back of the house, where he had been feeding his two greyhounds in their shed.

'Whit's the barney aboot?' he asked.

His mother-in-law rushed at him, shaking her fist in his face.

'He's kilt Tibby, that's whit! If ye don't dae something aboot him quick . . .'

Pate shoved her towards the house. 'Just you try the polis caper again, my lady, an' see what it gets ye. There'll no' be a hen left at your back door the morn.'

She ran to the house then, shouting that he was a lazy loafer, a good-for-nothing poacher, a ne'er-dae-weel and a sheep stealer. Pate grinned. He called Blue Jim, put a rope on him, and led him to the back of the house.

There he beat him with a stick. At first Blue Jim accepted the beating, without cry or yelp. This seemed to make Pate angrier than ever, and he rained blows on the dog's ribs and rump. At last Blue Jim growled. Then he bit the hand holding the rope. Now Pate began to use his feet as well, kicking and booting unmercifully until Blue Jim gave up the struggle and cowered in submission.

He dragged the dog into a shed, kicked him into a corner, and slammed the door.

All that night, and the following day, Pate kept the dog in the shed without food or water. At darkening on the second night he brought him into the house, where he fussed him roughly and gave him milk and raw rabbit. Blue Jim responded by wagging his tail effusively and patting Pate's face with a friendly paw.

Pate now thought it was time to begin training Blue Jim. Though he had no clear idea how to go about it, he had it in his mind that the dog would make a lurcher, so the first thing he did was to give him a kill.

To 'give a dog a kill' means, in greyhound-racing language, turning loose a captive, healthy wild rabbit and slipping a long-dog after it, the release and the slip being timed to give the rabbit no chance of escape. Pate, besides being a poacher, had the rabbiting of many farms up and down the country, so he always had plenty of rabbits for the kill. Some he kept for his own dogs, but most of them he sold to other greyhound owners. He bagged three from his hutch to give Blue Jim a worry.

He turned them loose, one at a time, in a netted enclosure from which they could not escape, and there Blue Jim killed them. He upended twice before he caught the first one, which screamed like a handled piglet before he broke it up in a corner. The other two he killed quickly and cleanly, with inherent skill.

Blue Jim proved he was a natural poacher. He was quick and eager to learn. Pate's schooling was a beating when the dog did the wrong thing, and a rough pat when he did right, and Blue Jim quickly learned what not to do. Most of his accomplishments were the result of his own natural flair. He had the knack of doing something on his own while Pate was still thinking how to make him do it.

He quickly learned to take cover at the sight of a policeman. He learned, at a word from Pate, to make his own way home, across country, from any place within eight miles of the house. He learned to leave Pate's van at a word, course a hare in a field, and return with it to the van. He hunted mute without teaching.

Pate often took him in the van when he was going out for a night's long-netting at a farm. He didn't use Blue Jim at the netting—no good long-netter uses a dog—but in the morning, on the way home, he would sometimes slip the dog at rabbits in a field, often adding a dozen to the night's bag in this way. One morning Blue Jim killed twenty rabbits in an hour.

The dog had speed, tremendous speed, and stamina. Pate also discovered that he had a nose. In threshy pastures the dog could rouse rabbit after rabbit from their lies and chop them almost as he put them on foot. After that he learned to mark silently at burrows when Pate was purse-netting with ferrets, lying quietly back while the rabbits filled the nets and Pate tomahawked them deftly with the edge of his hand.

Not until his second autumn did Pate run Blue Jim at hares on the high moorland pastures. It was then he discovered the big dog's true speed. He was the best hare catcher Pate had ever seen, and probably faster than any long-dog in the neighbourhood. Pate had an ethic. It was that he didn't poach the ground of people he liked, unless they refused his price for their rabbits. He poached the ground of those he didn't like without making them any offer first.

One of the farmers he didn't like was Winter of Winnock. Two or three times a year Pate raided Winter's hens, taking a dozen or so each time and selling them to poulterers who asked no questions. He poached Winnock as regularly as he would have worked it by arrangement.

The Winnock, which included a wide area of rough grazing, had always plenty of hares, and one morning at tattie time, just after sunrise, Pate took Blue Jim there to try and catch six hares for the poulterer. He parked his van off the moor road, in a bay used for road grit. In it were twenty-seven pairs of rabbits netted earlier that morning by Pate and his brother.

Blue Jim had caught, and brought back, three hares when Pate saw Winter's car leaving the farm close, about a mile away across country and nearly two by road. He hid the hares, ordered Blue Jim home, and returned to the van. He was starting the motor when Winter arrived.

The farmer was squat and bald, with three days' stubble on his chin. He stopped his car in front of Pate's van, blocking his entry to the road.

'The polis'll be here ony meenit,' he said, 'so ye micht as weel save your petrol.'

'Polis?' said Pate, switching off his engine. 'Whit does the polis want?'

'Don't try an' fly-man me!' Winter shouted. 'I've been watchin' that bliddy dug o' yours for the last hauf hoor, catchin' hares an' rumblin' up the hoggs!'

'Dug? Hares? Whit dug? You'll be seein' elephants next! Hae ye been on the batter again or whit?'

The farmer grew red in the face. 'Aye, hares!' he roared. He moved towards the back of the van.

'Jist try an' open thae doors,' Pate said, 'an' I'll crown ye wi' a spanner.'

'Nae matter,' said the farmer. 'The polis'll see. Here they come noo!'

The village constable arrived from one direction in a police van; the prowl car, with a sergeant and constable, came from the other. Pate took the offensive.

'Mebbe *you'll* get this man oot the road, sergeant. He's . . .'

'That'll dae, Pate,' the village constable interrupted him. 'We ken whit's whit!'

'It's a guid job somebody kens,' Pate grumbled. 'This eejit's haudin' me up.'

The sergeant said to the farmer: 'This man . . . whit's his name . . .?'

'Pate,' said the village constable.

'This man Pate,' the sergeant continued. 'His dog was on your grun'? Caught twa-three hares? Coupit some sheep?'

'That's right,' the farmer said. 'I seen this dug coursin' wan hare, then it got in among the hoggs.'

'Did you see this man wi' it?'

'Naw! I seen naebody till I got here. But the dug—it was a big dark-coloured greyhound—made for the road here. This yin,' he indicated Pate, 'was the only wan on the road. He'll hae the hare in his motor.'

'Dug! Dug! Dug!' Pate broke in. 'I've seen nae dug! Where is this dug?'

'You shut up!' warned the sergeant.

'Belt up yoursel'!' said Pate hotly. 'I've done nothin'. An' you're no' the Gestapo, mind that!'

'I tell ye,' the farmer said. I seen this dug. It came ower here. It canny be far away.'

'Where's the dug?' the sergeant asked Pate. 'And what hae ye got in the motor?'

'Rabbits!' said Pate.

The sergeant's eyebrows went up. He looked at the village constable, who grinned wryly, and said: 'That'll be right enough.'

Pate ran to the back of his van, and threw open the doors. 'There,' he said, 'tak' a look. Twenty-seeven pairs o' rabbits, caught legal an' papers tae prove it, every wan cauld as a stepmother's braith an' deid since before daylicht!'

The sergeant and his constable rummaged among the rabbits in the van. They removed the front seats and looked under them. They looked underneath the van. They found no hares, and no dog.

'Nae hares,' the sergeant said to the farmer.

'Where's the dug, Pate?' the village constable asked.

'Ma dugs are a' at hame. It's somebody else's dug you're looking for. Noo, if

119

ye've a' feenished, I've thur rabbits tae clean an' deliver before nine o'clock.'

'Just a meenit,' said the sergeant. 'Hoo mony dugs hae ye got?'

'Fower,' Pate said wearily.

The sergeant turned to the village constable. 'Sam tak' a run ower tae the hoose an' see. Ye should be back in fifteen minutes. I'll see whit's whit here.'

When the constable had gone, the sergeant said to Pate: 'Hae your smoke while I look aboot.'

'I don't smoke.'

'Well, just wait.'

Pate watched the two policemen and the farmer searching up and down the roadside ditches, looking for hares. They left the road and searched among threshes and heather. They found nothing.

Twenty minutes later the village constable returned.

'Well?' said the sergeant.

'The dug's there. The wife says Pate's been away since eleven o'clock last night. The dugs have never been oot.'

The sergeant said to Pate: 'Right Pate, you can go.'

The farmer fumed. Pate, starting up his van, called to him: 'Next time you're on the batter, see elephants. They tell me the Zoo's short.' The sergeant covered his face with a mittened hand.

Next morning, the farmer reported to the police that twenty-five of his Light Sussex pullets were missing, and that three hoggs had been worried by a dog. At the time he didn't know that two hoggs were missing.

Pate sold the hoggs to a butcher in the city. The poulterer took the pullets.

In those days there were two Sunday schools about a mile from the village: the pontoon school and the tossing school. They met in a ruined steading. The bobby-watcher, an unemployed man, was posted in one of the attics. He was paid by a levy on every pontoon and on every pair of heads tossed.

Pate lost his van when Irish, the kerb dresser, who was on the pennies, headed three pairs in a row. But he wasn't out of business. On the following Tuesday, he won £33 off the greyhounds, and on the Thursday bought another van for £18. He gave his wife £7 and held on to £8 for the schools on Sunday.

He arrived at the steading when the church bells were ringing in the village. The shot-firer was on the pennies, and Irish was squatting with over £40 at his feet.

'Ho-ho!' greeted Irish. 'Anither wan o' the faithful. Hullo, Pate! How's the dug?'

'Fine. I'm for the pontoon the day.'

There were six men at the pontoon school, with four greyhounds bellied down in an outer circle. Pate hunkered down in a gap with Blue Jim at his side.

'For a haun', Pate?' the banker said. 'The kind man's in!'

'That'll be right,' Pate replied. 'I see ye skint the van aff Irish. Dale me in.'

He won £1 on his first bet, which gave him £9. The next deal brought him a deuce, on which he bet £2. His second card was also a deuce. He bought for £2 and got a third deuce. With expressionless face he bought for £2 again and got a five. Holding the cards, he put down another £2 and got a king. He threw his cards, face up, to the banker.

The banker grinned. 'It's your lucky day, Pate.'

'We'll see,' said Pate.

The banker turned up a pontoon, and Pate was left with a single note.

He played for an hour, up and down between £5 and 5s., and then he was out.

'That's me,' he announced. 'How aboot runnin' up the dug?'

'Nothin' doin',' the banker said. 'I'll tell ye whit, though; I'll gie ye a straight bet for the dug. Ten poun' against the dug on the next deal.'

'Right,' Pate agreed.

Three minutes later he had Blue Jim and ten pounds. In ten minutes he had £22 and the bank. An hour later he had cleaned up the school, and had £68 in his hand. Then he went to the tossing school.

After half an hour at the tossing school he had £93, and was on the pennies. On the ground there were £36 in single notes when he threw the coins. The men looked up, then down.

'Head two!' said Irish. 'The man's anointed!'

A low whistle came from the bobby-watcher in the window. He shinned down and hurried over to collect his money. 'The polis,' he announced.

The schools broke up. Cards and money disappeared into pockets, and fifteen men and seven greyhounds and a lurcher filed through the hedge into a rough field, where a youth was working with an old bicycle, a drum of ham twine and a ba' hare.

Irish and Pate helped the youth to rig up the bicycle. They fitted the drum of twine to the sprocket, and turned the bicycle upside down, planting it firmly. Once the twine was paid out to its full length, two men, ca'ing the pedals by hand, could wind it in at speed. The ba' hare was attached to the end of the twine, and a man carried it 200yds. down the field, unwinding the twine as he walked.

'Right!' said Irish.

Seven men walked seven greyhounds to the far end of the field, 250yds. away. Pate followed, leading Blue Jim. Two men took their places beside the bicycle pedals, and Irish stepped clear, removing his cap from his head. When he dropped it, the men would wind and the dogs at the far end of the field would be slipped.

'Here's the polis,' said one of the men on the pedals. 'Who's the civvy wi' him?'

'The C.I.D.,' said Irish. 'Wi' a hat big enough tae shade a town!'

The policeman stopped beside the bicycle.

'Training school, eh?' said the constable.

'That's right,' Irish said. 'Whit did ye expect?'

'A tossin' school.'

Irish laughed. 'On oor wages?' Noo, if ye'll staun' back, I'll wave on the dugs. We maun keep them in shape, ye ken.' He raised his cap aloft, held it for a moment, then threw it to the ground.

The men on the bicycle pedals began to ca', and at the other end of the field seven greyhounds were slipped. Almost on their heels, Blue Jim broke away from Pate, trailing his short leash. Despite the trailing leather, he lengthened out into his tremendous stride, and quickly closed the gap. At 150yds. he was up with the bunched dogs. At 200yds. he had headed them, and when he crashed into Irish at the other end, sending him on his back, he was three clear lengths in front.

'Did youse see him?' exclaimed Irish, sitting up. 'He came through them like a dose o' salts!'

Later, when the police had gone, Irish said to him: 'That mongrel's just bate the best dug oot o' Ireland! That brindle o' mine coast a hunner poun' and this yin left him a park length. He should be racin'.'

'Blue Jim? Wi' a hide like that?'

'Shave him,' Irish said.

Next day Pate blunted three razors and broke two pairs of hair clippers smoothing Blue Jim down, and at the end had a greyhound standing 29in. tall, weighing 74lb. and nicked in nine places. A few days later, running under the name of Honest Heck, Blue Jim qualified at a flapping track, his weight and time kept up with a bellyful of porridge.

The night Honest Heck won his first race—seeing the hare home while the second dog, the favourite, was still five lengths behind—he broke the track record and the bookmakers' hearts. They had to claw the caup to honour the bets which had been placed, unobtrusively and simultaneously, by six men with six different bookmakers just before the start of the race.

The six well-put-on, gainfully employed who came to collect had been deputised by Pate to spread his money so that he could make a kill. The kill was £360, and that in the days when the pound sterling could still look the dollar straight in the face. Pate gave his adjutants four fivers among them and drove home with Blue Jim.

He took the dog to other flapping tracks and at the end of three weeks was more than £700 in pocket. Blue Jim had won more than that, but Pate parted with many tenners on other racing certainties. Inevitably, the time came when the dog was bet off the boards. Pate discussed ways and means with Irish.

'He's been back marker three times an' won in a canter,' Pate complained. 'The bookies'll hardly write his name.'

'Here's whit we'll dae,' said Irish. 'We're in the same race next week. Wi' Blue Jim in the race I'll get threes aboot Parnell. If we stuff Jim and send on Parnell he could win. But ye'll hae tae take ten yairds aff Blue Jim. Can ye dae it?'

Blue Jim had already beaten Irish's brindle dog Parnell by five lengths when giving him two yards. The other dogs in the following week's race couldn't catch Parnell.

'In fact,' said Irish, 'it'll be a two dug race at best, an' if ye can stop Jim we're there. I'll see the others aboot stoppin their dugs, an' we'll let them in on Parnell.'

'Dae that,' said Pate. 'I'll get Blue Jim stopped.'

On the night of the race Pate found Irish in a dark corner pouring stuff from a bottle down Parnell's throat. The other owners were standing by with their dogs. The arrangement was six £25 bets on Parnell, placed simultaneously by the six men.

'You're sure aboot Blue Jim?' Irish asked.

'He's stopped,' Pate assured him. 'I raced him at five hares in a row, after they had a good start. He's stood a' the way here in the van. An' he had a basin o' parritch jist before he came in. Oh, aye, an' last night he had castor ile.'

'That'll dae it,' Irish said.

They took their dogs in and waited. At the parade Blue Jim's tongue was showing. The dogs were put in the traps.

Blue Jim trapped first, bounding like a cheetah. Parnell held him off until the last turn, then Blue Jim flashed ahead of him. Pate swore, Irish swore. The others accused Pate of shopping them.

Blue Jim won by two lengths.

Pate was warned off every track at which Blue Jim had raced, because, they said, the dog wasn't a greyhound. For a spell, Pate put him to catching rabbits again. Then Irish came along with the idea of the drive to the east.

They drove through and qualified him, then entered him for a race under the name of Tricky Jim. They entered him in the name of a local man, a relative of Irish.

Because the opposition was strong they bottled Blue Jim before the race. The bottling put five yards on him and he won by three lengths. Pate made £80 from the trip, and booked the dog ahead.

'Maybe we can mak' a kill through here,' Pate said.

'He should be good for two races yet afore he'll need shavin' again,' said Irish.

'The opposeetion's good, but. We'll hae tae keep bottlin' him.'

In his next race Blue Jim took the lead at the second bend, but, when the nearest dog growled at him, he turned his head. He lost balance and was bumped out of his stride, and Pate groaned when he saw the dog run wide then fall head over heels in a fankle of legs. Blue Jim limped off the track, carrying his near hindleg.

When the veterinary surgeon saw the dog, he said:

'Fractured femur. He'll not race again.'

'You mean you canny sort him?' Pate asked.

'Oh, I'll fix him all right, but he won't win another race.'

'Sort him, then,' Pate said. The veterinary surgeon set the leg up in plaster. Pate paid him and drove home.

He shot Blue Jim in the morning.

Billy Butts In

The big thin-furred fox—no lank halflin, but a robust himself probably pre-Watergate—went over the high fence with the greatest of ease, which any fox with more than a distance piece between his ears can do at any time, as those who know anything at all about foxes will tell you. And that was him in the big field where dwell our wild goats—William the billy with his two wives and two this-year's-vintage daughters.

Luckily I was sitting in my van on a vantage point from which I could view nearly the whole field, seeing the fox without being seen, and well out of his nose. As I say, he was a dog fox, a fact he made clear to me by exhibiting his diagnostic features.

Was he in merely to cross the field? If he was, why cross it at all when he could as easily have gone round it? Was he interested in the goats? The kids, I thought, were maybe too big for him, but I had to bear in mind that we lost a very small one two years ago, without ever being sure about what had happened. It could have been taken by a fox.

Anyway, here was a fox now, and I was in a position to watch.

So was William. And in case you've forgotten about our William I'll tell you that he came to us as a gentle and genteel wee character whose main interest in life was a milk bottle. And he grew up at my back door, playing with all and sundry, or one and all if you like, and taking milk from anyone who cared to offer it to him.

123

Then he grew up, and up, with his horn growth keeping pace, and in the end he was placed in his two plus acres paddock with females for company. Last year he fathered his first kid; this year, back in March, he fathered two female kids. And that is our herd.

And now here was a fox visiting at seven o'clock in the morning, with the sun playing hide and seek among racing clouds, and the wind spitting rain from the west.

The two nannies stood, with kids at flank, watching the big tod, more in curiosity it appeared than in fear, while our William, who could and would beat your brains out when he's in the mood, just watched, turning slowly as the fox crossed his front. And I thought to myself—the tod is going on.

But no. He changed direction, and began to pad round William, maybe thirty yards from him, with his tail twitching now and again, just like a cat's. William stamped a foot, then began to come forward. At ten yards he turned side on, and minced alongside. And the fox kept turning in.

By now the nannies were hard against the rocks, with the kids close against them. The fox stopped, lifted his head, lowered it, then sat down to ponder. William replied by rearing tail, and butting the air with his armoured front.

What a show he gave, up and down, tall as a man, skewing his head, and butting the air, right and left and forward.

The tod decided to come on, keeping close to the rocks. William stopped upping and downing and watched. Then, when the fox was five or six yards away, he had had enough. He went in, with head down, and the reynard turned about smartly and ran. But only a score of paces. He stopped when William stopped, and did some more pondering.

He decided to try again, and inched back at a crouch. And then our William boiled over. He went in hard, and the fox fled, and our William followed him right to the fence, stabbing up at him as he went over the top. The fox kept on going, and William took out his spite on the fence, ramming it hard and hooking the strong mesh with his horns. I drove round and called him across, and he came, and I went in and congratulated him, whereupon he reared and thumped my palms with the flat of his horns, almost knocking me over. Then I scratched his neck and he subsided, and walked away to graze as though nothing had happened.

I doubt if the fox will try again.

How Scotland Got Its Wild Goats

From the time I was a small boy at school until I was nearly a man grown I was friendly with a big, lean Irishman by the name of Cornelius O'Donnell with whom I shared the distinction of having a stove-in nose. He kept goats: a hirsute, sheemach crew in all shapes and sizes. If you asked him what kind of goats they were he'd say they were Irish goats.

Con (we called him Con, what else?) lived in a small cottage, with a bit of garden that wouldn't have maintained one goat in affluence, let alone the dozen or so he had at any one time. Opposite the cottage was a whinstone quarry and a pond where frogs spawned: behind it was a strip of deciduous woodland and six acres or thereby of rough, hillocky ground. None of these exists now, but that is neither here nor there.

The goats fascinated me. Over the years there were piebalds, skewbalds, black-faced silvers, greys and blacks: every permutation in fact except green-and-white and orange. But their colourings and markings interested me less than their habits. They were free rangers, oot ower a' at every opportunity, or as Con would have said: 'mair oftener oot an aboot than in an aroon'.

You would find them in the quarry, jumping from rock to rock, or getting the kerb dressers' goats by stealing the woodbines and bread-and-cheese from their jacket pockets. You would find them in the wood stripping bark. You would find a nanny kidding behind a hillock. Or a billy wandering into town.

But nobody bothered very much about such goings-on. The world was a more easy-going place in those days. Once in a while there would be a bit of a splore when a Con goat got into one of the big hoose gardens a quarter of a mile away; but it wouldn't last. Con would tether the goats for a bit, mostly on the rough ground which didn't belong to him. Then, after a bit, the *status quo ante* would take over.

The collars Con used would have tethered the *Hood;* the leather and buckles were good in these days. One August a roebuck tangled with one of Con's tethered billies (I used the incident in my book *Six Pointer Buck*) getting an antler under the goat's collar. They were in serious trouble, and if anything was going to break it wouldn't be the collar. When I reached them the billy was near strangled and the buck likely to break his neck. Instead, the antler broke, and the beasts were free.

Con's goats started my interest in goats in general, and I kept a pair of white milkers myself when I had my little farm. But it was when I began looking at the wild goat herds of Scotland that Con kept coming to mind. The wild herds looked just like Con O'Donnell's goats. And, in fact, they all originated from domestic stock gone feral. The Cornelius O'Donnells of centuries ago produced the ancestors of the modern wild herds.

Goats, as Con's animals showed, have a natural predisposition to take off, and will quickly live wild if allowed to do so. There is no such thing as a truly wild goat in Scotland.

The ancestor of the domestic goat breeds was the Persian wild goat, and domestic goats were the ancestors of the modern so-called wild goats of Scotland. Feral goats are like Con's 'Irish goats', and come in a variety of colours and markings: black-and-white, brown-and-white, black, brown or white, or even cream. Kids are often beautifully coloured. Adult billies can grow a routh of horn

125

like the true wild goats.

The wild goats of Scotland are born survivors, even in the most shriving environment, but mortality among kids can be high. They are born in February/March, when the weather can be severe. But this makes for rigorous selection. Apart from this culling by weather, so to speak, there is some predation on kids by eagles and foxes, sometimes even by wildcats. But the herds survive, as they have done for centuries, and heft themselves so well in some areas that they have become a nuisance.

Where they are not disturbed wild goats can be very approachable, to about a hundred yards or so. But where they are being hunted they can be as wild as the wildest red deer on the hill, and have to be as carefully stalked. Like the red deer they move up and down through the contours, and will cling to the highest ground and the wildest screes when under pressure.

During the first week or two of its life the wild kid stays close to its mother, then it begins to play with others of its age. In almost every way the behaviour of goat kids is like that of lambs of the same age.

Litter Louts and the Trigger-Happy

It was one of the warm, sunny days earlier in the month, and we were sitting in armchairs, he at one end of the bed and I at the other. We were facing the Campsies, conversing between lunts of our pipes.

'The furniture's better arranged this year,' I said. 'It gives a better view.'

'That'll be right,' he agreed.

'Last year it was facin intae the wid. But mind ye I saw a squirrel wan day, jinkin aboot a tree ower there. Watched it for quite a spell.'

We smoked and watched the peewits wheeling and strumming in the field across the road.

'These are mair comfortable chairs,' I observed.

'Aye! I prefer this mocket stuff. It's warmer. Last year it was yon rexine stuff, a bit cauld on the backside.'

'An nae cushions,' I added.

We smoked in silence for a minute or two.

'The trouble wi this place,' he mused, 'is there's nae privacy.'

'An it's draughty,' I carried on, 'an gey weet if it rains . . .'

'And bloody scandalous!' he exploded.

A lorry came round the tight bend, its wheels spurting up a cloud of dust in our faces.

'An nae windaes,' I said. 'Hoo long has this place been furnished?'

'Oh! aboot six weeks this time.'

I rose and turned over the bed. It was heavy with wet and the grass pressed and pallid underneath. The ground was grooved with vole creeps and as we looked a fat vole scurried from a ball of dry, shredded grass into a hole in the bed.

'It's an ill wind,' I said, probing for the vole. I caught it and it nipped, drawing a droplet of blood from a finger.

'That'll larn ye,' he laughed.

'Not many of them bite,' I told him, as I wrapped the vole in a handkerchief.

'It's amazing what a man finds when he turns ower the bed,' he said, shaking his head.

We stepped on to the road, then turned about to look at this one of many outdoor sittingrooms, dumped there by some litter lout at dead of night, or perhaps even in broad daylight because the littering habit is no longer exceptional.

'You should take a picture o that,' he suggested.

'I've taken dozens o them,' I replied. 'They're auld kale noo. Refurnishing the countryside is a growing habit. We ca doon wids and grub oot hedges and replace them wi auld chairs, bedding, prams and what-don't-you-want. The County people will shift this lot one of these days.'

'There'll be birds nestin in the chair stuffin before that,' he countered.

'It has happened,' I agreed. 'A pair of blackbirds nested in an auld chair at anither dump like this. Even the badgers hae got in on the act.'

'D'ye tell me that?'

'Aye! Somebody dumped an auld pouffe in a wid aboot a mile frae here. That was last year and the badgers ignored it at first. Then this spring, when they were takin in bedding, they ripped oot the stuffin and balled it away intae the den.'

'Jeez!' he said. 'Marks & Spencer furnishing badger dens.'

'Naw! This was badger brand.' He thought the weak-kneed joke was funny. 'They're clean beasts and like a clean bed. That's why they left the pouffe to air for nearly a year before they used it. And come to think o it; they cleaned up somebody else's litter. We should use the badger in a *Keep Scotland Tidy* campaign!'

'Well use it!' he said.

I have just used it.

There is another side to this abuse and misuse of the countryside— uncontrolled, slap-happy shooting. Nothing gets any peace any more. The woods crawl with people with guns every day, including Sundays, right in the middle of the breeding season. Some shoot anything that moves; others kill what they are pleased to call vermin, which means anything they don't like.

In the past fortnight I have found a dead tawny owl, a dead kestrel, a dead domestic cat, a dead fox and a dead weasel—all shot. That doesn't take account of the magpies, rooks and crows. Last night I had a Larkhall man on the phone telling me he had found eleven curlews shot dead on a local patch of wetland.

It seems that democracy isn't doing any better with wildlife than the worst of the old-style gamekeepers.

Vanishing Wildlife

On Arthur's Seat there once lived Artaxerxes—not a Greek general, but a butterfly. The Artaxerxeans lived there, fluttering about the crags above Duddingston Loch, long before the New Town was an idea in anybody's head, and for a century after it was built. It's extinction was not brought about by the building of the New Town. Nothing had changed on Arthur's Seat to cause it. The butterfly was wiped out by collectors of a type we still know today.

In case it is thought that a local extinction such as this was no more important than, say, the dispossession of peewits by the new town of Cumbernauld, it has to be explained that Artaxerxes was a very special butterfly. It was first discovered on Arthur's Seat, and there the first description of it was made. At that time, and for a long time afterwards, Arthur's Seat was its only known habitat. When it was exterminated there that, so far as anyone knew, was its final demise. Edinburgh had lost the only known population. That it was dicovered elsewhere later on does not lessen the impact or the lesson of the story.

Arthur's Seat then was as world famous for its butterfly as St Kilda is today for its wren. By 1844, according to Ritchie, collectors had preyed so heavily on it that all the English cabinets and most of the foreign ones had been supplied with specimens from Edinburgh. The collectors gathered not only butterflies; they gathered pupae and caterpillars, and the rock rose on which the butterflies laid their eggs, and stocked the cabinets of the world, leaving Edinburgh with nothing.

You may feel—well, after all, Artaxerxes was only a butterfly; its disappearance didn't really affect anyone, except perhaps the specialist lepidopterist. If you feel that way, then let us look at a food species and by all accounts, an economically important one—the oyster.

Lowther could write in 1629: 'Hard by the town of Leith be oysters dragged which go to Newcastle, Carlisle and all places thereabouts; they be under 3d. the 100.' In 1778, according to *The Old Statistical Account*, 'there were 8400 barrels of oysters exported from the city's fishing grounds. This trade was increasing so much as to threaten the total destruction of the oyster beds. The magistrates have therefore prohibited the exportation, and even the fishing of oysters under a certain size.

As late as 1794 the oyster beds at Prestonpans were the main fishery for the port. But the oyster beds and the oysters have gone: an old version of a now familiar story—over-exploitation of resources and no conservation.

The Old Statistical Account is notable, so far as Edinburgh is concerned, for its failure to mention even the barest details of the fauna and flora of the area, although it does have the interesting note that whereas public cockfighting was unknown in the city in 1763 it was flourishing by 1783, and continued to do so in 1790. The *main,* as the cockpit was called, was a place 'where every distinction of rank and character is levelled.'

It is a far cry from 1416, when a pair of White Storks nested on St Giles Cathedral, to 1730 when the cry of *Gardy Loo* was still familiar in Edinburgh, but the pace of change in wildlife numbers and species was slower in the three centuries before 1730 than in the two centuries since.

In eighteenth century Edinburgh dogs, cats and pigs were still common scavengers, but so were birds like the raven and crow. The raven, indeed, used to breed on the Castle Rock and was nesting on Arthur's Seat as late as 1837. There would no doubt be kites, too, because they were common in Stirlingshire in 1795, and still present in East Lothian about the same time. Nowadays all the British kites are Welsh, and you could number them on the fingers of your hands.

It is difficult to appreciate today that the polecat, now said to be extinct in Scotland in its wild form, was common in the countryside about the New Town. Some time before 1880 it disappeared from Midlothian.

The pine marten, highly prized for its valuable fur, was once a common mammal, too, still to be found in numbers as far south as Kirkcudbright as late as 1796, and classified as 'rare' in Stirlingshire at the end of the eighteenth century. Facts like these are no less surprising today than the nesting of storks on St Giles in 1416. The badger, although killed to some extent for its skin, was hunted for a variety of reasons—sport, baiting, food—and at one time was considered rare. Edinburgh had badgers in the eighteenth century and still has them on Corstorphine Hill.

Duddingston Loch is today a haunt of wildfowl. Two centuries ago there were others, the Nor' Loch, now Princes Street Gardens, and that other, drained and planted by Thomas Hope, and now his memorial—the Meadows, which the people of Edinburgh take for granted. And there was the morass that became the Mound, built of 1,305,750 cartloads of earth, dug to make the foundations of the New Town.

What a wealth of wildfowl must have been there, in the heart of modern Edinburgh.

An Obscenity Called Velveteering

If there's a fast buck to be made out of exploiting some animal in some way, you can be sure somebody will get around to making it sooner or later. The newest kinky one is making the fast buck out of the red stag's velvet (no pun intended).

It's the old, old, story. You do something questionable that earns you a buck. Then you justify it on the grounds that it is profitable. Then the practice, in terms of profit and economics, becomes its own justification. And if somebody tries to stop you that somebody is accused of taking away your livelihood. Not to mention being accused of putting animals before people.

Maybe you saw the discussion programme on TV, where a deer farmer tried to make a case for velveteering, while the British Deer Society's consultant veterinary surgeon made a very poor showing in opposition. Why is it that considerate humanitarians so often allow themselves to be maneouvred on to the defensive? It seemed to me that the interviewer missed a great opportunity to score some waterline hits on both with heavy shell.

Avoiding all double think and newspeak let me say that the red stag's velvet, like rhinoceros horn, is considered to be an aphrodisiac by certain jaded orientals itching to become crowing roosters again. That's the way it is. That's why the rhinoceros was hunted almost to extinction. That's why they want the red stag's velvet.

129

It doesn't matter a damn whether the stuff is an aphrodisiac or not, and it's totally irrelevant whether somebody has proved that it is or it isn't. To subject the red stag to the brutal business of having the living velvet stripped from him to pander to such a market is close to being an obscenity. So we'll anaesthetise the stag they say. Shall we anaesthetise our consciences too?

The buck, the buck; it's always the buck. We did it when we put hens in battery cages, pigs in sweat boxes, baby calves on an iron-deficient diet to make their veal white. And then animal loving Britain is accused by the Germans of cruelty to hens.

Not so long ago many farmers would have liked to see the red deer eliminated because it competed with sheep and cattle for the grazing. Then the deer farming thing came in instead. Deer became big money.

I can think of no good reason why deer shouldn't be farmed if it's done properly. But to subject them to this because of somebody's waning sexual libido seems to me to be beyond the pale. And I don't care how many New Zealand veterinary surgeons have proved this or that.

Then there's the hypocrisy bit. They're talking now of the alleged medicinal value of red deer velvet. Medicinal my sphincter! That's the PR boys and the euphemismerists at work to con the big sillies—US. The velvet is wanted as an aphrodisiac: full stop.

And I say to hell with the Oriental nuts who want to flog their waning potency. There are plenty of hookers around who would do the job for them.

My own feeling is that enough is enough, and I would oppose any attempt to legalise velveteering. Let's be clear about one thing. Nobody is in the business of velveteering because they want to help suffering humanity by providing a magic medicine. They want to make a buck out of pushing an alleged aphrodisiac.

When it was the horn of the rhinoceros many of these same people laughed at the quaint Oriental notions. But now it's red deer. In abundance. So let's get tore into the bonanza.

The red stag's growing antlers are living bone, covered with living tissue. When the blood supply to the covering (velvet) stops, the velvet rots, and the stag rubs it off. His clean antlers are dead bone and can be cut off without pain. What we're ettling to do is to strip the living velvet in the search for the fast, very fast buck. This one says no.

Of Buttons and Bucks

This sapling roebuck (1981 vintage, and well forward at half his twelvemonth) came out of the young pines upwind of me, a clean, slim cut-out against the cranreuch, and when I put the glasses on him I could see he was wearing his buttons.

Buttons . . . Now there's a word I don't fancy much to describe what a six-month-old roebuck wears on his head. They're more like cones, little hard horny cones, from half to thimble size, and they're the first thing he sprouts on his pedicles: a sort of John the Baptist before the real antlers.

Pedicles . . . There's another word, a right enough word if not very vernaculary, if you know what I mean. People often mistake pedicles for antler buds, which they're not, although they are like enough antler buds. In fact they're bony protuberances from the skull, providing the bone-forming material of which antlers are made, and from which antlers will eventually grow.

So this halflin I'm telling you about had buttons on his pedicles. No antlers. Yet . . .

These buttons, cones, horntips, John the Baptists—call them what you will—appear on the young buck in November or December. Unlike the pedicles, which are furred, the buttons are as bare as a cow's horns. As far as I could judge, the young buck I was looking at had buttons about the length of my pinkie nail, and maybe a little broader at the base.

The young roebuck doesn't wear his buttons for long before he loses them: in late January to late February or early March, depending on whether he grew them in November or December. You can't make rules about them, although many people try to, just as they do with antlers. In my part of the world February, mid to late, is the normal time.

When the buttons fall they come off cleanly sliced from the pedicles, like a candle sliced with a cut-throat razor. Then the antlers begin to grow and, unlike the buttons, they're furred over (in velvet) from the first sprouts. Just what sort of antlers the young buck will grow first time round is not predictable. A lot depends on his condition, the food supply available, and, no doubt, his genetic make-up.

The way I learned it as a boy was that the young buck grew single spikes in his first year, a forked head in his second and the full six points in his third. It's a neat caper, typical of the way we like to make rules for animals. But it is demonstrably untrue. It can happen, but as often as not, probably more often than not, it doesn't.

I had no way of telling how the young buck I was looking at would turn out. But I know how others like him have turned out. I've seen spikers in spring, forked heads in spring, and from time to time full heads in spring, albeit the six-pointed yearlings weren't all that remarkable: four good tines and two draidlets. But six there were. Our favourite roebuck (the late Bounce) put up six points as a yearling, and they were good.

A mature roebuck will cast his old antlers in November/December and his new ones will be complete and clear of velvet in late March or early April. That is just when the yearling's first antlers are beginning to grow, or are at an early stage of growth. The yearling will, therefore, have antlers in velvet long after the mature bucks are clean: that is in May, or even into early June in the case of backward beasts.

Come November again the old beasts will cast their antlers, but the young beasts, the rising two-year-olds probably won't do so until December. But a lot depends on how forward the young beast is and how backward the old ones are. You can't make rules about it.

Unlike the bucks, roe deer does do not grow antlers except in rare cases that nobody has yet been able to explain. An antlered doe (and I've seen only one) grows simple spikes that remain in velvet. It used to be said, and probably still is, that an antlered doe is barren. Barren such a doe may be, but suckle a fawn she can. I know, because I've watched one doing it. And she was, as I've just said, the only antlered doe I have seen.

By the way, at this time of year when mature bucks have lost their antlers, and yearlings haven't begun to grow theirs, you can always tell buck from doe in the field. Does grow an anal tuft that looks like a down-pointing tail. Bucks don't have this winter adornment.

On a Cold and Frosty Morning

One day it's spring, you think, with the lark up crossing the T of a jet vapour trail, and getting a hearing despite the engine shout-down, and the next there's cranreuch and a crackle, with grizzled fields and grey-haired trees, and the sky blue with cold, and the dispersed birds making back to the bird table again, like a film running in reverse, and instead of the dozen mouths of yesterday there's five dozen again, and a hurried scraping of the barrel to find enough for them to eat.

You'd wonder what the rooks get to do out there on the frosted field, with the grass bitten down to the bone by sheep, but they soldier on rookfully, stabbing and dibbling, glossed like beetles in the sun, swinging their ragged kilts with inimitable crow-swagger, their eyes like ripe elderberries peering over wrinkled yashmaks of grey lizard-skin; gey birds and gaucy, as much a part of the vaporous blue and silver morning as the shepherd and his two dogs or the tractor chugging along the woodside with bales of hay for ponies and sheep.

The woodpigeons are down, too, pecking frosted green, bosomy and big-bellied, legless you would think in the way a shuffling hedgehog looks legless, round and buxom, with smooth pates, and not a feather out of place, well tailored compared with the ragged rooks, dabbing with yellow and vermilion beaks, like mechanical toys almost the way they dip and lift, white-necked, pomaded blue and purple in the sun.

There are starlings scattered thinly about, like fleas in a blanket, moving dots, nondescript, until you zoom into them with binoculars, then you see them for what they are, some glossed purple, others spangled like costermongers, waggling about puposefully on copper legs, all savoir faire and sang froid—vulgarians you say? proletarians? commoners? Listen to the chap on the chimney pot with the wolf whistle and the castanets; he'll play you a drum roll, or pop corks, or act greenfinch or even peewit depending on how he feels. So he jaups the jimp in Glasgow? So he does, but this isn't Glasgow, and a chimney pot without a starling is like Nelson's column without Nelson.

More starlings come in—the overspill from the city?—and form cloud, and then there's a display of formation flying that makes the Red Arrows look like beginners, and then they come around again and down like a buzz of blowflies settling on a carcase, and the grizzled field is stippled with restless, bustling dots, all seeking the same apparent nothing as the rooks and woodpigeons.

The dung spreader arrives, an ugly, clanking intrusion, but what it scatters is as much a part of the countryside as the birds it scares to one side; dung is life to the soil and hides life in it for the birds, and it smells good to a nose brought up to it, and what euphemiser would dare call it manure? The machine clanks away to reload and the birds come back—the rooks and starlings to the feasty spread, the woodpigeons to the unbrowned silver green.

132

Now come two black crows, dark silhouettes, slow flapping, and swing down to the spread; no ragged kilts here, or yashmaks, but alert, wary, bright-eyed caterans who can tell a stick from a twelve-bore at 100yds. and recognise the difference between goodwill and damn you! which is why they have lived so long. They pitch on the brown, purple-black and not a feather awry, and probe and prise and pinch out the crawlies with shining ebony beaks. It's an ill wind, and they know every wind and what it brings.

Then there are five magpies, suddenly, and here is bounce, elegance and swagger on light feet; hop, skip and jump, with long rainbow tails flirting and scissoring, their white waistcoats snow-white against the grizzle; nimble lightweights compared with rook or crow whom they presently join in the dung search—lifting, laying, turning, probing, teasing, extracting and swallowing.

Lesser fry now arrive—itinerant sparrows, a pied wagtail with nervously twitching tail, a pair of blackbirds, a mavis. The skylark goes up from the field to sing while the others work. Again one wonders what so many mouths can be winning from a close-eaten, lightly frosted field.

Brief Encounter in the Moonlight

The big reynard came foxing into the haugh, picking his way where the rimed grass was short, and putting his feet down delicately, the way a walking fox does, overprinting his fore-pads with his hind.

The goblin's mirror wasn't limning him much, but enough, and now and again there was eye-shine to betray him. It wasn't all that clear, like looking out through a dirty window, and I felt like wiping the air clean to get a better view.

I was on the opposite side of the burn, below the badger sett, well in the shadows and not sending him any alarming scent signals, so he was at ease, and taking his time. I was at my ease, too, and not going anywhere.

He would be halfway across the haugh when he stopped, sat down, and clawed at an ear; then he shook himself, like a dog just out of water. Clouds came over the moon's scliff, and I lost him for a moment, but they passed quickly and there he was again, as shadowy but as recognisable as before. But now he was stalking.

From tussock to tussock he moved, slowly, lifting and placing his feet slowly and precisely. The hind feet came forward to take place vacated by the fore. Every now and again he froze. Then, after bellying down, he leaped up, formed curve, and came down forepaws first—the classic up and over curve of a fox pouncing on mouse or vole.

I am sure he caught something, because he turned about and stood as though swallowing. It was at this point that the roe came slouching into the haugh.

Roe and fox meet often enough; I have seen them meet often enough. And you can never be sure what is going to happen. Sometimes they pass the time of day; sometimes the roe, if it's a buck, will posture in threat. Sometimes the fox will put on a show of aggression. One never knows.

The roe was a doe; I could see the white tuft down-pointing from her white seat. The fact that she had no headgear proved nothing because the bucks on the ground were now without their antlers. Anyway, I proved it afterwards with a quick flick of the big torch. The fox didn't like that much, and quietly disappeared, but that was later, maybe three whole minutes later.

At first the doe ignored the fox, and I thought she was going to mince right on and over the burn into my lap. Instead she stopped, turned about, and lowered her head. The fox sat down and looked at her. They would be about ten yards apart.

She began to walk round him, still with her head down, and he rose to go round with her, on a lesser circuit, the one fixed on the other. The fox took a few paces towards her, and she retreated. She took a few towards him, and he retreated. Then he sat down and the doe reached down to nibble perfunctorily at the frozen grass, turning her back on him.

She nibbled towards me until she was at the edge of the burn, then turned her head to look at the fox. He was now sniffing tussocks again. At this point I flashed the torch at her, quickly, on and off. She didn't react at all. But the fox did. He disappeared as though into a tunnel; in fact he was running along a drill.

Now I was left with the doe. She stood looking in my direction for a bit, then turned aside and followed the edge of the burn, not hurrying, until I lost her among the bushes. And I had the night to myself again.

Brief encounters like this have always fascinated, and frustrated, me. They must happen often, yet one sees so little of them, even when one is a confirmed night bird like me. I was surprised at the doe taking time off to have words with a big fox. A buck doing so would not have surprised me at all.

Some years ago my son and I watched a fox putting a roebuck on foot. The buck ran for a bit, then stopped, as though saying to himself 'what the hell'. He turned about and almost prodded the fox into the ground. The fox didn't take any time off to dispute with him.

The Sorting Out of Barney

It's all houses now in Maukinshaw—the big field where once dwelt Barney— and the only danger to life or limb is when a motor car nicks too smartly round a corner, or you're knocked off your bike by a dog chasing a cat. Or the like . . .

It wasn't like that after the Kaiser's war; then it had a northern frontier of thorn and dogrose, and a bramble slope to the south, and in summer it was patchy with ragwort, and there were always fat bumblebees lumbering in the clover, and as likely as not there would be a hare there, and sometimes leverets as well.

Maukinshaw was a great favourite with small boys who went there, armed with jam jars, to catch bumblebees. Each jar was lidded with paper, held in place by a string round its throat, and the lid had a finger hole in the middle for putting bumblebees in, and a dandelion head as stopper to keep them from getting out.

Hares could go into the field any time, but small boys, like big men, could do so in safety only when Barney was absent at work, or housed where he worked. For Barney was a man-hater, not to say a man-eater, with a kick like an Argentinian mule, and a closed mind on the subject of hominids—all except one. And that was old Vincent who worked with him.

Who was Barney? He was a horse—a big Irish gyp: a horse among horses, and the whole Apocalypse among men. Except when he was in harness. For Barney had his rules: quate in the shafts and all hell in Maukinshaw.

The few old hominids still around—the real old ones: the ones with scars and hoasts and glass eyes and memories of the Somme and Paschendaele—could always find the extra breath to speak about him, mostly with a kind of pride, and even affection, for in their microhabitat of this great biosphere the gaunt, thrawn Irish donnybrooker had the status of Pegasus, or the Byerly Turk, or the Darley Arabian, or the Godolphin Barb.

A dram set before any one of them—thank ye kindly; the blid's thinner than it was—would start him off on Barney. That was if you mentioned the name.

'Noo, man, I mind the day . . . Be Jasus, now there was a baste that if Billy had been up on him he'd have landed in the bliddy Boyne.'

They would brulzie over Pope and Dutch William, and swap flowery insults, but they were unanimous on the subject of Barney. Of course, they remembered other horses; like the braw American trotter the lawyer used to drive into town; the butcher's pacer; the hawker's mustangs; the snorting Belgian Blacks, foaming against the bearing rein, that pulled the hearse, giving a man time to look, say a word in God's ear, and doff his cap.

But Barney was special: the horse who could put the fear of God in men. They could call him bad names, mind you, the names depending on which foot you kicked with, as they used to say and still do. But whether they called him an Irish get, or a tarry-hoofed pape, or the seat of an Irish horse protestant, they all meant the same. He was all horse: ever feared and dearly loved.

None ever really insulted him, the way yon Charles V of France did with his horse by speaking to it in German. In fact, if you'd searched their souls, and they'd kent how, they'd have spoken to Barney in Spanish, the language which that same Charles said he used when speaking to God.

Old Vincent, rest his sowl, laughed at them all, and agreed with everybody. He knew all about the *plus ca change* thing, although he'd never heard of it.

'He's a bould bad boy when youse tant him,' he used to say, 'but a grand baste of a horse if youse leave him alone.'

In those far off days—they were good old days to me, although I'm told such days are a myth—colliers and quarrymen used to use Barney's field as a shortcut to and from their work. They were all right from Monday to Friday as a rule, when they could trip heavy-booted through the clover and ragwort and buttercups and dandelions without fear of equine retribution.

But on my Irish granddaddy's gless-and-pint day, as on the Sabbath, you'd have needed a choir of Mons angels to sing you through. On those days Barney was almost always in residence; and believe me, he invented the territorial imperative before Robert Ardrey was born.

In Barney's neck of the woods—if you'll forgive the cliche—the commonest words that floated on the zephyrs, from mouth to eager ear, were: 'A'll bet you're feart.'

Such was the ritual challenge from those who knew Joseph (Barney) to those who knew him not, or only by hearsay. People being what they are, and were—all feart but none admitting—the challenge was usually taken up: to cross Barney's field while his hippo majesty was in residence. You could call it a kind of local Russian roulette; it was certainly like running the wild bulls in Pamplona.

Vincent excepted, Barney couldn't thole anyone of species *homo sapiens* in his field: brown, black, yellow, white, Cromagnon, Neanderthal or whatever. Five days did he labour and do all his work, and on two he was a one-horse executioner among men. He would have hoofed you from Maukinshaw to Edinburgh (a tidy step) or hefted you over the hedge. A collier once said he'd have bited your ears aff.

'Haw Vincent!' a fancied runner cried out, having beaten Barney to the hedge by a short nostril: 'That horse could hae bate The Tetrarch. He damp near seen me bits aff.'

The wise ones used to lie up behind the hedge, waiting for the next runner of the gauntlet to crash home. One day, with watch ostentatiously in hand, a collier greeted a short nose winner with:

'A good rin man. But ye hinna bate the record.'

That was Barney—who chased and snorted and thundered; who kicked, and sometimes bruised, and infrequently bled. But serious Maukinshaw casualties were nil.

Vincent himself was a quiet man, barely literate, but nae mair barely, as a quarryman once put it, than some o the wans ye fa ower nooadays wi the letters efter their name. One thing he did have was a magical rapport with his horse. And two others—a fear of no man, and a reputed fear of his missis.

Though a bantam, he had the bantam's ramgumption. He could put the clamihewit on the biggest with his tongue, and sned the gawsy like thistles. Yet he offended none, for his wasp stings were skaithless.

'Hello man,' he was greeted one day when, arms behind his waist, he was walking like a small guardsman with a loaded breadboard on his bare head. 'Hello yersel,' he replied, 'you're speaking to the only level-heided man in Scotland.'

His favourite jousting mate was an Irishman the height of a smoke stack, with shoulders like a hoose-end, and nieves ever restless to be performing feats of strength, especially when he could talk God into being his witness. One day Vincent was backing Barney up to a railway wagon when a drum fell off a lorry, and burst, oozing a delta of thick green paint into the loading bay.

'Look at that, Barney,' Vincent said, 'big Mick's jist had anither haemorrhage.'

That was Vincent, who always reckoned that if he had ever been able to find anyone who could stay long enough on Barney's back the horse would have won the local Derby oot the park.

Missis Vincent, as they called her, was a gran wummin—ye hardly ever hear her cheep in public—who was almost a legend, and whose panacea for everything was a cup of tea, the like o whit they dook the whin metal in.

'Mark ma words,' she was fond of saying. 'Wan o thur days yon horse is gaun tae rin some poor eejit intae the grun, then ye'll hae tae gether the bits, and that'll be nae laughin maitter.'

'That wull be right,' Vincent would answer. 'They torment him and get his wild up. But he's a fine baste o a horse if they leave him alane.'

'Weel,' she would comment. 'Aw A ken is that A widna tak a meal bag o sovereigns and gang intae that park while he's there.'

The school summer holidays came round, and for a week of the time Barney was on holiday too, which meant he was in his field from the Friday until a week

136

the following Monday. It was a good old summer of the good old days, and Barney grazed, slept, rolled, galloped, trotted spring-footed, snickered and whistled. He spent a lot of time, too, looking over the boundary fence, and there were those who said that maybe he was lonely.

Well, on a warm day of butterflies and bee-buzz and sleepy trees two small boys entered Maukinshaw by the thorn hedge, armed with a ham net and a jam jar with dandelion stopper, and began to stalk among the clover. Barney was lying down, near the brambles, at the other end of the field, and didn't immediately notice them. Likely he was as drowsy as the trees.

But they hadn't got far before he saw them, and sat up to look. They were in his field all right, and that, as the fencer put it later, wid get his arse aff the grass. Up he jumped, and shook his head, and whistled piercingly. And suddenly he was thundering towards the two small hominids; seventeen hands of long-legged, raw-boned, sinewy horse.

The boys called out, and began to cry, but they didn't try to run away. They stood as though nailed to the ground. Barney, who must have looked the size of a locomotive, came on at speed then ploughed to a halt about five yards short. He snorted; he snickered; he shook his head; he kicked his heels; he scraped with a forehoof. The boys screamed.

It was then Missis Vincent appeared, drying her hands on a hessian apron. She ran up and fronted the boys. She took off her apron. She waved it at Barney, who stood there blustering but not coming on.

'Get oot o there! Gaun noo; get oot! Ye big Irish galoot, fleggin the bairns wi your cairry on.' She struck out with the apron and Barney backed away rearing and grumbling.

'Come on noo. Oot! At wance, A say. Or A'll tie your mane an tail thegither an wrap you roon a tree.'

As they tell it, Barney just stood there, while she led the boys away. Explain it? Maybe he was at a loss in a new situation for there had never been a woman in the field before during his reign. Or maybe it was because it was Missis Vincent.

When they were ragging Vincent about it afterwards, especially when they went on about the way his wife had taken him verbally apart afterwards, he silenced them.

'Youse are a feart as hell o Barney but the wife sorted him oot. Take a gander at me,' he went on, 'for A'm the wan that bides under the same roof as the only bliddy human being that put the fear o God in your bogeyman.'

Mobile Homes

The tractor has been lying idle for a couple of weeks or so and when the farmer goes to start it up he finds a pied wagtail nesting on the cylinder block. She flies off when the farmer lifts the bonnet but is back on her eggs within a minute or two of his closing it. And where do we go from here? the man says to himself.

He further says to himself that she'll go when he starts up the engine, but the engine didn't bother her none as he said to me later, so there he was in a situation that isn't novel but which is always supposed to happen to somebody else. You expect to read about this in the papers was what he said to himself.

Anyway, he needed the tractor, and when he streamlined its programme, allowing for this and that, and bringing in his other machine, he found it still had to work for an hour a day, half an hour out and back morning and evening. And that was the programme for the few remaining days to hatching. The bird took the half hour off while the tractor was away and went back to work when it came back.

Once she hatched out the tractor was left idle except for the occasional unavoidable use. The birds accepted the interruption to their feeding routine, and reared their brood successfully. The man lost a lot of tractor hours, and had a lot of pleasure, which you can't put in the bank; but as he said there's mair than the fast buck or the slow wan. And that's about the whole of philosophy when you come to think of it.

Mobile nests are not all that uncommon, but they're the sort people tell each other about, or write to the papers about, because they're not all that common either.

A mistle thrush built her nest one year on the jib of a huge crane in a scrapyard. The nest was built, and the eggs laid, while the crane was idle, but the bird had to incubate her eggs, and the pair had to rear their young, while it was working. Hour after hour the bird sat while the crane swung this way and that, lifting and dropping the great bomb of metal used for smashing scrap.

The sitting bird ignored all the noise and activity and sat right through. Unlike the tractor the crane had to work its full shift, even after the young hatched, which faced the adults with a feeding problem. It's one thing sitting on a nest and being swung about; it's something else to fly after a moving jib with a beakful of food. The birds solved the problem by coming to one spot and sitting there until the jib swung round to them.

Then the operator gave them a few seconds to stoke the gebbies of the young and clean the nest. It all worked out. Nobody lost any time and the men had a lot of fun.

I knew a pair of blackbirds that had their young on a digger in a sand quarry. Unlike the mistle thrushes they didn't wait for the nest to come to them at feeding time; they flew to wherever it happened to be in the quarry and fed their young on the move. That was a big adjustment for the birds to make, for a nest is a very restricted spot on the map and birds have a very definite restricted mental map of it.

A priest came to me a while back to say he had a problem. A blackbird had her nest on the engine of his curate's small car, and was now sitting tight on four eggs. The curate was due back the next day and would certainly want his car again. So what was he to do?

I told him there was really no problem. It was a matter of simple decision. The bird and her nest could be thrown out and the car made available for instant use; or the bird could be left sitting her time, and then allowed to rear her brood, which would mean that the car would be out of action for about another three weeks.

The priest looked at me—he was old, and Irish, and the look was old fashioned, if you know what I mean—and then made his decision.

'Och, well,' he said, 'the walk will do him good.'

I came away wishing I had said it.

A Villain on Four Legs

The black cat, with a white face and an eye patch, a real villainous cartoon-type baudrans, was bellied down on the bank, beside the exposed roots of a tree, with his eyes fixed on a wee hole under a tussock, patiently waiting, maybe not realising that if it was a woodmouse's burrow he was watching he could be there until moonrise at least, the woodmouse being nocturnal, and not even liking bright moonlight in its eyes when it peeps out at night.

All this went through my mind as I approached the *grand chat* who was a stranger to me, and very likely one that had become a renegade from man's fireside, trying his fish-hooks in the free life of the woods, a not-so-warm life these nights with the frost chirping and the wind with snow on its breath.

Friendly-like I greeted him: 'How goes it puss? What hae ye there, a moose or maybe a wee weasel at the moosin?'

He—he had to be he with the ba face he had—turned the pirate's patch on me, and stared. The eye patch was a bit of a confuser; it gave him a fierce look, although it didn't follow he was a fierce cat. But with cats you don't probe for ferocity; you wait and see.

I waited, but he just stared at me, or rather glared at me, aloof, not afraid, not obviously hostile, yet not forthcoming, not saying a word in fact. I spoke to him again, inching nearer, wishing to make his better acquaintance.

'Come awa puss,' I said, hunkering down with a hand held out in friendly gesture. 'If it's a moose ye want I hae hundreds o them, and could spare wan for a freen. Come on, gie us a bit shake o the fish-hooks.'

He glared at me, then turned to stare at the mousehole, then switched his eyes back to me. And yawned, showing canines like dirks. Then rose, and stretched. Then came to me with his pole up, with his bagpipes at full blast. The purring of him was about as loud as the drumming of a woodpecker. In a moment he was rubbing against my legs.

So I stroked him, the signs being unmistakeable, and was astonished at the muscle on him, and the sleek good coat of him, which looked as though it had been groomed half an hour before. He was neither haggard nor hungert, so he hadn't been long on his own, or he was an expert feral feline of long standing.

When I poked at the mousehole, his head followed my fingers, rubbing against them. He was an inveterate head rubber, and followed my fingers as though the head were attached to them. This cat, I said to myself, didn't quit any home, and I came to the conclusion that he had been ditched by someone wanting rid of him, probably dumped from a motor car and left to sort himself out.

I was surprised when I walked away that he didn't follow me for more than a few paces. Watching him over my shoulder I saw him stop, and turn about. I heard him mew, peradventure in feline farewell.

Whatever it was he then stalked with considerable dignity back to his mousehole, and crouched down just as I had found him. He was determined to have a mouse-eye-view of moonrise.

Ten minutes later I was home, and still thinking of him. So making one of those instant decisive, daft decisions for which I am duly famous, I collected a fat mouse from the freezer, and set out on the short journey to the mousehole. Eye Patch was still there. And when I wheedled he came to me again, with the bagpipes playing as before.

'Here,' I said to him, 'I've brought ye a bit that'll maybe keep ye going till ye hae your right dinner on Apodemus.'

With these words I ceremoniously produced the mouse, which he snatched up at once, frozen though it was.

And at that moment my braw, gentle, genteel baudrans became a very wildcat. He shot three feet up the tree, clawing to a knuckle on the grey trunk, and moaned and groaned and wailed at me, even fizzing when I tried to reason with him. He went higher up the tree, and girned on. There were no bagpipes now. Should I try the hand of friendship again? I did. And a lightning left hook almost did some surgery.

'Ye've a nice way of saying Thenk ye,' I said to him.

He was all cat, like my own aged Teuchter.

He, Twee and She

A man, looking at a mated pair of my weasels, said to me: 'Which is the She?' And I replied: 'You mean the bitch?' And he said: 'Yes, the female . . .'

Why is it that so many people avoid the term bitch as though it were a four instead of a five-letter word? They will avoid it, even for the opposite of dog, although it is in universal use and sanctified by the Kennel Club. There are those who use the odious and ridiculous term Lady-dog for a bitch, and there's nothing more twee than that.

I loathe the term Lady-dog, although I have known bitches whom one could have called, justifiably if anthropomorphically, ladies. I have one ancient of my own of whom I frequently say: that old bitch is a perfect lady. But Lady-dog—God help us!

I also loathe She used as a noun. Is this a She? It was a She. I'm sure it was a She. Who wants a She? Anent a bitch puppy a man said to me once: 'Won't a She be a lot of trouble?' To which I replied: 'She could be.' That puts she in her place. The indefinite article is the trouble.

She as a sexer is about as bad, although we have grown accustomed to it in certain contexts. She-bear isn't at all bad because we're used to it. I had a teacher once who called Bloody Mary a she-tiger, which didn't sound so wrong because he was right. But who ever heard of a she-dog, a she-wolf, a she-horse, a she-cow, a she-stoat, or whatever? They've got names.

The curious thing is that the same run of people, almost, do not hesitate to use the words bitchy, bitching or bitchiness, yet when these terms are used in the context in which they are usually used they are almost invariably wrong. When I reflect on the she-men to whom these terms have been applied in my hearing I have been unable to find a characteristic that would apply to the bitch dog, bitch wolf or bitch fox.

We take a human behaviour pattern, name it after a she-dog, then blame the bitch for being like the person. Which the bitch isn't.

Anyway, there's nothing wrong with a bitch if it has four legs, and it is a proper word to use of dogs, wolves, foxes, stoats, weasels, ferrets, polecats, without any nuances of bitching or bitchiness.

The language abounds in proper names for this and that, age-wise and sex-wise. Of goats, a male is a billy, a female is a nanny, and the young a kid. Is there a more apt, delicious, or baby-faced word than Kid? Kid also applies to roedeer young of either sex, although fawn is more usual; the adults are buck and doe. With such names who would want to call a roebuck a he-deer or a doe a she-roe?

Of foxes, dog is used for the male, vixen or bitch for the female, and cub for the young. Yet the young of the domestic dog are called pups, and pups is usually used of young wolves as well. Pup is also used widely for young seals, although the adults are bull and cow.

Depending on your taste and your upbringing you will call a male hare a buck or jack, a female a doe or jill, and the young a leveret anywhere. Buck and doe are used for male and female young rabbits. But what of the young rabbits? The term kitten is widely used, and there's nothing wrong with it as a word. But I believe that rabbit (rabet) was originally used for the young, when the adults were known as conies (in the Game Acts for example).

Badgers are commonly called boar and sow. Some people call the male a dog, but I have never heard the sow called a bitch. Young badgers are cubs. The term kits is customary with such beasts as stoat and weasel, but young otters are cubs.

The sexes in red deer are stag and hind; the young are calves, never kids or fawns. The reindeer male, on the other hand, is a bull, and the female a cow.

But the euphemisers need not worry. Even in scientific circles there is a reluctance to call things by their proper names. I once wrote a paper on stoats, and the general editor cut out every He and She and substituted It. I dislike that almost as much as Lady-dog.

Give a Fox a Bad Name

Thou shalt not suffer a fox to live! might well have been the Eleventh Commandment if Moses had come down from Cruachan instead of Sinai, and the indications are that this Commandment would have been fervently and single-mindedly observed, for we are a peculiarly dedicated nation of fox-haters, thirled to an anti-vulpine Solemn League and Covenant.

That is fair enough. A man has a right to covenant, or hate, or to be solemn. The trouble is that if one singeth not the Hymn of Hate, because of misgivings or reservations about its validity, one becomes like the Civil Rights worker in America who is branded nigger-lover. These are irrelevancies. So are cleverisms. I was once asked to make up my mind whether I wanted a pound of fox or a pound of mutton off the hill, which was a pure sillyism because you can't have a pound of predator per pound of prey, or a pound of predator instead of prey, anywhere in the wide world.

From grass roots up to grass eaters, and the things that eat the grass eaters, right up to the fleas that torment the foxes, you can build a pyramid of integrated numbers, with the carnivores a tiny part of the structure. A fox has to have on his range more than he could ever eat if he is to be sure of eating every day. He may eat 2000 voles in a year, but he has to have many times 2000 voles on his range if he is to survive on voles alone.

Foxes are a problem of hill sheep farming, and the best way to look at the problem is to look at foxes. It is surprising how little foxes have been looked at, how little is really known about them, and how much remains to be discovered. So let us look at foxes, forgetting the anathematising and mathematising, the pontificating, the intuition and the second sight, and sticking instead to the few facts that are known.

Fact number one is that foxes are not responsible for the general malaise that afflicts present-day sheep farming. At least I have never heard it suggested that a new Golden Age of Hill Sheep Farming would dawn if all the mountain foxes had their teeth pulled.

Fact number two is that we know that an adult fox requires between 12oz. and just under 16oz. of food per day, depending on age, size and sex. Cubs require proportionately less according to age and size. We know that a fox on ground heavily populated by voles will eat mainly voles, even up to a dozen or more a day. We know that a vole, weighing an ounce, will eat two ounces of grass a day, and Lockie has shown for Wester Ross that voles eat 23lb. of grass each in the winter months, taking about 2cwt. from each acre.

Fact number three is that foxes bolt voles like a man bolting porridge, so there is hardly ever any evidence of vole remains at a fox den, unless the vixen brings in more than the cubs can clear at once, which is not very often. Evidence of predation on voles comes from the fox's droppings, which contain the indigestible remains. This can show vole as high as 80 per cent in vole years. Lockie, working on sheep ground in Wester Ross, found droppings contained 31 per cent vole by weight in summer and 26 per cent in winter.

Bigger prey is not totally consumed. Cubs can't do much with the leg of a stag after they have stripped the meat from it. Remains of hare, rabbit, capercaillie or lamb are also obvious, and lie there to give a false, or exaggerated, picture of total predation.

Fact number four is that remains of prey found at a fox den are not an indication of predation unless it can be shown beyond doubt that they were killed and not taken as carrion. The same is true of an eagle's eyrie, especially where lambs and deer are concerned. Twelve lambs found at a fox den may mean much or nothing, according to whether the fox killed them or scavenged them.

Fact number five is that some foxes kill lambs. Fact number six is that many don't. Fact number seven is that many foxes eat a lot of carrion. Nobody knows how many lambs are actually killed by foxes, and consequently nobody can put a value on the lambs killed by foxes. It would be interesting to learn the hard facts about the cost of killing foxes in the Highlands, and the man-hours expended, against this unknown quantity.

Fact number eight is that we have some idea of the amount of carrion eaten by foxes in a given situation—it isn't a great deal of information but it is hard information, and much better than the intuition. Lockie found that, in Wester Ross, foxes ate 63 per cent carrion in winter (25 per cent adult sheep and 38 per cent adult deer), and in summer 5 per cent adult sheep carrion. In the same place, during summer, 15 per cent of the fox's food was lamb, dead or alive.

Available evidence suggests that foxes eat more lamb on poor sheep ground, or in years of bad lambing, than on good ground or in years of good lambing. During a fifty-one day period I found that neither the eagles nor the foxes on a good sheep hill had lamb in at all. But the eagles were killing fox cubs.

The fox, as we have seen, is a considerable carrion eater, and it is the presence of carrion that makes possible the density of foxes (and eagles) in certain parts of the west where natural wild prey is scarce. A fox that is there to kill lambs in spring has to survive on something else for the rest of the year. In many places what carries it through is carrion.

One hundred years ago Osgood MacKenzie was killing big bags of hares, grouse and ptarmigan on western hills where hardly one of any of them is to be seen today. Yet there are plenty of foxes on the ground, two pairs to every 16,000 acres by recent census. What do these foxes live on? It has been shown that the average yield of lambs is 65 per 100 ewes in good years, and as low as 30 per 100 in poor years, with many ewes dying in addition. This provides a bonanza for a carrion eater like the fox, but if all the dead animals were buried, as the law requires, the fox population would be cut in half at least. On such ground therefore, it is not too much to say that the fox problem, such as it is, is the direct creation of the interest that suffers most from it.

I was appalled, some years ago, to find lamb mortality as high as 40 per cent in one place, but not surprised to find that eagles were present in high density. For the mainland substitute foxes and the picture is the same—an abundance of carrion providing a persistent or carry-over food supply for carnivores adapted to exist on carrion. And the carrion holds the predators on the ground, so that they are there when the lambs are born.

The rights and wrongs of not burying dead sheep, or the kind of management that leads to large numbers of dead deer on the hill, are not my concern here, although both are entirely relevant to this discussion. But when one finds five dead sheep on three acres of hill, or twenty-three in an hour's walk, it should be no surprise to anyone that foxes, buzzards, ravens and hoodie crows are thick on the ground. The blowflies buzz where they have something to buzz over. By this yardstick there can't be 'too many' foxes; there is too much fox food.

The same carts-before-horses thinking comes out in statements about foxes now hunting in packs, as though singly they were not worrying the sheepman enough. Five foxes don't have to combine to pull down a dead stag, or course a dead sheep already rotting into the ground. The five foxes, or seven as I once saw, are not a unit but an aggregate of individuals, attracted to the same place by the same thing—food. On the day when foxes have to combine to find a living, every hoofed animal will be on its hooves, running away. You don't have to combine to kill a corpse.

It is unfortunate that foxhunting in the sheep areas is based entirely on lamb losses (dead or alive) without other factors being taken into account: the fox as a predator on voles, or as redder-up of the hills. The suggestion made some time ago by Lockie, and now supported by the Nature Conservancy, that only cubs should be killed, while the parent foxes are allowed to go on with their work of killing voles, has never been tried. Yet all the evidence suggests that adult foxes, without a family to feed, are little problem. Unfortunately, too, foxes are still killed between lambings, when they can do no harm. Killing cubs might help but in the long run the answer is to reduce mortality among sheep and deer to a minimum, and to improve management.

The fox is a pest animal some of the time, but not all of the time, and there are times when it is entirely useful. Since, after years of ruthless war against them, foxes still remain numerous, it seems to me that the time has come for some rethinking on methods of 'control'. We have not reduced foxes. It might even be argued that present methods have made the world safe for foxes, by pruning them down annually to a nice relationship with their food supply. And at the end of the day we don't know how many lambs they kill. At such a stage of unknowledge surely castrating dog foxes to run barren races with vixens is again putting carts before horses. Or is it that we must have a scapegoat? The facts available do not warrant the kind of constant costly war that is waged.

In their book *The Highlands and Islands,* Fraser Darling and Morton Boyd write: 'We live in an era of admitted over-population of the Highlands by deer, yet the persecuted fox is the surest check on the calf crop, failing adequate stalking of the adult stock by man. Highland sheep farming has been one of the depressants of the biological condition of the wild lands and the intense persecution of the fox which does so much else than eat lambs has been one of the factors putting the Highlands out of ecological equilibrium.' It is a pity that a serious ecological problem should almost invariably be reduced to a For or Against the fox. One is a partisan only of the facts, which have yet to be discovered.

The fox problem, because we believe it to be what we will it to be, gives rise to certain abuses and malpractices. On the strength of official attitudes to foxes we have retained the gin trap, long since outlawed in England, where they also have foxes and sheep.

The argument in favour of the gin is that Scottish foxes are a special problem. The snare, another cruel instrument, is also widely used. So is poison, which is illegal. The most humane way to kill foxes is the way practised over much of the Highlands—the use of terriers at the time when vixens are lying up with cubs. But one wonders if it is all necessary.

I quote without comment from Duncan Ban Macintyre, a great man in his day, despite what he said about foxes, and still a great man, except for what he said about foxes. He wrote:

'My blessing be upon the foxes, because that they hunt the sheep—
The sheep with the brockit faces that have made confusion in all the world.
Deeply do I hate the man who abuses the foxes,
Setting a dog to hunt them, shooting at them with small shot.
The cubs, if they had what I wish them, short lives were not their care.
Good luck to them, say I, and may they never die but of old age.'

I doubt if anyone would dare to take that line today, but I quote it because it is as expressive of a built-in attitude as the opposite views of today, and certainly with more reason on the basis of known facts then or now. The war against foxes had begun in Duncan Ban's day, and we are still using the same slogans. Couldn't we, to misquote Cromwell, beseech ourselves for a wee while at least that we might, just possibly might, be mistaken? That for the past many years we have all been aiming at the wrong target?

In the meantime, if more men carried shovels instead of guns the fox would quickly feel the pinch, and we would come still nearer to understanding the true

144

status of the fox if more of us realised that a fox's dung is a more revealing thing than a dead lamb, and a dead sheep a direct invitation to lamb killing.

Vermin Vocabulary

A mentor of young hominids, looking at my wildcats, Teuchter and Pibroch, said to his brood: 'Wildcats are vermin.' And I, like the late Herman Goering in another context, did what I usually do when I hear the word 'vermin': I reached, not for a gun but for an epitaph. For the word is slow a-dying . . .

'What is vermin?' I asked, with that sweet air of tolerance for which I am justly famous, and, unlike Pilate, I waited for an answer. I got it.

'Cats for a start,' he said. 'And hawks, stoats, weasels, foxes . . .' I finished the list for him, because it's been a sort of game preserver's catechism since the nineteenth century, and I learned it by heart as a boy.

'Why?' I asked.

'Why what?'

'Why are they vermin?'

He gave what I can only describe as a pitying grin and said: 'They kill birds—grouse, partridges and whatnot—and hares and rabbits . . .'

'So do people,' I said. 'As well as killing each other. Vermin,' I went on, 'is a totally inadmissable term in any serious discussion on wildlife. It means anything you don't happen to like. Politicians have been known to use the term to describe other politicians. You shouldn't be serving up this arcane muck to children at this time of day.'

'What would you call them?' he asked.

'Predators,' I said. 'Which is what they are. Like you and me. The term 'vermin' is meaningless except to describe something you don't happen to like.'

This kind of thing happens all the time. The game preservers wrote the bible on wildlife relationships, establishing a sort of canon law that fooled even the industrial proletariat right down almost to the present day. So the sparrowhawk is 'vermin' because it kills wee dicky birds (the ones we all happen to like) while the kestrel is a good bird because it kills voles and mice (which people happen not to like very much).

The term 'vermin' is sometimes used by people who ought to know better. I've heard professional ecologists use it and, when I twitted them about it, the answer was that they had to speak to keepers in the language they understood. And here am I trying to change the language. The fact is that I know a lot of keepers, among the best in the business, in the real sense, and I've never heard them use the term. It never enters their heads to do so.

A few weeks ago, while lecturing, I was asked if eagles were 'vermin' and, if so, why were they protected. Well, of course, they're protected by the people who always protected them, but they're still knocked off by the people who always knocked them off. But they do it more circumspectly than they used to; because the canon is changing.

Furthermore, I made the point that if I were looking for a good grouse moor for somebody, and I visited five, all with resident eagles, and one eagle had a dozen grouse in the nest while the others had none, I would recommend taking the moor with the eagles that were killing grouse.

I knew one pair of eagles that had eighty-eight grouse during the rearing period of eleven weeks and four days; the guns following in August had a record bag.

This 'vermin' business is a funny one. Somebody, with no knowledge of wildlife, and not much more about the countryside, decides to take up shooting. Fair enough. But, suddenly, a strange thing happens. He starts talking about 'vermin' and before you know it he has fettered himself with all the old shibboleths, as though we were all back in Glen Artney in 1843.

In general, shooting men are the worst offenders in killing what they consider 'vermin', whatever the bird's legal status may be. In the particular, it is a matter of pride with many shooting men that they play the game according to the rules, and bodies like WAGBI are tough with the cowboys.

I don't care how many pheasants or grouse or whatever people shoot, and although I do not shoot I have always defended the right of other people to do so. Until they decide to kill whatever they don't like, protected or not. Then I part company. Dukes or dustmen they're crooks and it's time that word came into the canon.

And Behold, There Was Light . . .

Mull, as you probably don't know, had a blackout one Friday, and before you start saying, 'So what? Who hasn't? This is dog bites man stuff!' Let me go on.

They'll tell you—those who believe that man's chief end is to fill the unforgiving minute with ninety seconds' worth of slog and gather—that the folks in the West and the Islands are un-with-it dawdlers with less initiative than a toad in a hole. Well, now you can judge for yourself.

There I was on Mull on that Friday; scheduled (horrible word) to lecture, with lantern slides, at the village of Dervaig at 8p.m., and in the middle of my getting ready, garnishing myself into that state of sartorial splendour for which I am not noted, word comes through that there is to be a blackout from 6p.m. likely to last for four hours. That, said I to myself in one of the soliloquies for which I am famous, puts an end to the lecture.

But we decided to drive to Dervaig anyway. The folks would surely be there, and there is no difficulty in sipping malt by candlelight. So we set off in the mole's fur dark, along and over and round, but mostly over and round, and when we turned the last bend behold there was a beacon in Dervaig like yon star men still speak about, and it wasn't the moon, for the moon was at our back. And I said: 'Behold, there is light.' And it was the only light in the night.

And the light was outside the door of the village hall where I was due to speak. And I was sore amazed, and said so.

'Well,' said Auley MacInnes, organising secretary and retired sheep farmer, 'ten minutes after we got the word I phoned the Board and wass told the power would likely be off for four hours from six o'clock. And I said to Scrap (Major

Rupert Lyon Balfour-Paul, hyphen and all, to you) "What the hell are we to do? And David Stephen all the way from Cumbernauld too".'

And Scrap said: 'We can try Knight. He has an engine.'

So they went in search of the engine, and got permission to take it from the field where it had been lying, weathering and fading into the landscape, for three years past. And by 3p.m. they had the engine in Dervaig. But how to connect this generator up to the power system in the hall?

So they contacted (good word) the only electrician on Mull, Roger Felter, who, I am told, is a French Algerian, or Algerian French, or whatever, a man not noted, being overworked, for instant service. But he came as instantly as possible and connected the motor to the supply. But he couldn't get it to work.

'Well, to hell,' said Auley, 'there wass only one thing for it. Get McCrone.' But McCrone, whom I take to be some sort of magician, was on a roof at the time, aided by Pedro (Peter McLean) the salmon fisherman. But Auley unroofed them and got them to Dervaig. And the McCrone got the engine started, and behold, they had light in the Dervaig village hall.

And it was the right light. It lit all the standard fittings and it fed my projector and we got started at ten minutes past eight, and I doubt if anyone could have pulled that off in Glasgow or Edinburgh unless there had been a resident generator. We had a full house, too, and the electric was there for the tea as well. But what I forgot to tell you was that the power, the mains power, was restored at 8p.m.

'Well, to hell,' said Auley, 'we'll do without it now and run the show on the motor.'

Which is what we did.

Another thing sticks in my mind about that night. After I was finished, and had had a cup of tea, a beverage to which I have no particular objection, Tommy the fisherman asked me would I like a dram. Would I not, but the place was shut by then, and there was no bar at the hall.

He took me along the street about fifteen metres or so and pulled a length of guttering from a wall, then put his arm into the hole and pulled out a bottle of the best, which we sampled on the spot. We sampled it twice. And then he put it back, guttering and all.

And I hope this appreciation hasn't revealed his hiding place.

A Question of Cruelty

What is cruelty?

If Pilate had asked the question I'm sure he would, wise man, have made it rhetorical, and turned away dichting his hands, not waiting for an answer.

After a *Nationwide* programme, when Robin Page (author of *The Hunter and the Hunted: A Countryman's View of Blood Sports*) was interviewed, I said to my wife: 'Here goes. Should I leave the phone off the hook?' I didn't, and after that it was the deluge.

To first Person I said that, in the first place, I did not, as a countryman, accept

147

the view that there was anything special, *per se,* about a countryman's view on anything. I also said that I didn't hunt, shoot, or fish, but that didn't make me automatically an expert on the ethics of blood sports. Further I was not prepared, as a matter of course, to underwrite, unconditionally, anybody's pet hate, despite the fact that I have always opposed most forms of hunting, and said so unequivocally over half a century.

To another Person I said that what bothered me, as much as any incidental cruelty involved, was the psychology of those who hunted some animal to death for fun. Even, I said, if it could be unarguably shown that there was no cruelty in hunting this or that I would still wonder about the hunters.

What about otter hunting? To this Person I said that I could see no justification whatever for hunting otters with hounds, whether otters are plentiful or rare. It is a nonsense, and there is no doubt, in my mind, about the cruelty involved. It cannot he argued that the otter is a menace to river stocks, so there is no economic justification either.

Foxhunting? Well, what kind of foxhunting? If you happened to be a fox living in certain parts of Scotland you might pause before answering that one. I wish the fox could give his point of view. But since everybody, nearly, goes after the persons on horses let's look at them.

They hunt the fox in the way they do because they like it, and I wish they'd come out straight and say so, instead of using mendacious arguments about helping farmers or controlling foxes. They crop foxes on the one hand, and ensure a supply on the other. They're a bunch; but I'm still not certain that the fox, if given the choice between being trapped, snared, poisoned, shot or whatever, would choose one of them rather than the big bow-wows and their holly-red followers. If foxes have to be killed, I'm not sure that hunting with hounds is the cruellest way of doing it. I'm not sure.

Which is no justification of the person on the horse, who, if he or she wanted to kill foxes, might well consider getting off the horse and hunting some way that would receive popular approval, even if it were more cruel, as snaring, gin-trapping and poisoning, are. I know the gin trap is illegal but that doesn't stop it being used.

Snaring is slow strangulation; trapping means a lingering death; gunshot wounds can leave a beast to die of blood poisoning. I wonder what the fox would say. Would he, perhaps, ask if much of the opposition to the persons on horses wasn't opposition to top-drawer chinless wonders? I don't know.

As for the Devon and Somerset Staghounds, I don't like what they do. But surely the first point is that the hunt is arranged by man for his amusement? As far as the deer is concerned the dogs might as well be wolves, and deer were born to be preyed upon by wolves. But wolves take the poorest specimens, not the best. And is it really true that the rifle can't be used on Exmoor? I doubt if any Highland stalker would agree. And would the deer on Exmoor really vanish if there were no hunting? Maybe they would, with all the cowboys after them. I don't know. In any case, I have always been reluctant to interfere in the internal affairs of another nation.

An old person brought me a very obese wee terrier to have its nails clipped. 'Do you like your dog?' I asked. 'I love him,' was the reply. 'Then why do you fatten him like a pig and make him uncomfortable and shorten his life?' I don't think this is an irrelevance.

Miss Wise and the Magpie

When about 3000 people visit a place in a weekend, and you happen to be in the middle of it, it's more than likely that one or more of them is going to hail you with do-you-remember? Which brings me to the subject of Miss Wise.

Once upon my time there was this delightful old maiden lady called Miss Wise, whom we all loved and respected in a special way that those of us who are left still talk about, which becomes seldomer and seldomer because there aren't so many of us left. The love and respect we had for her had nothing, or very little, to do with the fact that she had a Jenny-a-thing shop, that she gave us sweets from time to time, whether we swept her floor or not, or whether we carried in boards of pies for her, or did nothing, or anything.

For some unaccountable reason, probably Freudian, of which I knew less than anybody, she called me John till the day she died. And I grew up accepting the fact that I was John to her and myself to everybody else.

By the time I was nineteen, and no end of a worldly wise man, having lived in Spain and travelled in France, for God's sake—that was when Wall Street was falling apart—Miss Wise still called me John.

By then she was a very senior senior citizen, a bit thin on top for sure, but still hirsute, and still with the old grocer's trot, and smart and neat as they come. And she used to curdle my blood with her stories about treacherous Spaniards with knives in their belts.

Came the day when she passed out in her shop with a severe nasal haemorrhage, and she called out on recovery 'get John', and I rushed into the shop, and did what needed doing, and she was extremely grateful, and I had tea with her, and then I said to her: 'Miss Wise, my name is really David.' And she said: 'Yes, John, and thank you.'

And she was dead for several years before I knew her first name, which was Mary. Can you believe such a thing? It's true. She was Miss Wise, and first names didn't seem to enter into it.

Well, I've come the long way round—she believed every word in the Bible, literally; she believed that no foreigner could be even half good; she believed that Spaniards were always ready to stick knives into people. She even believed that I was a good boy, which proves how way out she was in her notions.

But I loved her.

And I humiliated her.

I had a tame magpie at the time and, like me, he grew to like Miss Wise, and if I couldn't find him I knew where to find him, if you know what I mean. He would be in Miss Wise's shop, mooching tit-bits, usually bits of Fray Bentos, or Dunlop cheese (where is the Dunlop today?) or boiled bacon trimmings.

Well, one day he was standing on the counter, with Miss Wise feeding him and talking to him, when in came another maiden, a decade younger than Miss Wise, with an outsize Airedale dog. Dog barked at magpie, and magpie flew on to Miss Wise's head. Dog continued to bark, and magpie took off, and out of the door, and on to the nearest roof.

With Miss Wise's wig in his talons.

Leaving her bald-headed. Yes, yes, she was as bald as a coot.

149

And leaving me with a most difficult, and the saddest public relations job I ever had to do.

But she was a brave old girl. Saw the funny side of it. And said so. And laughed about it later.

But I never guessed she had a wig. If I had been made of less stern stuff my psyche would probably have been irrevocably damaged.

When I apologised, what she said was: 'Thank you, John.'

Comic Crows

Some birds have what, for want of a more definitive term, can only be described as a sense of fun.

Take a pair of well fed ravens seeing a fox bellying up to the carcase of a deer from which they have just eaten their fill. Do they ignore him? Sometimes they do, of course; probably more often than not; but they are just as likely, in the right mood, to intervene and make life difficult for him.

One bird flies down and struts around with that semi-circular crow swagger so familiar to anyone who has taken the time to watch corvine cussedness. With head feathers raised and beard quilled out, the strutting bird taunts the fox. If this gets no reaction the bird works in closer, with wings half open, and says this and that in raven language.

At this point the fox usually stops eating to watch out for raven nonsense, and while he is engrossed with this raven the other one very likely skips in to pull his tail. The fox turns on the tail puller, and the first raven will now try to be the tail puller. And so it goes on, with the birds having a great time and the fox a frustrating one.

That is when it is fun on the birds' part. Sometimes they go beyond being funny and become aggressive, and then they can hurt more than the fox's feelings. One would like to know much more about such relationships. Seton Gordon used to refer to the raven as the eagle's cross, because it tormented the eagle so much. The same might be said of ravens and foxes.

Carrion crows will sometimes behave in the same way, although they are not so bold as the bigger ravens. I know a pair that used to fly down to a fox den to pick up bits of carrion, and there were times when they would chase after the cubs, pulling their tails and pecking at their noses. But they did not try it with the vixen.

Tame ravens and crows carry this kind of behaviour to the limit, playing with, or becoming aggressive towards dogs, cats and people. A friend of mine had a raven that played all sorts of games with his Labrador bitch. There was the usual tail tweaking and nail pecking but no aggressiveness, and the attitude of one to the other might have been described as *bonhomie*. The things they did together were more than fun; they were funny.

When I had a raven some years ago it was friendly with my Labrador and terriers, but aggressive towards an Alsatian we had at the time. The Alsatian got the fox treatment (aggressive); the others got the fox treatment (fun). The bird would gently peck their teeth, nibble their nails, tweak their whiskers, and sit on their backs when they were lying down. But between him and the Alsatian it was armed truce.

Many of the crow family become aggressive towards cats, and this applies to some wild individuals as well. My raven, crows, magpies, jackdaws and rooks were all friendly with the household cats, but they carried war to any strange one, and I had a magpie once that blinded a cat in one eye.

This is all fun, or whatever, at the expense of another species, but the crow family finds amusement in other ways; often, as far as one can tell, to relieve boredom.

Two years ago I was looking at a pair of ravens who spent all their time pecking scliffs off the branches of a dead tree and watching them drop to the ground far below. The dead tree was on a steep slope, and the nest was in the nearby crag. The birds were feeding their young on mutton from a dead sheep lying a little distance away, so it was easy for them to keep the family satisfied and they were left with a lot of spare time. They spent the spare time dismantling the dead oak, studying each beakful, then tossing it into space and watching it drift, spiral or drop down according to its weight.

It isn't unusual to see magpies or crows playing with such things as old fungi, or golf balls, or clothes pegs, or pebbles, or anything that takes their fancy and can be safely pilfered. Crows can be a great nuisance on a golf course.

Baby deer sometimes have a lot to put up with from crows. I don't know how often it happens, but I've seen it happen twice. The crow pecks the youngster's rump, or pulls its ears, and such foolery can quickly become aggression. I saw it happen with a wild fawn, and later with a tame one I had at home.

The Survival Stakes

Asynchronous hatching means that if you're the one out of the first egg you have a survival edge on the next one by up to forty-eight hours, on the next one by four days, on the next by six days, and so on. And, of course, the second out has a similar edge on numbers three and four.

But it has no edge on you. It is behind by up to forty-eight hours. So, unless you take sick, or are injured, you are going to finish ahead of it in the survival stakes.

Tawny owls are like that. The hen owl lays her eggs at two-day intervals, perhaps longer and, if she lays five eggs, as she sometimes does, the interval between the first and the last is considerable. That interval shows at hatching time because she begins to incubate her eggs as soon as she has laid the first one. The effect of this is even more pronounced in barn owls, in some years, when you will find half-grown young and half-clocked eggs in the nest at the same time.

Let's take an actual tawny owl nest that had five eggs. The first egg hatches on 30 April, the next on 2 May, the third on 5 May, the fourth on 7 May, and the last on 9 May. Chick number one is ten days old by the time chick number five struggles out of the shell. A ten-day-old owlet is a gey bird: a newly-hatched one is a helpless mite.

In those early days the cock owl brings in all the food; the hen is occupied in brooding and feeding the chicks. If food is plentiful the hen manages to fill every craw in the nest; if it is at all short, the smallest owlet is missed and goes hungry that time round. That puts it still further back in the growth, and therefore the survival, stakes.

If prey continues scarce, or if the nights are so wet that the cock can do little hunting, the smallest owlet will almost certainly die. If it does it will be torn up by the hen and fed to the others, if number one doesn't swallow it whole on its own. In any case, that leaves four owlets where there were five.

If food supply improves, the hen will rear the four, as she would have reared the five. But if it continues to be scarce, the fate of number five will be the fate of number four and it, too, will be cannibalised by the others, or one other—the big one. There is no share and share alike, no sentiment, only the fact of life. The food is to the strong, and the strongest is the first one out of the egg.

In bad years, only half a brood may be reared. Or even less. In good years the whole family will be reared.

Asynchronous hatching ensures that, in competition for food, the weak will die and the strong will live. Unto them that could do without for a day will be given.

I've known a hen harrier rear only two out of six chicks. The smallest chick got so little that, by the time it was a week old, it looked two days old, and might well have been mistaken for the young of the oldest chick in the nest instead of a brood mate. It died, inevitably, and was torn up by the hen as food for the others.

Away back in the forties, when I was working a lot with tawny owls, I used to keep broods alive by going round the nests, either putting prey into the ones that were short, or robbing Peter to pay Paul. Half a mile could make a lot of difference to the success of the birds on the ground. Of course, helping birds out like that was a pointless exercise, but it seemed a good idea at the time.

Predation on Pigeons

A lot of things can happen to a woodpigeon's nest when it is still only a pad of twigs in some high tree or low bush. For one thing, it can be dismantled and carried away by rooks as building material for their own bulky nests. Or a squirrel may use it as a place to sit on while it is stripping pine cones. In either case, the woodpigeon ends up laying her first egg on the ground and leaving it there.

Squirrels sometimes eat eggs and even nestlings. So, if the woodpigeon happens to lay an egg when the squirrel is somewhere else, the squirrel is more than likely to eat the egg when it comes back to use the nest as a feeding platform.

Woodpigeons sometimes make their nest on top of a bulkier structure. More often, what they build is a mere fretwork of twigs through which a man can see the sky when he is looking up from the ground. In some nests, even the eggs can be seen through the weave.

A lot of things can happen to woodpigeon eggs long before they are ever filled by chicks. This is especially so in the earlier part of the year when the trees are bare and white eggs can be easily seen by a bird flying overhead.

Carrion crows spike and suck woodpigeon eggs, especially in the early part of the year. Magpies do the same. In fact, magpies will do the rounds in woods with a lot of woodpigeons. Nests near the ground, for example in a rhododendron bush, are sometimes emptied by foxes, and the fox, being a good climber, will take eggs from nests up to ten feet from the ground.

The stoat is also a considerable climber and will empty woodpigeon nests at ten, twelve or fifteen feet from the ground, especially if the nest is in an ivy-clad tree. Pine martens will also empty woodpigeon nests.

A lot of things can happen to young woodpigeons, apart from growing up to become big woodpigeons. When very small, they may be killed in the nest by crows acting in consort and pushing the adult pigeon off. A stoat will climb and take small squeakers as readily as it takes eggs. The pine marten, presumably, is able to do the same. The fox can certainly do so, and I have known a fox take woodpigeons out of a nest fourteen feet from the ground.

Woodpigeon squeakers, half-grown or fully fledged, do quite a bit of jostling in the nest and sometimes one jostles the other right out of it. If it isn't caught up by foliage, for example in a thick spruce, it falls to the ground and all it can do is lie there if it is too young to walk about, or walk about if it is not able to fly.

A grounded woodpigeon fledgling is thus open to attack by fox, badger, stoat, crow, magpie and hawk. It stands little chance of survival unless it is strong enough to flap into a low branch and hop and flap its way to some height above ground.

Adult woodpigeons have their range of predators, too. Man is probably the most important and many woodpigeons are killed annually by shooting. In the old days, when corn was stooked in the fields, it was often possible to kill up to a dozen woodpigeons with one shot, the birds offering a mass target on the stooks. More accurately, one should say that it was possible to bag up to a dozen with one shot, because one shot was unlikely to kill them all outright.

The peregrine falcon is a predator on woodpigeons, killing them by stooping at them in flight. The sparrowhen is capable of killing fully-grown woodpigeons and some sparrowhens become specialists at it. I knew one that killed thirteen in the nesting season. Golden eagles also take woodpigeons and, in some eyries, the bird figures quite prominently as a prey item.

The Wa-Gang of the Futtrats

September is one time you're likely to see them, if they're there to be seen— one here, one there, maybe two or three together, maybe even a family of mother and five or six—marching feet on their own compulsive *Drang nach Osten,* for futtrats are cater-cousins, and the young have to hive off from the land of their father when his sons, having already outgrown their mother, are ettling to reach his size.

A futtrat, or whittret, or whittrock, however you care to spell the name, is a weasel, and a master dog weasel is a landowner—a thrawn landowner, dour in defence of his frontiers, which he patrols against invaders and signposts with his scent, scats and urine. Of course there are infiltrators, the landless who slip through an unguarded sector, and poach behind the owner's back, then move on. And there's the mate, lesser in stature and status, a second class citizen, as it were, confined to a sort of weasel ghetto until emancipated by maternity. Then her authority burgeons, and she can hunt vole and mouse and rat and bird in every corner of her lord's domain. Her lord will yield ground to her, fear her even, while her kits are still dependent on her. But in the end they have to go, unless the lord dies, and there is a son strong enough to take the territory and hold the frontiers against mature and powerful neighbours.

The old dog weasel of Mossrigg held twelve acres of ground near the farm—hedgerows, fields, a young planting, and a sunken lane with verges smothered in brambles, dogrose, ferns, raspberry canes and matted grass. His mate had reared a family there in early summer. Now she was leading away her second rearing of six, perhaps to escort them to the frontier, perhaps even to leave the old ground with them. Zoologists have not so far evolved a technique that will explain intentions.

Anyway there she was at twenty-five minutes past noon on a breezy September day, after a short sweeping on-ding, peeping on to the road from the shelter of tall fireweed stems, with a bourach of six kits in the ditch at her back. The wet grass had left brush-strokes on the white fur of her neck and chest. She peered up and down the verge. The clouds were thinning like smoke. Light sklented on the loch, whose surface was now mirroring blue pools of sky. The watergaw was melting away over Cumbernauld. The futtrat drew back and bounded along the verge under cover, her kits following. When she looked out again the sun was kindling pink flames in the fireweed.

The roadmen in their plastic igloo, smoking and studying form after drinking black tea, saw her across the road, sitting up under a spreading ragwort, displaying her white neck and waistcoat. 'Jesus!' said one. 'Look! A wheesel!'

They scrambled out of the hut. The bitch futtrat pulled back like a light switched off. The men could see the stirring of other bodies behind the ragwort, betrayed by flashes of white.

'There's mair'n wan!' said the smallest man. 'There's dizens o' the bliddy things!'

All four looked and looked, but could see nothing stirring near the ragwort. They withdrew about twenty paces and stood in the middle of the road. Whether they were real scared, or half scared, or just a wee frichtit, or merely uncertain, would have been hard to say, judging by the mockery and jest they could bring to bear on their retreat.

'That's me lowsed,' said one.

'You're dead right,' said another. 'Yous'll no get me back there, even for danger money.'

'They're vicious bliddy things,' said another, 'an' ye can include me oot o' mixin' wi' them.'

'Ach!' said the fourth. 'They're only bliddy whittrets!'

'Ho! Ho! Listen tae brave Bobby!' the first mocked.

A farm worker, wiring a straining post along the road, stopped when he heard the loud voices, and laughter, and badinage, and bawdry, and hyperbole. Then he downed tools, stamped the dark earth from his boots, and came along to join them.

'Whit's a' the donnybrook?' he asked.

'Wheesels!' said the small man. 'Dizens o' the bliddy things, right where we were haein' oor piece.'

'Wheesels?' the farm worker said, and began to walk towards the hut. The roadmen followed a little distance behind, advising as they walked.

'Mind they don't sook your bluid!'

'By here, you better pu' your socks ower your troosers!'

'That's Dracula in there!'

But there was nothing at the roadside. The farm worker trod down ragwort and thistle, fireweed and scabious, but no weasel was seen. The men contented themselves with tossing stones and clods into the thickets, then all went back to work.

The bitch futtrat made her crossing round a bend, out of sight of the men, not bounding, but caterpillaring at speed, so that her belly seemed to scrape the road. Five kits followed her, bounding. But the sixth, a muscular dog, sat up in the middle of the road in classic listening pose, and was flattened by a car coming round the bend. The car drove on, leaving him ankylosed to the tarmac.

In the hedgebottom the bitch futtrat stopped and sat tall, after snarling off a dog kit who, for some reason, rushed at her and upended her, snapping at her throat. Then she darted to the road, and snaked to her kit, bowand, high-rumped, a fingerling, unwary, owning with her nose the onion taint of crushed glands. She gripped the crushed body by the neck, and bounded, trailing it, back into the hedgebottom. And there she stuffed it under a thorn root, and scraped leaf flinders and grass fluff over it.

The field was sun-bleached barley, waving bristles dried out by the sun. The endriggs were cut and combined and the kits began to play under and over the swathes of straw, poking out, peering, grappling, and cavorting like cut earthworms, their movements bewildering in their intricacy, speed, sinuosity and apparent abandon, yet without accident or collision, as precise as flocked starlings.

While they played, and tired, the bitch futtrat dozed in the sun, coiled like a caterpillar handled. And when they lay down she rose and poked, rump-shuffling, into a tussock. She stopped to listen, long-necking, and when a weasel stretches its neck it appears to begin in the middle of its body. The vole moved, hidden, but the bitch pounced with the unerring skill of long experience, and it squeaked once, and died, and when she slithered from the tussock she was holding it aloft in her jaws.

That waked the family to fury and excitement, and they leaped at her, bearing her back into the tussock. The biggest dog kit mouthed at her, savaged her, and tore the vole from her, and she leaped aside, leaving him standing over it, with teeth bared, shrieking at his brothers and sisters. The bitch kits rippled round him, not challenging; the dog kit challenged once, then cringed back in fear of the power of possession.

Four kits now bounded to the mother, crooning to her in the friendly fashion of happy family. But she had nothing to give, so they rumpled around, under and out of the swathes, following noses hopefully.

A bitch kit, hardly thicker than a man's thumb, and barely the length of a hand, suddenly left the others and bounded into the hedgebottom, pressing through the grass, sniffing up and down, bending and slithering over roots and obstacles, her rump arched and stiff unlike the snake-flex of the body.

Near the field gate lay a square of wire netting, held fast in the hedgebottom by grass matted in the mesh. Instead of running over it the bitch kit, weasel-like, went under and surprised a vole in its ball nest of shredded hay. But the kit's bite lacked grip and the vole, half her weight, broke bleeding away. In the open the young futtrat pounced again, and hunter and prey rolled on to the swathes in gymnastic grapple, the vole squeaking and the futtrat scratching with her hindfeet.

The scuffle would have been missed by human ears on the other side of the hedge, but to the ears of the old bitch futtrat, many yards away, the message came loud and clear, and was instantly interpreted. She bounded to the spot, thrust her daughter roughly aside, and took hold of the vole. Her bite on its neck was deep and lingering, and when it was dead she rolled over on her side, still holding, and treading with her hindfeet, an atom of boundless energy, her foot strokes almost too swift for the eye to follow.

Suddenly she released her hold, raced three lightning cartwheels, then stopped dead facing her kit, standing the height of her legs, with head up, nose twitching, and flanks heaving. Her eyes were dark and bright as ripe elderberries; her chestnut upper fur sheened in the sun. Thus she stood, watching, while the kit snatched the vole, and ran off with it, half dragging it, to the cover of the netting in the hedgebottom. Soon she could hear the smack and crunch of chewing coming from under the netting. She relaxed, and presently darted away in rippling run to hunt another part of the hedgebottom.

The futtrats stayed in the barley field until far into the next morning, eating and sleeping in the hedgebottom. Owls were calling, and it was still dark, when they left.

In the next field the corn was stooked, and hosts of old moleheaps were dotted about the stubble. The futtrats darted in and out of stooks, and climbed them, chattering and crooning, before fanning out over the field. They hunted mouse burrows and mole tunnels, flashing white signals from neck and chest each time they surfaced. The old futtrat killed a woodmouse and carried it under a stook where she ate what she wanted and curled up with the remains.

The family frisked around for some minutes after the old bitch disappeared; then all activity ceased abruptly, as with a time switch. Each sought a stook, and burrowed into the top among the ickers. They slept apart. The family ties were loosening, and hostility was growing.

They came out to play in the bright warm morning, while the old futtrat, yawning, watched them from the top of a stook. They ran under and out of the stooks, playing follow-my-leader. They climbed up and fell down. They birled and criss-crossed. They back-somersaulted and leap-frogged. They nipped tails and grappled, and now and again there was a clash of teeth, followed by snarls and screeches. The two dog kits were rough with their sisters, but their more meaningful bites were still for each other.

They were still on the territory of the old dog futtrat—their father, whom they had seldom seen because he had been kept away from them by their mother's aggressive displays. But their mother's authority was waning. They were now trespassers, with no guarantee of safe passage if they met him.

They met him.

He came from the sunken lane—a master male, heavy as a female stoat, bigger and more muscular than his sons, more than twice the size of his mate. Guided by the faint futtrat play-talk he bounded across the stubble, ferreted through a stook,

156

then stopped to view, framed in the dark triangle below the end shafes. The kits were too engrossed in their chases to notice him. But his mate knew he was there, and raced down to him, perhaps to greet him, perhaps to chivvy him off.

The moment she appeared he rushed at her. He gripped her by the neck and shook her. Then he rolled over and over with her in a convulsive embrace, still gripping her neck with his teeth. Her shrieks of protest, perhaps fear or anger, sent the kits leaping under cover. The big futtrat appeared to be killing his mate, yet when he released her at last, she was able to run on to a stook, unhurt and not bleeding, but touzled and wet with saliva. From there, when he didn't follow, she sneaked away, and escaped to the sunken lane.

The big dog now rumpled around in search of the family. They were all under one stook, their kinship restored for the time in face of the paternal peril. But there was no security for them there. The old futtrat burst snarling into the furry jostle and scattered them. They broke out of the stook, and went leaping like grasshoppers across the stubble, joining up in flight instead of separating, as though glad of a kent face in the hour of danger.

Without knowing it, they were heading in the direction of safety, because the next hedgerow marked the frontier of the old dog's territory. Once they were through he would ignore them. He was no practitioner of genocide. He was the territory holder, the landowner, straightjacketed by the rules of weasel behaviour. The rules said one territory one master male. He was a master male and he was on his territory. So the other feet had to march. His kinship with the young futtrats was forgotten, if he had ever been aware of it.

The last kit in line was the smallest female, and he caught her before she reached the boundary hedge. He sunk his teeth in her and rolled her over, screaming in his face. She tugged and twisted to escape his grip; she kicked her hindfeet in the air; she kicked his face. But she could match neither his strength nor his savagery. And presently she lay still, dead, a tiny creature, smaller than two woodmice laid nose to rump.

The four surviving kits, two dogs and two bitches, spent the day lying up near the boundary hedge, recovering from the shock of their fright. They rested unmolested. Either there was no master futtrat there, or he was in some other part of his territory.

At dusk they hunted, and when they had eaten their fill, and lapped water from a ditch, they moved out instead of lying up to finish off the prey at the next hunger. The urge to keep moving was now strong. For the moment they were transients, nomads, in search of a territory, the dogs to take over if they found one untenanted, the bitches to settle on sufferance.

They ranged the strange ground for most of the night, exploring, hunting, and resting, and met no strange futtrat. Nor was there the scent of one. Nor the usual signs to signify a master at home. Towards dawn the two dogs began bickering, and later they flew into a clinch that became a battle for supremacy. The struggle ended with the defeated kit running off with his sisters. The victor stayed behind, presumably to take over the ground.

Three futtrat kits now moved in line abreast towards Mossrigg Farm, within calling distance of each other, perhaps still held together by the last strands of kinship. They killed and ate, being hungry, because they had left the remains of their last prey behind and carried nothing with them.

The dog kit killed a small rat about his own weight and dragged it under a tree root on the edge of the farm stackyard. Later, having eaten, he was sitting outside in the first of the light, licking his fur and scratching at fleas on his flank. The scent of fox cut short his toilet and brought him upright, wrinkling his nose in distaste, and presently the fox appeared, coming at an easy walk. The kit had never met a fox before, but he knew at once that the beast was dangerous, and bolted under the tree root.

The Mossrigg dog fox was big and wise with years. He was also happy that morning, with half a rabbit in his stomach and the other half in his jaws. He was also a great fox for a ploy, and one of his favourite ploys was chopping weasels. When he smelt the young futtrat under the tree root he stopped and laid down his rabbit. It was early enough yet and he could take time off to investigate.

Although the futtrat kit knew nothing about foxes he was all weasel. The hole wasn't deep and he had little room for manoeuvre, so when the fox's face filled the entrance, blotting out the light and almost suffocating him with breath, he didn't cower back. He struck like a snake, and recoiled. The face withdrew at once, and the fox ran his tongue over the pinpricks on his lip. But he had found out what he wanted to know. The hole wasn't deep. A few scrapes and it would be open, and he could paw-whip the futtrat for the chop.

He began to scrape, but each time he put down a paw the futtrat dart-dart-darted, nipping and screeching. The fox pushed his face in again, perhaps to find out if he could reach the futtrat with his teeth. He couldn't, but the futtrat reached his lip with his. The fox yelped and withdrew. He pondered. He looked around as though measuring the light. It was coming up fast. He sniffed longingly at the hole, keeping out of reach of the futtrat's teeth. Then he yawned, picked up his half rabbit, and left.

That day the futtrats lay up with their kills, the dog under his tree root, the bitches together in an old rabbit burrow on a knowe near the farm. At darkening one of the bitch kits turned back the way she had come, and disappeared, maybe to settle on the range taken over by her brother. The remaining bitch travelled to the stackyard where she soon met up with her brother from the tree root. But their hunting there was interrupted by the arrival of a strange dog futtrat who was the resident landowner. They climbed on to the pighouse roof until he killed a mouse beside a stack and went on his way to his den in the glen.

The dog kit climbed down backwards and padded away to hunt mice in the stack bottoms, but the bitch explored the corrugated roof, padding with nimble forepaws and shuffling her humped muscular rear. Twice she fell over the edge, and twice she managed to hold on with her forepaws while she climbed back with thrusts of her hindfeet. Then she fell a third time, and lost her grip, and fell into a barrel of rainwater.

She swam, of course, round and round, but the water level was four inches below the rim of the barrel and she could not reach it. In a panic now she began to swim faster, and faster, until she was fighting against a belth of her own making. She broke clear and tried again to scramble out, but her efforts were becoming weaker, and assuredly she would have drowned.

But suddenly she was gripped by the scruff of the neck and hefted clear on to the roof. The rescuer was her mother! Why she had come there was a mystery. Or was she still keeping in touch with her dispersing family? She was two hundred and fifty yards from her home ground, whose outer limit marched the Mossrigg weasel's domain on that side. The futtrat kit, half drowned, snapped at her mother, then lay panting on her side. The old bitch climbed down from the roof and went her way.

The kit began to shiver, for she was soaked through. When she felt strong enough she climbed down, falling the last two feet. She bored into warm hay in the hayshed and slept the night through, and the following day. When she came out at dusk there was no sign of her brother. She killed, ate, and moved on.

The *Drang nach Osten* was not yet over.

In the West, a Flutter of Hope
for the Eagle

Well, we've had our teuchat storm with its bitter skite and sleet drench and glaur and frost and snaw bree, depending on where you happened to be, and to think it followed a day of nodding daffodils and lark-song and whooping peewits, with wind flaffs and freshets nudging gold dust from the chandeliers of catkins on the hazels.

In the hazel thickets every stone and boulder is cocooned in moss, under which lurk, insulated, fat white copper-nosed chafer grubs that lie like stranded whales when their rooty cover is peeled away.

Under one gnarly root is a dry cavity where exploring fingers find a handful of nut shells, part of last year's great harvest, neatly holed and toomed by woodmice—the shy, night lovers that like to gnaw and chisel in such dark secret places. The woodmouse is a craftsman, whether handling hazel nuts or snail shells, and treats each in much the same way.

Five oyster catchers, stiff-winged, like birds in a cartoon, come jerkily over the hazels and pitch on the shore where a solitary whaup is slow-walking up and down the terraced wrack, probing and turning, lifting and laying. Above, the creeping seagulls are tossing like blown paper. A hoodie on a rock is busy running his beak down the black darts running from his chest to his belly.

High above the rocks a pair of ravens are playing, masters of the heights of air, and higher still, great dark shapes against the blue, is a pair of eagles, drifting and riding the ridge tops, not a reassuring sight at a time of year when the hen should be tight down on egg or eggs.

Ravens and eagles have the upper silence to themselves; below, a kestrel is flickering along the face of the rocks. This is the line Concorde is due to travel soon, and if the boom doesn't start a big rock fall it will certainly make the big birds wonder what is happening to their wrapping of air.

Over the hills, where one can see a fleet of islands at anchor on the ultramarine sea—Lunga, The Dutchman, Fladda, Staffa and the rest—another eagle is wheeling with blue sky between the fingers of its wingtips, black in silhouette, wide-winged, alone. And a lone eagle, taking time off, is as reassuring at this time as a pair is not.

Reassuring, too—not spectacularly but significantly—is the evidence of greater breeding success among eagles in those parts of the west where there has been a decline in recent years. But it would be over-optimistic to read too much into this at present.

There is plenty of evidence of deer in sheltered hollows below the snow-line, and here, too, can be heard the lonely whistle of the golden plover. A dipper flies from beneath a waterfall and follows the burn, keeping close to the surface. A pair of hoodies takes off from a scrawny tree where there's an old nest, and a pair of buzzards flaps up from a rock face to rock and balance and drift overhead until we are clear of the gorge. Then they drop back again. They will be prospecting.

In the rain of the evening, when we were down again, we stopped by a field to watch a buzzard perched on a moleheap. The bird appeared to be doing no more than sit there then, suddenly, it came to life and began running this way and that, long-legged with wings half open, dabbing at the ground and swallowing. And the prey was earthworms.

It picked up seven, eight, nine, ten: then it lofted a big black slug and down it went too. I had 12x glasses on the bird, so had a ringside seat. It showed not the slightest interest in the car parked at the roadside, but went on with its high sprinting, more like a secretary bird than a buzzard.

One worm appeared to be partly anchored and this one the buzzard stretched until it broke with a high-tension splash. Then it was swallowed in two pieces.

At this point a second buzzard arrived, flying right over the car and down, and right away it began sprinting and dabbing without pausing to discover what was on. It obviously knew. The pair was still worm hunting when I drove off.

Of course, the buzzard is a considerable eater of earthworms, at times, as the badger is at most times. It takes molluscs as well, and beetles and grasshoppers, and sometimes a lot of them. It will even, at times, raid the nests of small ground-nesting birds like pipits, reedbunting and yellowhammers, taking chicks only a few days old.

The author in an eagle's eyrie with a golden eaglet. This eaglet was reared on rabbits and hoodie crows. The eyrie was on sheep ground but no lambs were killed.

A hen eagle at her eyrie on Mull. The hen eagle is bigger and more powerful than her mate. This eyrie had one eaglet which flew on 14 July.

A cock eagle brings a rowan branch to its eyrie with twin eaglets in the Forest of Atholl. Eagles carry in fresh material to the eyrie throughout the nesting period. The twin eaglets from this eyrie flew on 11 July.

A badger emerging from its sett in the Scottish Lowlands. When the soil is easily worked, established setts become like a series of small quarries and the mounds like a line of fortifications.

A boar badger at a Highland sett, 1500ft. above sea level. In remote areas badgers come out early, and may be found lying above ground during the day.

A hen capercaillie at her rest in a Highland pine forest. The chicks are hatching. The rest is usually at the base of a tree or a stump, and on a slope.

A cock capercaillie displaying during the breeding season. Like blackcocks, the cock capercaillies have a special display ground. The most active period of display is from 2.00a.m. until daylight.

The ring ouzel is the mountain blackbird and a migrant. Both sexes have a white crescent on the chest, more prominent in the male. Here the hen is feeding her young at 1500ft.

A redshank on her nest incubating. Redshanks are waders, like the curlew. They nest in rough pastures and moorland. Most rests are well hidden: this one is on open ground.

Red deer calves are born spotted with white and do not run with their mother until a week or so after birth. The spots begin to fade in August.

Red deer herd by sexes (stag groups and hind groups) outside the breeding season—September to October. After the rut the stags form companies again.

Roe does usually have twin fawns, but sometimes only one and occasionally three.

A roebuck in hard antler. Roe are woodland deer. They come out to feed from dusk to daylight.

A roebuck feeding on brambles. Roe eat a wide variety of food and they are especially fond of brambles.

A yearling roebuck in winter coat, showing the twin white flashes on throat and gullet.

Carrion crows can often be seen feeding on road casualties, and can be brought just as readily to a bait. A bait of dead hare brought this one to the camera.

The black-headed gull breeds on inland moors and marshes and is a familiar species near towns. The hood is brown, not black, and disappears in late autumn.

The kestrel is the mousing falcon, and probably the best known bird of prey. It's the familiar hoverer, common in the countryside but also to be seen in open spaces in town and city.

Dog weasels are more than double the weight of the bitches and can therefore hunt and kill bigger prey. They are territorial and may spend all their lives on a few acres.

The mink found in Scotland is an American, and the wild stock of the descendants of escapees from mink ranches. They have colonised most of the country and are breeding successfully.

Bitch weasels in Scotland vary in weight from 2½-3½oz. and require about one ounce of food per day, almost a third of their weight. They breed twice a year, exceptionally three times.

The original stock of true polecats was killed out in Scotland, but there have always been feral polecat-ferrets. The pure polecat is again present because many people have reintroduced it.

Ailsa Craig—Paddy's Milestone—is one of the great gannet breeding stations in Scotland, where 13,000 pairs have been recorded nesting. Dr. J.A. Gibson does an annual census.

The fulmar is a pelagic sea bird that has spread round the Scottish coast in recent years.

Shags are like cormorants. The puffin is like no other sea bird. He is often called the 'sea parrot'.

The heron is a fisherman, with long legs, a neck like a coiled spring, and a javelin beak.

The mallard is the well known 'wild duck' the ancestor of our domestic breeds.

Red grouse in winter. This species has been more closely studied than any other game bird.

Modern farming has not been kind to the partridge and its numbers are declining.

The red-throated diver of the North-west and the Islands prefers small lochans to large stretches of water. It nests near the water's edge, but hunts for fish in the sea.

The blackcock (the male black grouse) has a special spring display. The display ground is known as a lek, and on it the birds challenge and posture throughout the night in spring.

Loch Druidibeg of S. Uist, which is a nature reserve, is the main breeding place of the greylag goose in Scotland, but now it nests in many other places including Duddingston Loch in Edinburgh.

A dog stoat in summer coat, hunting for mice. Stoats of all ages are playful animals. When they are not hunting or sleeping they play singly or in a family.

A bitch stoat changing her coat to winter ermine. The change can take place in seventy-two hours but the usual period is two to three weeks. Bitches are smaller than dogs.

The adder is the only snake found in Scotland, and it is venomous. Therefore, all Scottish snakes are venomous. The snake strikes rather than bites.

The only lizard found in Scotland is the viviparous, or common species. It gives birth to living young because the eggs rupture at the moment of extrusion.

The slow-worm is a legless lizard, and totally harmless. Unlike snakes, and like other lizards, it can blink its eyes. It is often killed in mistake for the adder.

The toad does not spit fire although it has a tongue like a flame-thrower. Unlike the frog it has a dry warty skin. Again, unlike the frog, it can be easily tamed.

The great spotted woodpecker is common in many parts of Scotland. Like other woodpeckers it drills out its nesting holes in trees, often no more than eight or ten feet from the ground.

Blue tits can be easily persuaded to use nesting boxes in the garden, and will do so year after year.

A female whinchat carrying food to her nest on moorland.

A female grey wagtail feeding young at the nest behind a waterfall. Grey wagtails like clear unpolluted water, in and near which they find most of their food.

The woodmouse, a night hunter, is Burns' 'wee, sleekit, cowrin tim'rous beastie'.

The mole is well known but seldom seen. It feeds mainly on earthworms.

The water vole is the biggest member of the vole family and is found on the banks of burns and rivers. Although it is as big as the brown rat, its life is as short as that of smaller voles and it rarely sees a second winter.

The field vole is a grassland species notable for periodic explosions of population.

A leveret of the brown hare. Leverets are born fully furred with their eyes open.

The grey squirrel is an American and has colonised many areas once held by the native red squirrel.

Although the red squirrel has disappeared from many places it is still plentiful in coniferous forests.

A tawny owl brooding chicks at night, in the old nest of a carrion crow. This species can be dangerous when it has young in the nest, especially after dusk.

The hen-harrier was once practically confined to Orkney, apart from a few in the Western Isles. After the Second World War it began to colonise much of the mainland. It nests on moorland and in young coniferous plantations.

A cock long-eared owl arriving at its nest with prey at 2.00a.m. This is the most nocturnal of British owls.

A female sparrowhawk in the nest with her young. This hawk feeds mainly on small birds. It builds its own nest, favouring oaks and larches.

Dog foxes (like dog wolves) are good fathers and vixens are good mothers. It isn't easy to sex foxes in the field. Can you tell a bitch dog from a dog dog in the field at 25yds?

They'll tell you that the Scottish wildcat is untameable. Not true. Like the Scot it can be tamed easily if you get it young enough and again like the Scot it remains tame.

A bitch fox (vixen) hunting for voles. Foxes are shy, alert and wary; but you can get close to them if you take the trouble. It's a lot of trouble.

An Atlantic grey seal cow with pup, on Lunga of the Treshnish Isles. The grey seal comes ashore to breed in late autumn. The pups are weaned at the age of three weeks, or less, and the cows are mated again.

My daughter Kathleen with my tame dog weasel Tammas whom I hand reared from the age of ten days. He was beaten to death by vandals when he was three years old.

My German Shepherd bitch Lisa fostering my fox cubs Glen and Fiddich to whom she remained a mother figure for many years. She was the greatest fosterer I have ever known.

Lisa took wolf pups in her stride. Timber wolf Marquis was one of her protegees. She was still the unquestioned pack leader at her death when Marquis was five and a half years old.

'Give it to me' might well have been Lisa's slogan. She cared for this leveret until it was able to fend for itself. She mothered weasels, ducklings, geese, wildcats, polecats, blackbirds and my own grandson.

Most of the fawns I have reared were brought to me by people who thought they had been 'abandoned'. This is criminal. All young wildlife should be left alone. Roe deer do not abandon their young.

Glen was probably Lisa's favourite fox and she let him, literally, walk all over her. Not once over the years did she growl at him or show him the colour of her teeth.

Jess Stephen with seven and a half years old timber wolf Marquis, whom she brought up in the house until the age of four months. He played and slept with Lisa and was mothered by her.

Lisa

On November 1, 1980, in *The Scotsman*, in this column, I quoted my postman as saying: 'I wish all the houses I delivered to had dogs that gave me such a welcome . . .'

Three days later he was a welcome short, and a considerable chunk of what makes me tick went into the ground to stay there.

For Lisa, the *schwerpunkt* of his welcome, was dead: my timber-wolf-size Lisa, my gentlest of the gentle, charismatic German Shepherd, was in the past tense.

The first letter I opened that sad morning was from a woman who had met Lisa six or seven times. It was a totally unexpected letter, and I'm not even going to try to resist quoting the final paragraph:

'In her dark no-place she will not remember exuberances or battles: quietness now possesses her, your duchess, your gentle lady, keep clenched in your mind her love that was requited, her faith that was well-placed.'

For nearly eleven years she was my shadow, my adjutant, my rearer-of-all-things, my carer-for-all-things, the mother-figure, lover of children, favourite of the handicapped, who only twice in her life used her weaponry, and that at my express command.

We knew near the end that she was ailing, and probably beginning to feel pain. So we let her go for surgery, and she came back in the past tense. I keep thinking now that I should, myself, have taken her trusting, gentle life. With that compassionate treason I could have felt more like the person she thought I was. Instead I let her die, distressed, alone, in the clinical hands of strangers.

They keep telling me it was all done for the best. But one of us knows better. The other, had she been capable of thought, would have known too. In the end I failed her.

Looking at the empty cavern in the kitchen corner that her giant frame used to fill we are all agreed that she will be remembered best, by most, as a mother-figure, for although she never had a family of her own she treated every young thing in fur or feather as her own puppy, and that includes the Labrador and the Jack Russell who are no longer puppies.

As a puppy herself she took charge of the Siamese cat Skipper, and as she grew up she became his hatchet woman, ready to take a door down to get to him if he squalled outside. Skipper is now fifteen years old, and that relationship remained unshakeable until the day she died.

Then she had to put up with the weasels, Tammas and Teen, running over her like ants or slyly nibbling her nails. The most they ever got from her was a gentle *woof* of reproach. She was the same with the wild kittens, Teuchter and Pibroch. They liked the warmth of her. They also liked to tread on her belly with their hindfeet, and sometimes they pricked her enough to draw blood. The most they ever got for that was to be pinned down with a monstrous paw until they behaved.

She fostered the fox cubs, Glen and Fiddich, who slept, ate and played with her. She was their vixen, and she was vixen to them. A few years later she played wolf to the wolf cubs Marquis and Magda. For four months of their lives they slept with her, roistered with her, ate with her and obeyed her. I think her wolf days were among the happiest of her life.

When I was bottle-rearing my roe fawns, Bounce and Sanshach, Lisa took charge of them in the garden. She played with them, mothered them in sleep against her ample flank, and protected them with a consuming jealousy. When they went into their two-acre enclosure she played with them there. When Sanshach had her first two fawns, twins, she allowed Lisa to lick them over without the slightest expression of alarm or distrust.

One of the funniest things I can remember was the day she met her first wild roe deer: truly wild but with a growing tolerance of people because they were never under pressure. They ran about fifty yards, then stopped and looked back. Lisa trotted out towards them, and they ran another fifty yards and stopped. She tried again, and they ran again. She came back to me with a look on her face that was asking: 'Why are they running from *me*?' I knew how she felt. She could never understand why wild roe and foxes ran away from her.

My badger cubs Spick and Span were probably her roughest assignment. Badger cubs play exceedingly boisterously. She accepted all the gurrying, all the bear-claw grapples and brulzie, with tolerant dignity that was never ruffled. When it went too far she used the mighty paw to pin them down. Never did she utter a growl or flash brief ivory.

Once, when my terrier growled at Marquis (understandably because the horsing around was getting a bit much for her) Lisa rushed from her basket and dunted her aside. From that day my wife called her the referee. She would not stand for any aggressive display, any grumping or moaning or girning. She waded in and showed the iron fist inside the velvet glove. Her motto might well have been: if you can't stand the heat stay out of the kitchen.

She was well aware of her size, her weight and her fist-feet, and it was something to see her trying to cat-foot across the floor avoiding leverets, rabbits, polecat kits, blackbirds, owlets, goslings or whatever, while the Siamese cat, on top of the heater, looked down at her as though saying (which he wasn't): 'For a disgustingly big dog you might yet graduate to becoming a cat.' If Siamese cats could speak that's probably the way they would speak.

Little William, the wild goat kid, she loved, and he followed her around like her shadow. Cassius, my big dog polecat, she loved, and if she wanted him in her basket he had to go, by the scruff of the neck if need be. And Polar, the greylag goose, she loved, and would try to make room for him in her basket, even when he was fully grown and flying free.

My wife always addressed her as The Greatest, and Lisa could never resist the rhetorical question: 'Who's The Greatest?' That always fetched her for due commendation. She was the greatest, and she knew it. We did too.

Now she's dead. So what the hell? After all, she was only a dog.

Portrait of a Feathered Hoverfly

The Lordly Ones—the eagles, the falcons and the hawks—once the pride of kings, princes and noblemen, became the lumpens of the nineteenth century, when the modern fowling piece did them out of their jobs. Gunpowder took over from beak and talon. The Lordly Ones were not only unwanted: they were 'vermin'. And the killing times began.

Like Chesterton's donkey they had had their day, and it is a measure of their status that, in the days of Jamie Saxt, a pair of goshawks could cost you £1000. In earlier times twelve Greenland falcons could ransom a Crusader Prince. Savage laws were passed to protect the nests and eggs of hawks and falcons, and Henry VIII made it a felony to take nestlings or eggs. The 1954 Protection of Birds Act wasn't so tough after all.

Falconry is having a bit of a vogue again and, while I have no special quarrel with it as a sport, I often wish the vogue would die the death, because it has led to all sorts of hamfisted people putting on the style and making misery for many birds of prey.

The common mousing kestrel (hardly top drawer among the Lordly Ones) is a particular sufferer, and ever since the film *Kes* I've been inundated with escaped kestrels, wearing all manner of leg-gear, that were snagged up on power cables, telephone wires and fencing wires.

At which point I should give credit to the South of Scotland Electricity Board (unfashionable perhaps, but abundantly justified) because they once put a whole chunk of Cumbernauld out of action, and sent a team of men to get me a trapped kestrel down from a high tension cable. The bird had the usual clutter of leather round her legs, and was a bit under the weather, but we got her fit and hard set and released her to the wild.

I have a long standing liking for the kestrel, since 1936 in fact, when I first sat in a tree five feet from a pair feeding young. When I get to know a pair of birds well I often find myself giving them names, and in this case I called himself Yellowfoot and herself Kree. I knew them for many years, and if you ask me how I knew I was working with the same pair for so long I can say only that I know. I knew when the new female came, as I did when Yellowfoot was replaced.

I ringed Kree's nestlings in 1947, and in 1952 I found one of her daughters nesting near my farmhouse five miles from where she was born. Needless to say (but I'll say it) I called her Young Kree. And from that day her mother became Old Kree.

I got to the stage with Old Kree when I could stand on a branch four feet below the nest and from there, towering two feet three inches above her, talk to her at a range of two feet. She would be lifting, with the heaving of chicks below her, and sometimes she would give me a muted *kek-kek-kek,* and I would go into my hide where she couldn't see me and speak to her from there. But I reached the point where I could stroke her on the nest, and that was victory enough for me.

I always had plenty of warning when Yellowfoot was on his homing flight, because he was picked up and chivvied by other nesting birds as he flew over their territory. The whaups took him first, then the peewits, then the jackdaws; and syne he would be in the tree at my back, *wree-wree-wreeing* to Old Kree on the nest, and herself *wree-wree-wreeing* back. Then he would fly in and pitch on the edge of the nest with vole or whatever in his beak. Mostly vole.

He was a smart little falcon, well tailored always, and, as my mother used to say, as neat as Katie's leg. He was a good provider and Old Kree never lost a chick she hatched. A lot of people imagine that birds of prey of the same species are all equally efficient hunters, and that they always kill what they strike at. Neither is true. Some are better than others.

Yellowfoot was good. In the June of 1947 he brought 214 field voles, a mole, a weasel, a number of shrews and two baby rabbits to the nest. There could have been more because I wasn't always there at first light. Old Kree reared six young that year. Out of six.

I once rescued Yellowfoot when he became snagged up in a coil of old barbed wire in deep grass. When the farmer and I found him the barbs were through the webs of his flights and the feathers of his trousers. We took some time to cut him loose, whereupon he sank two sets of fish-hooks in the edge of my hand. Farmer said: that's gratitude for ye. I said: that's anthropomorphism. But we understood each other.

Most people know the kestrel; even town and city people. It's not only a common falcon; its hunting style draws attention. Although it's sometimes dubbed sparrowhawk by those who don't know any better, it's still a kestrel. The sparrowhawk's style is totally different.

The hunting kestrel hovers on flickering wings, with tail fanning or scissoring, and head down-pointing while the eyes scan the ground below for vole, or mouse, or shrew. The bird looks as though it were suspended from an invisible wire. I've called Yellowfoot a feathered hoverfly. Then there's the pounce, followed perhaps by a catch; or the change of hunting station following failure to find, or an unsuccessful strike.

Kestrels frequently fly up to the nearest tree or telegraph pole after a catch; thus many people see the bird with a kill. I imagine more people have seen kestrels kill than they have any other bird of prey. I'm one of them, and I've seen a lot of kills by a lot of species over many years.

The kestrel has come off better than most birds of prey because so many people consider it a 'useful' bird. Even some gamekeepers spared it, although most killed it, as one would expect. But the kestrel is just a kestrel, and if it happens to do something of which we approve (like killing voles and mice) fair enough. But that's not why it is there. And the notion that it 'keeps down' voles is a nonsense. If the kestrel keeps down voles I'm Pope John Paul.

Terror From the Nettle Patch

A man is always getting surprises with one bird or another, and this time, in my case, the bird's a partridge: the little brown couthy game bird of the farmlands and the moor's edge, which one might well be forgiven for calling the dearly beloved of the countryman and the sportsman.

Do you remember the lines of Charlie Murray in that wonderful poem of his,

'It Wisna His Wyte'?

Astride on a win-casten larick he sat
An pykit for rosit to chaw,
Till a pairtrick, sair frichtened, ran trailin a wing
Fae her cheepers to tryst him awa.

What countryman hasn't seen the pairtrick running low, cooryin doon, with wings half open, or one trailing? And which of us hasn't said, at some time: that's her pretending she's hurt so that we'll run after her and overlook her cheepers?

Modern zoologists have a different explanation of this display, just as some of them have cast doubts on the partridge's marital felicity, or rather fidelity; but by and large the pairtrick is the model sitter, the model parent, and one of the bonniest and best liked birds of the countryside.

How canty, gleg and homely it is! And how pleased any of us is when the partridge we have been seeing running with her mate for many weeks decides to nest with us—in the garden or close by in a hedge bottom.

My partridge disappeared, as all hens do when they begin to sit, and when I saw the cock on his own, in the hayfield, I said to myself: herself is sitting somewhere near at hand.

Two days later, at 3.30 in the afternoon, I saw the cock running through the trees from my garden gate. I walked to where I had seen him and worked my way down to the gate, searching every bit of cover.

And, in a clump of nettles, I saw the little brown bird, all dumplingy and streaky and motionless, crouched down on her nest.

And from that moment I was a worried man!

Would the Sussex cockerels running on the ground molest her? Would the pullets decide to trample lanes through this nettle clump as they had done with others? Would the magpies, who were always stealing the pullets' food, find the eggs?

The following day, at 3.45p.m., the hen partridge went off to feed, and I saw her in the grass park with the cock. I looked at the nest and counted the eggs.

Nineteen! A good cleckin for the partridge if they all hatched, and a grand breakfast for the 'pies' if they found them beforehand.

Almost every time I opened the garden gate there were magpies at the pullets' food, 15yds. from the brooding partridge. And the cockerels were showing a lot of interest in the nettle clump.

So . . .

We caught up the cockerels and shut them in a big house. We put up a piece of hessian on the fence to make magpies wary. And hoped for the best.

The pies stopped coming to the food trough. Each day I moved the hessian, and changed the colour of the piece I had pinned to it, and each day there were no magpies near the place. But I knew they would get used to it all in time. And I was sure they knew where the partridge was sitting.

I surprised them in the tree right above her one morning and tossed a stick at them. They flew away swearing, but didn't come back for two days. I converted the hessian to a hide and left it near the partridge. And every hour or so one of us made a brief sortie through the gate, up to the hide, and back into the garden again. The magpies stayed away.

On 10 June I got into the hide and took some pictures of the bird, tying back the nettles for the time, then closing them over the nest again. In addition I 'planted' extra baffles in front of the nettle clump.

The pullets then took a hand. They decided to sun themselves beside the nettle clump. And out came my partridge. She yoked on the lot of them, flying in their faces, and striking them with her feet, and drove them all right off the premises!

Three days from now the eggs will hatch. But I shall not be at home to see it. My wife will be the 'nurse' in attendance.

Danger, Naturalist at Work

A snake-hiss from a mousehole or a hole in a tree is a usual reaction of the blue tit if disturbed, and it's sometimes enough to put off a man the size of a door if the door-sized man isn't familiar with blue tits.

A young cuckoo in the nest can be very off-putting if you're the kind easily put off. Even young cushats, gentle young of the gentle dove, can deliver a tolerable right cross with a wing that has made more than one small boy fall out of a thorn bush.

The almost routine assault of the nesting mistle thrush, and the sometimes assault of the nesting mavis, are other defensive displays which most people take in their stride, treating them more as light comedy than serious threat, although one sometimes sees them reported in the buskins of hyperbole.

Probably the safest job in the world, if not the most comfortable, is the study of British wildlife in any form, unless, of course, you happen (stupidly) to fall out of a tree, or bust a bone falling off a rock, either of which you could just as easily do without being involved with wildlife at all.

But there is a persistent idea in some people's minds that wildlife is dangerous stuff to meddle with, and some newspapermen have the idea on the brain, so that wildlife is never featured as itself but as some Poe nightmare or Frankenstein creation. A year or two back, when I was supplying a picture of a Hooded Crow for a London publication, I was asked to send one that showed the hoodie looking 'as evil as possible'.

Somebody showed me a cutting the other day where the hoodie was being chastised for having no moral standards whatsoever. And what do you think of that, eh?

When the wildlife gets the treatment the barrel is scraped for the adjectives with the mostest. A weasel becomes a sinister, murderous, vicious, bloodsucking whatever-you-like to call it. The stoat is the torturer of innocent bunnies, the sparrowhawk the slayer of dear little dickie birds. The otter is a devil, a fiendish slayer of the lordly salmon which, one assumes, would rather be caught dacent on a fisherman's fly. Even old Brock is often portrayed as a kind of ravening Appolyon.

The dangerous elements in wildlife are few and seasonal, and do not, in any case, always run to form. The tawny owl with young can be dangerous when defending them, because she strikes silently in the dark and may claw one on the face. The bite of the adder can be dangerous. So can a wasp's sting.

A bonxie striking at an intruder can draw blood, but shouldn't really be taken seriously. The hen-harrier also puts on a defensive assault, but the display is one for admiration rather than awe, although a bird will occasionally rake a man's scalp with her fish-hooks. I had it the other day, but, curiously enough, I have lived to tell of it.

Many young birds of prey strike aggressive postures in self defence, and I have heard them called vicious when doing so. I doubt if the word vicious could reasonably be applied to any species apart from man. It certainly can't be applied to helpless young birds trying to ward off an imagined attack by striking an aggressive posture.

Young kestrels, peregrines and harriers will turn over on their backs when meddled with, presenting claws like barbed wire and fish-hooks to the foe. But you can stand them up on your bare hand the next moment. The apparent ferocity, and the threat, are really nothing, although the claws will take hold if one handles carelessly. But even a friendly kitten will claw if lifted clumsily.

Generally speaking, wildlife is safe when it is wild, unpredictable when tamed. More dangerous than any wild animals are the tame ones—the pet lamb grown up, the dairy bull at the age of cussedness, the billy goat, the bantam cockerel, the spoiled dog, the intellectual parrot. Dangerous wild animals are deer when hand-reared . . . because they are then tame animals that have no fear of man.

I'd gladly hold speech with twenty bayed wildcats rather than face one mature tame roebuck with his dander up. There is nothing about the wildcat that anyone needs to be scared of. The problem is ever seeing a wild wildcat at all.

No professional zoologist would ever dream of asking for danger money because of the non-hazards he faces in his work. The fear seems to be confined to people who hardly ever meet a wild carnivore, and to certain mass media that love catering for such fears. Wildlife-wise, Scotland is really a very safe place.

The Grim Reaper

The big field, being rumped for silage, was pallied where it had been cut. The great red mechanical monster, so like a dinosaur as my daughter put it, trundled slowly along, its head and neck showing above the hedge while it grazed the field into the ground. A few rooks, ever opportunists, were down on the pale rumped green, dabbing at exposed crawlies.

But my eye was taken by three other birds, for whom the dinosaur meant trouble—a pair of skylarks and a cock partridge. The larks were flying to a point on the bare cut, hovering for a moment, like kestrels, about three feet above the ground, then turning away in a wide half-circle, and returning to the spot to hover over it again. Both were carrying food, so it was obvious they had had a nestful of young in the field.

When the monster was at the bottom of the field, both larks pitched in the grass, walked a few feet, then began seeking at a featureless spot, which was probably the site of the tussock in which their nest had been. I couldn't see even a smudge when I trained the glasses on them. Their chicks, I had to assume, had been gathered for silage, like other chicks, and eggs, before them, and there was nothing anyone could have done about it.

Then there was the partridge. He was walking, sometimes sprinting, up and down, travelling about the same distance on each side of a low-spreading dogrose in the hedgebank. But he had less to worry about than the larks. His cover had gone, but his mate was still securely down on her nest under the dogrose, and, with luck, would get her brood off safely.

Field operations, especially the cutting of grass for hay or silage, inevitably cause some degree of upset to wildlife at this time of year. In a hayfield the most likely victims are pheasants, partridges, larks; perhaps young redshanks or curlews; and roe deer fawns. But it is surprising how many are missed, because some field workers have a flair for picking them out in time.

I know one farmer's son who usually manages to spot every peewit nest in his path, so that he can lift the eggs and put them back in a remade nest. This year, when working potatoes, he wrecked a nest, not noticing it till the next time round; but he put the one intact egg into a made-up nest and had the satisfaction of seeing the bird make do with it.

In my part of the world roe deer quite often drop their fawns in a hayfield. At haytime, the little beasts are usually spotted, and driven out, but sometimes they are cut up badly and have to be killed. I have seen a number of fawns with two, or all four, legs cut off, and have had to destroy them on the spot. A doe, put out of hay with her fawns in the evening, will often have them back in the morning, and this is when accidents easily happen.

I have seen a few foxes put out of hay at cutting time, but I can't recall ever seeing a fox cut up by the machine, although I have seen them knocked over by shot when they broke cover.

In the old days I used to run my Labradors through the hay on my neighbour's farm once the cover had shrunk to a stand in the middle of the field, and in this way we saved many a brood of pheasants or partridges, once a cleckin of mallards, and now and again a roe deer fawn. But I have seen a lot of mangled limbs, and a number of bloody messes, at haytime.

Farmers don't like cutting up anything (unless maybe a fox, and this hardly ever happens) and do their best to avoid it, but in such cover it is impossible to see everything ahead.

I remember one farm worker who was supposed to have a heart of concrete. The boys used to say he hadn't been born, but quarried. But I shall not easily forget his face the day he cut the legs off a roe fawn. He had rescued one, and was very pleased with himself; then he hashed on and cut the legs off a second one. I can see the wet glitter in his eyes yet as he said to me: 'There was twa o them!' They were five of the most expressive words I have ever heard.

Wildlife a Multiracial Society

Scotland's wildlife is a sort of multiracial society, and it comes as a great surprise when we first learn what a great amount of importing, swopping, and integrating has been going on over the years.

If we look only at mammals and birds we find there has been a startling amount of immigration and integration, although a man is not aware of this merely by looking. Integration has taken two main forms: the deliberate or accidental establishment of species that were not native, and the re-establishment, from foreign stock, of native species that had been lost.

In the first category are many common species which we tend to look upon as true natives, more Scottish than many present-day Scots. They are Scots by long residence, but exotics nonetheless. The house mouse and the rabbit are two such.

The house mouse was originally Asiatic, but has been carried all over the world by man, including here, where it is a commensal of man. The rabbit was brought to England by the Normans, and reached here much later to become a widespread and thriving pest. Before myxomatosis, something like 60 to 100 million rabbits were killed in Great Britain each year.

Some parts of the Highlands didn't have rabbits until the nineteenth century, and it is difficult for us to realise that generations of Highlanders knew the wolf but had never set eyes on a rabbit.

The brown rat, a great cosmopolitan universally loathed, came to us from the East, presumably via Norway, in ships. It reached the British Isles in the early eighteenth century, and immediately began to oust another alien, the black rat, that had arrived in the Middle Ages.

The American grey squirrel, often called the tree rat, was first introduced into England in 1876. There have been many other introductions, and the species has spread rapidly, replacing the red squirrel. The grey squirrel is now found in at least eight Scottish counties, and is a major pest.

Two of these five were deliberate imports: the rabbit and the grey squirrel. The Japanese Sika deer was a third, and this species is now found in strength in several Scottish counties: Argyll, Fife, Caithness, Inverness, Peebles, Ross and Sutherland. (England has other established alien deer: the Chinese water deer and the tiny muntjac).

We almost had the muskrat in the thirties. This species (it provides the musquash fur of commerce) was introduced in 1929, and soon there were escapes. One colony became established in Scotland in strength but was successfully trapped out. All British colonies were exterminated.

The big coypu, now well established in East Anglia, has failed here. I have seen a few beasts on the loose but they do not appear to have been able to gain a lodgement. The American mink, now breeding in several parts of England, has one establishment in Scotland, and has been recorded, and trapped, in other parts. It looks as though we shall have the mink on our list of natives eventually.

The reintroduction of former native species is in a different category from these. The reindeer was a native of Scotland which became extinct towards the end of the twelfth century. Now it is back in the Cairngorms, built up from a nucleus introduced in 1952.

The red squirrel came close to extinction in Scotland, but was kept going by introductions—from England, and from Scandinavia to Perthshire in 1793. Although we have never been without the fox there has been a lot of shuffling of stocks, and we have mixed Scots and Scandinavian blood. Then, of course, we have exported great numbers of foxes to England.

Among birds there are obvious examples. The capercaillie, a native, became extinct in 1760, then was reintroduced from Sweden in 1837 and has built up from there. The pheasant, brought to England by the Romans, is a long-established alien, still largely kept viable by artificial replacement rearing. Then there is the little owl, which has crossed the border from England, to which country it was deliberately introduced. The osprey, extinct as a breeder, has reintroduced itself.

We have had our losses, of course: the great Caledonian bear in the ninth or tenth century, the wild boar in the seventeenth, and in modern times (probably) the polecat. The wolf lingered long enough after our incorporation by England to become the British wolf once more.

It Was a Beautiful Morning . . .

It was a thistle-down morning of sun-glare and blue sky after a faint chirp of frost in the night. And the strands of spider silk, shimmering, finer than cat whiskers, stretched like a fragile system of communications from tree to tree.

The tall thistles, in the field along the woodside, were tousy and shapeless of head, like moulting deer's hair. Their white weightless seeds, like antennaed micro-satellites of spider silk, orbited the wood on flaffs of wind, and were lost. They were drifting in clouds dense as winter flocks of starlings.

Two rabbits, in fine bloom, ran without haste from the thistles to a slip-through in the drystane dyke, sat momentarily upright in the wood, then disappeared into freshly excavated burrows.

A few yards later, a third ambled aimlessly into the open, hopped in circles and figures of eight, then found the slip-through and its runway in the wood, and dived into a burrow.

This one had the disease, with callouses for eyes, and face puffed up. It was not seeing, but still eating, and navigating by its kinaesthetic sense—its ability to travel its own trails without seeing or thinking where it is going.

I turned into the wood, close to a clearing, and sat down with my back to a tree, and the sun on the side of my neck. Here the air was warm and flies were out. Magpies were yarring, crows calling, and woodpigeons flying overhead on whistling wings.

When I saw the foxy-red shape among the brackens, I thought fox then saw him for what he was—a slim roebuck, with finger-length antlers, full of the fun of a fine morning, with the spring in his legs and the itch in his head.

That was how I thought it, and that was how it was.

He hop-skip-pranced across the clearing, right across my front, then proceeded to dunt birches, bracken fronds and low branches with his antler spikes.

He would thrust, then jump back, as though evading a counter-thrust. He was

170

on a make-believe ploy, just playing, as so many adult animals do when there's nothing more serious on hand.

Thinking of other mornings, I waited with him, and let him wander off without letting him know I was there.

I followed a ride where squirrels had been shredding cones, and crossed the hardwood ridge to the burn. Now I was on the dark chill side of the wood. Here blackbirds were leaf-turning looking for creepies, and young robins, pale of breast, were flitting in the rhododendrons.

A kestrel flashed out from a burnside elm when I was near the burn, and a waterhen went squattering and *crooking* away when I made my step to the first stone. Then a pair of mallards beat up from the threshes on the opposite side, and swung across the blue arc of sky.

The morning was warming up, and I had wasp company most of the way along the cornfield hedge, with its stands of fireweed and bramble tangles on its south side. I found a seat giving me a view along the woodside endrigg, and lit my pipe.

I have always been a strong supporter of the inactive brand of activity, that is to say sitting down. A man, highly mobile, sees the sights. But there's no substitute for sitting when it comes to getting acquainted.

Five minutes or so after I had seated myself, a leveret hopped clear of the corn and came slowly towards me. It was half grown, an attractive beast, ruddy of coat, and with outsize whiskers.

It sat down a few feet from me and began to lick its flank and its crotch. Then it combed its whiskers, and grinned at me unseeing. My pipe smoke was drifting over my shoulder so its nose had no word of me.

It was the farm dog, barking on the road on its way for the cows, that frightened the hare back into the corn.

Perversely, I sat on, and for more than a quarter of an hour there wasn't a movement in the corn. Then, suddenly, there was. And this time it was a sapling reynard who came out of the cover.

He looked down the endrigg, then up towards me. Satisfied, he put nose to ground and began a walk-sniff on the line of the hare. But he didn't follow it back into the corn.

He turned aside, towards the wood, and followed some trail over the dyke and into the cover, where I could no longer see him.

I rose slowly, and put the field glasses on the spot, and there he was, weaving on the trail dictated by his nose, and taking plenty of time about it. I lost him at last in a rhododendron thicket.

Fifteen minutes later I was working at the porridge in its wooden bowl. It was still a beautiful morning.

Swedes on the Wane

The swede—Alas!—the swede. The root that once revolutionised animal husbandry in Britain has been displaced, like all revolutionaries, by still another revolution. Fields of swedes, except where sheep are to nibble *in situ*, are becoming things of the past, and a man has to go a long way to get his knees wet walking the drills of purple bulbs and green blades laced with chill morning dew.

Contemptuously, they'll tell you now that if you want to produce thousands of gallons of water to the acre—grow swedes! The bulb that once meant life and death is now no more than a skinful of water—so they say.

Yet a good swede is a brawsome thing, especially on a frosted blue morning, when it turns its round sheened belly to the sun and the partridges sneak under its armpits. Many a partridge, nosed up by dogs on a sharp September morning, has crumpled in a breath of gunsmoke and dropped to die under the green canopy, staining the purple with its blood. And one of the sights I remember well, for no special reason except maybe for the colours and the sun's glint and the crisp air, was a partridge perched on the hip of a coupit swede, rubbing an ear with the dew-itch in it against the knuckle of its wing.

The kale pot would never be the same without the swede. Nor the stewpot. And what would they have to go with the champit tatties at Burns suppers if the swede were no longer with us? It wouldn't be much missed for making Hallowe'en lanterns for hardly anyone bothers to make them nowadays anyway. Turnip heads, too, are things of the past.

I was thinking all these things aloud, and the lady said: 'I never imagined it was possible to reminisce about a turnip . . .' Yet a field of swedes has meant many things to me.

Many a juicy neep I ate raw as a boy, stripping down the skin with my teeth, usually as cleanly as a peeled banana, but sometimes chiselling the smooth flesh where a side rootlet was socketed. How much grit and soil, and how many soil animals, I ate with my swedes, I have no way of knowing. But I do know one thing: they grew a better neep than they do today. Which is true of just about everything that is grown today.

A field of swedes, when the drills had their hackle of green, was a source of extra cash, and extra cash was as useful to small boys then as it is today. I forget what the pay was then for singling turnips, but I know I travelled many a mile up and down the drills. My back couldn't thole the long rows now, but I can still manage my small stint on the drills I grow at home.

Farmers used to call out to us when, as boys, hungry after a long day's stravaiging during the summer holidays, we sampled his neeps of tennis ball size: 'If ye eat them the noo ye'll hae tae help shaw them later!' And, in fact, we sometimes did, which could be cold work, cruel on the hands.

There were years when the swedes were a poor braird, and had to be resown. And there were others when the woodpigeons hashed them about, leaving the drills scrawny and pookit. But nobody was ever short of turnips, as I recall, and you could take away as many as you could carry on your back for a shilling.

I spent many a morning then watching the woodpigeons eating the green shoots, tugging, shaking, swallowing, and stopping now and again to scrape soil particles from their beaks with a foot. The partridges would sometimes peck a few tops, but they were never dense enough in a field, and not fond enough of turnip tops, to do any damage. They were more interested in the weeds and clovers on the endriggs.

The partridges would sometimes peck at the purple bulbs in summer, but again they were mainly concerned with weeds, weed seeds, and the insects in the drills. Much the same applied to the pheasants.

Roe deer would visit a field, and scallop a turnip here and there; rabbits would eat their way into one; and you could sometimes watch a hare stripping down the peel to get at the sweet pulp, in much the same way as we did ourselves.

Several times in later years when I was walking the drills with my dog, I would put up a fox, as I also did in the potato drills, and I could only assume he was lying up waiting for unwary birds, or maybe after mice. I've certainly seen a fox running from a potato drill with a headless pheasant.

I never liked to see the swedes going in, because then we were surely on the threshold of winter, but many a cart I loaded as a boy, and many a big-footed, big-shouldered, willing Clydesdale I led into the farm close where they were being heaped to rone height to feed the dairy cattle through the long days in the byre.

Changing the Face of Nature

Pope it was (and I quote from fallible memory) who said:

Know then thyself
Presume not God to scan:
The proper study of mankind
Is man.

No-one would dispute the broad truth of this proposition. But, if the proper study of mankind is man, surely part of that study has to be an assessment of man's trusteeship of the universal environment; and examination of his role as the dominant world species; an appraisal of his management or mismanagement of the world's resources: soil, water, forests, minerals, plants and wildlife. That which is ecologically bad or unsound can hardly be ethically or morally correct.

No-one—and certainly not I, whose creed is to conserve by wise use—would deny man's right to exploit and use any resource, whether soil, plant, water, mineral or wildlife. But to misuse or abuse, to squander or destroy, arbitrarily and arrogantly—these are a betrayal of trusteeship, an abdication of responsibility.

Conservationists are often accused of being 'blindly opposed to progress' when, in fact, they are merely opposed to blind progress. All too often progress has its monuments in man-made deserts and dust bowls, soil erosion and exhaustion, river pollution and destruction of wilderness, not to mention the sordid and sorry story of the extermination of species, with complete disregard to their role in the fine spectrum of life, or their use, or their aesthetic value, or their right to existence.

Man is constantly changing the face of Nature, and very often such changes are necessary for his own survival; but there is no way that survival can be ensured by destruction of habitat, despoliation and poisoning of environment, or the elimination of wildlife. To paraphrase one infinitely greater than any of us: 'What shall it profit a man to make the fast buck if he loses his habitat and his life?' Long ago Aldo Leopold, the American, said it in other words:

'Our present problem is one of attitudes and implements. We are remodelling

Alhambra with a shovel, and we are proud of our yardage. We shall hardly relinquish the shovel, which after all has many good points, but we are in need of gentler and more objective criteria for its successful use.'

Some massive and monstrous surgery has been done to the earth since man took over, with power if not with glory, and very little of it was for the good of the earth. In the old days this was the result of ignorance; nowadays it is much more frequently the result of arrogance.

The surgery still goes on, mainly with the technologist wielding the scalpel, with little or no thought for after effects, or side effects, or the ultimate health of the patient.

The go-getters and the tear-aways and the fast-buck merchants all speak of conquering Nature, which is simply a euphemism for destruction and despoliation. The ecologist attempts to understand, and work in harmony with Nature, and to ensure that the world's renewable resources are kept viable and renewable.

God preserve us from the 'experts': the highway of history is signposted with their directions to destruction. In this country of ours, where we were brought up to have (rightly) a healthy scepticism of experts, we now accord them a status presently denied to God Almighty.

Conservation means waste not, want not. It means the husbanding of assets, including wildlife. It means the wise use of assets, including wildlife. It means respect for life, for the delicate equilibrium of Nature, and sound management of wildlife and other renewable resources, based on humble ecological understanding.

Conservation has taken a long time to graduate from the status of a dirty word to its proper status as a synonym for commonsense, and an ethic in itself. It is still, unfortunately, a mere whisper of caution and restraint amid the bawling of the technocrats, and my own view is that the whisper will be drowned out in the end. The late and much lamented Sir Frank Fraser Darling wrote some years ago:

'The battle of Conservation is *not* being won. We of the Conservation Army have available the best scientific knowledge in the world. Yet the best we can achieve is a costly stalemate against the horde of knuckleheads who are bent on driving us into the sea.'

Things haven't changed all that much since then.

Life Among the Tatties

A tattie park is a tattie park you might think, and nothing more; but like other man-made habitats it provides a niche for some forms of wildlife to exploit, and exploited it is before and after the official tattie gatherers have gathered— variously booted, in urban garb or garbed like scarecrows, with wire baskets armed—to pick up tatties like hens dabbing grain, and to run their losing race against tireless tractor and always-on-the-move trailer.

One such is the rook, bold, versatile and well informed, who flap-swaggers down to dibble where the endrigg crosses the Ts of the drills, and in this way

174

maybe a few sacks of potatoes are gobbled which, although they may not amount to much compared with, say, fourteen tons to the acre, are understandably grudged, so what are you going to do about it? Nothing likely, because the time and the shot would be uneconomic in terms of the percentage of rook gobblers put out of business.

Another one, not so loudly shouted about in Gath or headlined in Askalon, is the pheasant, the jewelled rajah with a licence to misdemean, who can do his share of sneaking up the drills, poking out and gobbling the near-the-surface smalls and sometimes probing deeper for more than pay dirt; but from whom the farmer can get some carcase return, as he can't with the rook, for up here we don't eat rook pie, only humble.

A great sight, if farmer you are not, is when an escaped grumphie forces her way through sagging wire fence, dragging her under-carriage of balloons, with eight, nine or ten eureka-squeaking piglets following the bulk tanker, to show how Nature's specialist with no more than specialised proboscis can root out tubers from man-made ridges, helped by lesser dig-noses, all flaking, chipping and crunching, and swallowing with much pig noise juicy starch and red skins and dirt and centipedes and worms and snails and maybe even a nestful of luckless woodmice.

Many and varied are they who eat the murphy, even today's soap-textured quick blackening breed.

The cushat takes the bool-size, and can pouch a boy's fistful without bosoming noticeably bigger. The black crow, better tailored than the ragged rook, although maybe not such a confirmed digger, can and does pouch the bools and skewer the bigger. Gulls are rather gleaners, mopping up the overlooked ware, or the scatterings, when they're often seen in the company of jackdaws. Even Charlie, carnivore by preference and breeding, will gnaw the raw tattie, ignoring the disproportion of carbohydrate to protein in his unfavour, and maybe even liking it.

Then there is the brock, mighty among diggers, and a great eater of underground storage organs, who, although he gleans surface potatoes, seems rarely to dig them out, which is not to say he never does, because he does sometimes. But loss to the badger is less than a lost chip in a fish and chip shop.

Then there are the woodmice who scurry out in the dark, or when the moon is hiding, to nibble the leftovers, and that brings the ghost owl from the farm doocot and the tawny from the wood, and maybe that highly mobile sinuous furred sausage the whittret, immaculate in tiresome-TV-white waistcoat—all three to eat mice not potatoes, and all three expert, especially the whittret who alone among them can follow the mice into mouseholes or through the fankles of withered shaws.

At night, too, the artiodactyls come to glean, but are of no importance, the roe deer taking what is already lost and of no account; but in other parts of the country the red deer can become tattie happy and make a nuisance of itself multiplied by gregariousness.

So a tattie park is a tattie park is a stage, on which a varied cast perform round the clock for a brief season before departing to play the same parts in a new script elsewhere. The script and the stage may change, but the players are type-cast.

The Threat in Tameness

An Ayrshire bull, hand-reared, petted as a calf, fawned upon as a yearling, and fussed over on the day he wins his first card at a show, is a potential killer when he is mature, whereas the red deer stag, maned up and roaring his head off in an October gale, is not. But hand-rear that stag, pet him from calfhood to the days of roaring, and you have a potential killer on your hands.

What makes the difference is the very fact of tameness. The tame wild animal, the hand-reared and imprinted, is the one that poses the threat. So the wild roebuck you hardly ever see, shy, elusive, wary, cunning, with a total fear of man, is safe so long as he's wild. Hand-rear him, pet him, make him a member of the family, and he's as likely to gut you as fuss you. That just happens to be the way things are. Roe are neither good nor bad. They're just roe deer. And if you take him as a pet you take him with his innate behaviour pattern.

The wild wolf, singly or in pack, is no threat to man. A rabid wolf will bite anything of course. So will a rabid rabbit. But there isn't an authentic instance of wild wolves attacking human beings, despite all the hair raising stories from the Balkans and elsewhere. Mowat in America literally stuck his face into a wolf den and all he got was a green glare.

For wolf read dog, and the picture can change because the dog is a tame wolf at several removes. The wolf's is a hierarchial unit whether family pack or bigger company. Each member knows and accepts its place. The boss wolf has an edge just as the territorial roebuck or stoat or grouse has an edge.

A dog, like a wolf, has to find its place in the peck order. In a house the dog should be the lowest member in the hierarchy, however highly each member of the hierarchy may esteem it. I yield second place to no man in my respect for my dogs, but they are last in the peck order. And of course they accept it. And this, so far as I can see, is the first arrangement one has to make with one's dog or dogs.

I have been in houses where I was asked not to sit in a certain chair because it was the dog's chair, and I would be bitten. It staggers me that some people accept such an arrangement, with the household dog the boss wolf.

Another thing that strikes me is the concept of watchdogs. A lot of watchdogs make their own running, and decide when and whom they are going to watch, and whom they will bite or otherwise scare. It seems to me that this is the owner's problem, and the decision should be his. An out of control dog, making up its own rules, isn't a watchdog; it's a menace.

Savagery in a dog is no asset unless the owner can turn it off and on like a tap.

My own view, for what it is worth is that a dog that is to become a guard dog should be taught everything else but in the first instance. Any other way is putting carts before horses. To take a known bad-tempered, savage dog in the belief that it will make a good watchdog is a mistake; it is more likely to become a problem, biting as it feels like it. A guard dog, like a gundog, has to be made; and no part of the early making includes the final act. Any big dog is potentially dangerous and making guard dogs is specialised business, best left alone by most people.

The dog that chooses not to hear when called, that goes this way when you want it to go that way, that disappears off the map as soon as it is let loose, is a potential danger. And the owner is entirely to blame. There is responsibility, as

176

well as pleasure in keeping a dog.

It is easy to teach dogs, any dogs, to leave livestock alone, to put up with cats, mice, birds, anything that moves. It is easier still to get them to hunt all these things. And this is the point. The dog has got to be taught that what it wants to do is of no account.

I've had lots of dogs, and my children grew up with dogs. With each new puppy I did the same thing. The children gave food and took away food, including meat and bones. Lots of people thought this was dreadful to put such strain on a puppy. But the dogs accept this. Children get their place, which is a niche higher than dogs. After all, in a pride of lions, the lionesses kill but the males eat first.

Left But Not Abandoned

The roe is cursed with what the old Hollywood moguls called star quality; beautiful in the morning of life, beautiful in its noonday—beautiful even in death when he lies grassed by a bullet. So people can't resist picking him as they would a flower, and he has about as much chance as a flower of surviving afterwards . . .

Behold then the fawn of the roe! He is the stuff of fairy tale when first he views the world with soft, violet eyes, glowing under long sweeping lashes. He is an elf—white-lipped and jet-muzzled—with wavy, snow-flaked coat, slim legs, and tiny polished hooves. As still as a carving he lies, chin to ground or muzzle to flank, without flicker of eye or twitch of ear. Such is the pattern of his baby life when his mother is absent, and he has to maintain it until he is fit to run with her.

So he lies where he is put down by the doe, and only the shock of alien contact—fox or dog or cat or man or bird—or extremity of hunger, will make him cry or put him on foot. Only the tread of his mother, or the assault of flies, will make him twitch an ear. And this is the way of him: whether he is born in the heather, on the open hill, or in some cool concealing thicket.

Then see him with his mother, on a June morning when the trees are sequinned with dew. While he nurses she is alert for danger, her ears restless, her nostrils flared. When he dunts her udder she is patient, and will change stance for him. When he *peeps* to her, bird-like, she will squeak to him, mouse-like. And when he has suckled, and is milk-happy curling his tongue over his blue-black nose, she is firm with him when she presses him back to bed once more.

Thus she trysts with him until his legs find their strength and he is able to run with her. A doe, dying, will tryst with her fawn who is alive: fit and unwounded herself, she will tryst with her fawn who is dead.

Into this age-old scene there sometimes walks man or woman, boy or girl, and one of two things happens. The intruder admires and moves on, not touching or interfering; or he (or she) carries off the fairy-tale creature and condemns it to a fate worse than death. And I use the cliche with forethought and hindsight.

The script from that point goes something like this: 'I have a baby deer I found abandoned by its mother, and I can't get it to feed: can you take it?' 'Why didn't you leave it where it was?' I ask. 'It was abandoned.' 'It wasn't abandoned.' 'Well there was no deer in sight. Anyway I can't get it to feed and I'm worried sick.

177

Could you not take it?'

Now I need baby roe about as much as I need two wooden legs, so I ask the person: 'Have you any place to keep this deer if it survives?' 'Not really.' 'Then why did you take it?' 'It was abandoned . . .'

Roe do not abandon their young, and the best thing to do with a roe fawn is to leave it alone. In any case a baby buck grows into a big buck, causing more problems for itself and its 'rescuer' than any two dogs. The elf at a year is more likely than not to become a dangerous savage, fit and willing to tear holes in a man or kill a child.

'But he's so tame!' He is indeed, and that's the problem. If he had been left wild he wouldn't have gone anywhere near a human being. Now he has to be shot, or penned out of harm's way—harm to you not to him.

A tame roebuck is a potential killer. That is not putting it too strongly. Some bucks are worse than others. Out of seventeen my wife and I have reared over the years (all but two of them thrust upon us by rescuers) only three have been potential killers. All the others, with one exception, were unreliable and unpredictable, especially during the rut in July/August.

But apart from this, it should be taken as a golden rule that young wildlife should be left severely alone.

Foxy Tales

The man comes on the phone and asks: 'Wid ye settle an argument for me an some pals in the work?' He says he's sorry to bother me and I tell him its nae bother at a'. I tell him I'll settle the argument if I can.

'Whit's the argument aboot?' I ask.

'Well sir,' he says, 'it's wan o' the mates like. He says that foxes can sclim trees, and that started an argument, an' then the poun' notes were flashin', an we agreed that we'd a' take your verdict for it.'

'You mean you were betting on it?' I ask.

'That's right,' he agreed. 'It looked like easy drinkin' money tae me.'

'Whit side were you on?' I asked.

'A said they couldny.'

'Well you've lost,' I said.

'Ye mean they can sclim trees?'

'That's what I mean, yes.'

'Well, ma argument wis that they had nae claws like a cat, an if they get up hoo were they gonny get doon.'

'Well,' I said, 'they can get baith up and doon wi nae bother. I've wan here at this meenit who thinks nothing aboot gaun up sixteen feet. In fact, he's a better sclimmer than I am.'

'Looks like A've lost then, whit?'

'It's mair than that,' I told him. 'It's a certainty.'

They'll tell you things aye gang in threes, and they're telling you right. Less than an hour later the phone went again, and this time it was a newspaper man to tell me he had a report of a fox fighting with two cats. He'd never heard the like.

The confrontation had taken place in a suburban garden. One of the cats had a slight cut on the face; the other was unhurt. The fox took off when the householder opened the back door.

I told him it wasn't news, but the fox must have been a 1981 model and the cats byornar powerful and bold. A big fox, meaning business, would have had at least one of them dead or badly mauled. In fact, I told the man, one of the quick ways to lose a pet cat is to let it run loose at night where there are foxes, especially where there's a vixen with cubs.

I've seen a lot of cats that were executed by foxes, and I've seen a few foxes executing. The impression one gets is that the fox does it for the hell of it, although a hungry fox will eat cat if it has to. It will eat fox if it has to.

A cat stands as much chance against a fox as it does against a dog big enough to play executioner. Most dogs play at chasing cats, and break off when they get a fizz reaction. But a dog that means to kill will kill. I remember one cat that specialised in chasing dogs, until it met a bull terrier whose favourite ploy was killing cats.

The third call, on the same day, came from a woman who had just seen a fox in her garden. The garden was in Glasgow's suburbia. She wanted to know what she would do about it.

'Is the fox living in your garden?' I asked her.

'Oh no!' she said. 'I've never seen it before.'

'Have you any rats about the place?' I asked.

She hadn't any rats about; nor were there rats nearby so far as she knew. I suggested that the fox was probably midden-raking, as urban foxes do. But apparently it wasn't that either. What she wanted to know was what she should do about it.

'I'm not very clear about why you should want to do anything about it,' I said. 'You may see it again, or you may not. If it isn't bothering you why do you want to do anything?'

'What about rabies?'

I should have guessed it would come. Like so many others she had the idea that foxes were carriers of rabies in a country where rabies is unknown.

'No British fox has rabies,' I reassured her. 'No British dog has rabies. No British anything has rabies. If we ever get rabies it will be because some stupid human being smuggles in some animal with the disease. You can forget about your fox.'

She accepted this assurance, although I could tell she was none too happy about it.

Portrait of Spring

Whether it comes in with a warm smile or freezing breath, or lurches in with a shake of fists, or gently steals with brush dipped in sunshine to paint gorse and coltsfoot and marigold and daffodil, March is the rabble rouser, and if it can stir up

179

a peck of dust along with the rooks the farmers become as active as the birds, because a histie tilth means seed-time, and seed-time gives the kiss of life to farming.

The rook is the mouthpiece of March; he can talk out the loudest-mouthed gale. A gey bird he is—an ebullient bird, a swaggerer, a comedian (and to hell with anthropomorphism); an auld-farrant bird, a knowing bird, bright-eyed, with a mask of crumpled paper, a leery buffoon; a colourful bird, resplendent even; a gaucy, gleg hallanshaker of a bird; a gabby roisterer, whose clishmaclavers have more fundamental significance than most of the talk in yon other place; a ragged-trousered bird who can ride the unbieldy top twigs like a trampoline, while the wind whips his feathers about his legs.

In the parliament of the rooks—and that is what their social organisation boils down to—they achieve a better distribution of birds to jobs, and a more realistic density of mouths to food supply, than civilised man has been able to do with all his miracles of technology.

Good farming and rooks go hand in hand; the rook is the camp follower of the arable farmer. You will find cultivated areas without rooks, but you'll hardly find rooks where there is no cultivation, and even where there is cultivation they need trees to nest in or they won't be there round the year. The distribution of rooks mirrors the distribution of farmers, and the birds live mainly off farms.

But do they live off farmers?

Farmers ken the rook fine, even when they call it a crow. Anyway the bird is zoologically a crow. I haven't met an arable farmer yet who didn't know the difference. Where I come from the rooks are called *craws,* and the carrion crows are the *big black craws.* The names are the same; it's the saying of them that makes the difference. Those who look upon arable farmers as purblind ignoramuses should tune in to nuances.

But knowing the rook is one thing; knowing about it is something else.

We all know that the rook feeds mainly on farmlands, and we know that it eats wireworms, leatherjackets and corn, among other things. As a result the bird has partisans, for and against, with one lot supporting it for eating wireworms and leatherjackets and the other lot damning it for eating corn. Dr W.E. Collinge, after examining 2000 rook stomachs, pronounced the bird harmful. Sixteen years ago I had my farmer neighbour shoot for me a baker's dozen of rooks off one field where, he was sure, they were eating newly sown corn. I found not an icker in one of them.

Doing little sums on the basis of such findings is a form of arithmetic as dangerous as the partisanship it leads to. How does one equate leatherjackets or wireworms with corn? Anyway, we are not dealing with fixed quantities of leatherjackets or corn, where every mouthful eaten would reduce the one or the other. We are dealing with variables, with living reproducing organisms, whose capacity for increase is greater than the capacity of their environment to absorb it. The increase is expendable, lost in one way or another.

It is axiomatic that predators generally do not reduce the population of their prey species; they live at its expense. They take the interest, not the capital. On this basis it is unlikely that rooks make any difference to the number of wireworms, or the weight of the corn harvest. If all the rooks disappeared tomorrow it is questionable if farmers would harvest an extra sack of corn, or wireworms thrive beyond the capacity of the habitat to hold them.

180

The rook, one might say, is of no account either way. Rooks take corn at seed-time, but far more grains are sown than could ever grow into plants. Rooks eat corn after the crop is growing, so the grains taken then would never become plants anyway. In September it has been shown that six-sevenths of the grain they take is stubble corn, which is lost anyway. Even where the harvest is untouched the farmer doesn't get it all in; there are incidental, and accepted losses.

The same is true of the rook's sometimes predation on small birds and their eggs, or on mice or voles. It makes no difference to the numbers of either, so there is no reason for the coy pro-rook partisan to make whispered asides like 'occasionally it stoops to the crime of robbing a small bird's nest.'

The rook doesn't need any such 'defence'. It is neither an angel nor a crook. It is a rook, and nothing more: a bird with a niche into which it fits well, and where it finds its living in the best way it can. That it is an opportunist predator on small birds and mammals is neither here nor there.

Even if the rook could be shown to be a pest of agriculture it is doubtful if that would solve anything. We should probably find ourselves spending more of our resources controlling it than the cost of the damage it had done.

This is becoming apparent with the woodpigeon, considered by most of us to be the greatest single problem of the grain farmer. With the most modern organisation and techniques the cost of killing a woodpigeon is a long way above the cost of the damage it does in a year.

The woodpigeon, or cushat as I prefer to call it, is a big-busted, buxom and braw bird, all hyacinth blue and salmon flush and flashing white, whose husky, sultry *coo-coo-coo-roo* is, like the rook's *caw,* one of the best known songs of spring.

Besides being big-busted the cushat is big-bellied and has a big gebbie. It requires up to two ounces of food a day, if it is eating all cereal grain, and more than four times that weight if it is eating clover and other greenstuffs.

The young, after they are past the pigeon's milk stage, are fed on seeds, mostly cereal grain, which can easily be felt by the fingers through the skin of the crop. So the woodpigeon lives off the farmer.

Cushats nest early and late, and a man doesn't have to look very hard to find fresh eggs from March to October. April is a fair nesting month, and it is then one can watch young birds in the nest being fed grain from the drills. But the main period is in July, August and September—a fact now well established—and only a minority of birds breed successfully outside that period.

There is a heavy toll of eggs between March and the summer leafing—by jays in big woods and forest, by magpies and carrion crows in spinneys, hedgerows and smaller woodland. Hatching and survival are much higher after mid-summer, not least because July and later broods are in the corn season. The very act of arable farming suits the cushat and its numbers go up as the scale of grain production increases. Predation, or the lack of it, is a negligible factor.

On a UK scale, whatever the cushat's cost to farming, the bird isn't going to put any farmer out of business; it levies a tribute. But damage by cushats can be severe in a small area, and even make all the difference between profit and loss. Assessment of cushat damage is based on crops lost that could have been harvested, not on crops that would have been lost anyway—such as stubble grain. I have been cleaned out of brassicas three times in the spring, three years in a row, and by one pair of cushats.

Unlike the cushat, the peewit is the farmer's protege. Its whooping by day, and its keening at night, are part of the fabric of spring, and there can be few farmers who don't express regret when it is scarce as it recurrently is. They like it as they like the swallow.

Big scale mechanised farming doesn't do so well by the peewit as the family, or small unit, farm where the men take a pride in saving peewit nests from machinery during spring operations. They lift eggs, and replace them in new nests, and the birds accept the change. There is a real association between farmer and peewit.

We have a tendency to divide birds into useful and harmful, and part of the peewit's popularity is based on the idea that it is a useful bird. It eats hardly any grain, but it does eat large numbers of noxious soil animals. Yet it is unlikely that, at the end of the day, it makes any difference at all to the final crop, and there is no reason at all to believe that it reduces the number of pests in the soil.

The peewit nests early and many of the early clutches are destroyed accidentally by machinery, or systematically pilfered by people, although bird and eggs are protected at all times by law. Once the main field operations are over the birds have more peace and, since they come on to lay after losing a clutch, the early loss makes little difference. But modern big-scale farming methods take little account of any man-bird relationship, and it remains to be seen what the final effect on the peewit will be.

The curlew, another firm favourite with farmers, is less affected by modern methods than the peewit, but the draining of wetlands and the reseeding of threshy pastures are gradually prising it from its niche on many farms. Some contrive to nest in hayfields but a hayfield isn't much use to them if there is no boggy ground nearby.

It is ironical that the farm, hitherto one of the great reservoirs of wildlife, is fast becoming a deserted place, despite the goodwill of the farmer. The modern byre doesn't suit the swallow; nor do modern farm buildings of corrugated asbestos or whatever. The modernisation of buildings, and new buildings, don't suit the barn owl. The destruction of hedgerows, which proceeds apace, has wiped whole plant and animal communities from the map. The bug of tidiness that has bitten so many county councils has made the roadside verges deserts. Wrote Nash:

Spring the sweet spring is the year's pleasant king,
Then blooms each thing . . .

One wonders for how much longer.

182

'Scarecrow' with a Stoop
and No Front Teeth

The tall, thin scarecrow of a man, wearing a clergyman's cast-off jacket and ancient knickerbockers that had once been the roadman's, stopped at the field gate when a loud voice hailed him from the road. The scarecrow had a stoop, a bad sniff and no front teeth.

'Noo, see here Greenbeak,' the farmer accosted him, 'A've telt ye afore aboot paradin ower ma grun wi a gun, and A'm no gaun tae tell ye again.'

The farmer was a short barrel of a man, round-faced but not jovial. He was wearing horsehide boots, with turned-up toes, and green with coo-wash. His elbow and knee-patches, in contrasting shades, reminded one of a prisoner of war.

'Gun Tam?' Greenbeak asked. 'A've nae gun Tam.'

'Aye, gun!' the farmer shouted at him. 'An don't Tam me!'

'Well, you ca me Greenbeak,' Greenbeak said.

'A'll ca ye what A damt well like. An don't you try tae fly-man me!'

'Naw Tam,' Greenbeak said.

The farmer lost his temper.

'They a think you're daft,' he roared, 'but you're no sae bliddy daft if they but kent. A see that wee .22 o yours, and A ken you've got it broke and doon the legs o your britches. So don't try an tell me ither.'

Greenbeak sniffed, a kind of snort in reverse, and stepped out into the road. He shrugged and leered at the farmer.

'There's nae law o trespass in Scotland,' he said. 'So you try and stop me.'

The farmer brandished his stick, and began to froth at the corners of his mouth.

'A'll trespass ye,' he bellowed. 'A'll get the polis. A'll interdict ye. A'll . . .'

'That's whit tae dae,' Greenbeak agreed, and walked away, giggling to himself and sniffing.

'Aye, on ye go,' the farmer called after him. 'An tae think if A gied ye whit ye deserve, a belt on the kisser, they'd likely pit me in jile . . .'

When he was a small boy, snorting and sniffing, they called him Greenbeak. The snort cleared up when he reached his teens, but the sniff stayed with him. And the name.

In his early schooldays they sent him to the parish doctors to be examined, thinking he was simple, and perhaps in need of special care. The doctors examined him, dicussed him, and referred to the categories of mental defect laid down in the Mental Deficiency and Lunacy (Scotland) Act 1913. But Greenbeak defied classification. The doctors couldn't find a niche for him at all. The legislators, it seemed, hadn't foreseen a Greenbeak. The doctors sent him back to school.

Psychologists might say that this event must have seared his soul or his psyche or something.

In fact Greenbeak would have told you that it was a great help in later years because, until people realised he was as cunning as a fox and as ruthless as a ferret, they let him off with anything—anything within unreason. Even those who swore they'd never be taken in again by his daft-man act relented, and called him a puir sowl, when they saw him in the village on a cold winter's day, eating a shafe of bread and jam, or carrying firewood to his single-end.

He accepted the role of eejit because it paid, but if anyone called him a fool to his face he would tell them that, although his heid might button up the back, he had at least been proved sane by two doctors.

'He's no daft,' the pig-killer would say. 'When he was at school he couldna hae tellt ye if twa and twa made six or pie crust, but noo he can bamboozle the bookies wi his mathematics.'

He should never have been in the Army, they said. So they let him out.

And for the duration Greenbeak helped himself to unkeepered pheasants, and long-netted rabbits, and gate-netted hares, and all the other things that most of the with-it couldn't do.

Wind-Fall

If the blast o Janwar wind that blew hansel in on Robin was anything like that we've had on some days the lad must have been born in a Concordengland of decibels. This profound thought came to me when I was driving out to look at our Highland cattle, with the on-ding of fine snow overtaking and passing the van at motorway speed; the snow flight was horizontal, the windspeed seemingly defeating gravity, and I found myself wondering if the spots ahead of me would ever make a landfall.

Well, the snow passed on, but the wind kept up its rant, and everything bowed before it, from the topmost twigs to mouse-high grass. I got out of the van, almost losing the door, and started to walk.

The terrier was blown over twice, and wouldn't go on, so I put her back in the van rather than carry her. The German Shepherd, her long tail blown sideways and her fur French-combed, plodded by my side. And we got to the Highlanders, which were lying against a bank, unruffled and cudding.

The next storm hit us when we were 400yds. from the van, and the hail hit the face like granite gravel. The big dog slitted her eyes. I turned my back to the storm and let her put her face against the front of my thighs. Twice I was almost blown on top of her.

We were halfway to the van when the dog turned across the track, with her nose reaching into the wind. She whimpered, flagging her tail—muscle this time not wind force. I left the path and looked around, and found the short-eared owl.

It was lying on its side, snagged in the heather, and stared at me wide-eyed, as owls always do. Thinking it was injured I reached down to free the snagged wing, and was met by clutching talons. Then the bird struggled clear, and took wing. But in a matter of seconds it was blown down again into the heather. This time I picked it up, and looked it over. There was no damage at all, so I tucked it under a heather clump and left it. It was simply unable to clear the heather in the teeth of such a gale.

I sat the next blast out in the van, near the edge of the wood, facing the loch. When the air cleared the wind seemed to reach an even greater pitch of fury,

184

tossing the gulls over the loch like bits of paper. A mallard coming in for its splashdown was caught at the last moment by a gust and went in wingtip first instead.

At my back a carrion crow was doing a kind of rigadoon on a patch of green, dancing sideways when the wind tipped it under a wing, or being caught under the tail and sent over on its face. In the end it did a fast sideways skip into longer grass where it sat down. And stayed.

I drove along the woodside to the big clearing, and stopped. All the small birds in the neighbourhood seemed to have gathered in it and around it: tits, goldcrests, goldfinches, chaffinches, yellowhammers. I was on the point of leaving when the big Sitka spruce began to go.

The clearing had been a stand of Sitka a number of years ago, but a storm had laid most of them low. Two years ago we cleared the windblows and left the area to the colonising birch, rowan and Scots pine. This big Sitka, up to the ankles in brushwood, was one of the few we left standing. And now it was leaning over again, sweirt to lie down.

But I waited on, and saw it teeter, then lean over at an angle from which it couldn't recover, and begin to fall, in slow motion. So it crashed in an explosion of branches, but the crash and the explosion were silenced by the wind. It was like looking at a silent film. And then surprise . . .

As the head hit the ground two roe deer burst from the brushwood at the base and came bounding towards the van. Within feet of it they leaped sideways, and stotted into the pines with their white rump fur spread wide in alarm. I looked at the nearest spruces and realised that, if one decided to come down it would come down on me, so I started the motor and headed for the house, which had its own share of decibels, double glazing notwithstanding.

Predator to the Rescue

One of the most futile rescue operations a human being can perform is hijacking a predator's dinner. Rescuing a blue tit from a sparrowhawk may give one a feeling of wellbeing, of putting the world to rights, but it would be as rational to rescue a caterpillar from a blue tit or a tadpole from a dragonfly nymph. A food chain is a food chain, and God Almighty's chain-making is a pretty good job, even if some of us reckon He might have arranged the links more to our liking.

Rescuing a piece of domestic livestock is something else. Then man, the predator, is fighting for his own back. But it seems to me completely irrational to rescue a rabbit from a fox or a woodmouse from a weasel.

A friend of mine once ran after a sparrowhen that had snatched one of his homing pigeons from the cottage roof. The weight of the pigeon brought the hawk down on a long slant, and my friend ran 400yds. for the touchdown, a remarkable enough performance considering he was over seventy years old at the time. The sparrowhen gave him no argument, and the pigeon soon recovered from its wounds. But the old man was in a state of collapse for the rest of the day.

Then there was the miller's wife who, hearing a cuffuffle among her hens at peep of day, ran out in her night attire and chased a fox across a field. She got the hen back, minus its head, and when I saw the body it was in the pot, so man, the predator, again got back his own. I have always regretted not seeing this piece of counter-predation, because a woman in a nightdress, chasing a fox at four o'clock in the morning would be an unco sight.

Sometimes man counter-predates when he has no right to, as when a lorry driver stole a hare from my old cream cat, the late Cream Puff, and left him running in circles at the front gate. The man in this case wasn't rescuing the hare, because it was already dead; he was just plain stealing it for himself.

My own feeling is that rescue operations involving a wild predator and its prey result from our predilection for the 'nice' as against the 'not nice' animals, for the 'gentle' as against the 'savage'. Hence the rescue of 'innocent' small birds from hawks and owls, of rabbits from stoats, of fledglings from crows. Or the rescue of 'lost' baby deer that are not lost until we rescue them, an act that makes us worse than the predator that might kill them.

Nobody would dream of rescuing a rat from a terrier, or a weasel from a fox, or a worm from a blackbird, or a snake from a hedgehog, or a butterfly from a swallow. We decide what are nice and not nice, and would like to regulate predation accordingly.

It isn't so very long ago that a lady asked me if I would come along and shoot a sparrowhawk that was killing birds at her bird table, where she was happily feeding mealworms to robins. I found it difficult to explain that the hawk was as much in need of its dinner as the tits and robins, that tits and robins were among its natural dinner items, that in any case I didn't own a gun, and wouldn't shoot the hawk even if I had.

I can think of nothing more irrational than for a man to rise from his dinner of lamb cutlets, or sirloin, or chicken, and rush out to the lawn to rescue a chaffinch from a kestrel, then, having rescued the prey, to try to nurse it back to health while the kestrel goes off to seek another prey, compelled to do so by the well meant interference of predatory, carnivorous man.

We have a sort of built-in hostility towards predators that shows itself against the predator caught in the act. We don't mind so much the osprey or the heron killing fish (fish are remote and cold, nice enough but not likely to kindle emotion), or the kestrel killing voles; but the falcon killing a grouse, or the sparrowhawk a thrush, or the eagle a deer calf, or the fox worrying a hare, or a crow drowning a duckling—these acts stir the emotions and we rush to the rescue.

Further hostility comes from those who look on predators as competitors with themselves for certain prey. Lots of people still shoot falcons and other birds of prey in the belief that this will mean more grouse or whatever, and the fact that it means nothing of the sort doesn't deter them. It is salutary that game fish are in decline, not because of otters, but at the same time as the otter itself is in decline. The peregrine falcon is in decline, while the grouse is getting on. The sparrow-hawk is in decline despite the multitude of small birds still about.

When you come right down to where you live, it becomes obvious that man is the most ruthless and wasteful of all predators, who has exterminated and decimated species in a way no wild predator can do. If we are going to become emotional we should become emotional about the species whose very existence we are threatening by direct predation and subtler means.

Not the Word for It

A woman, admiring my big German Shepherd bitch, said to me casually: 'What's its name?'

'It's name,' I said (riding the neuter) 'is Lisa.'

'But that's a girl's name,' she said.

'It is indeed,' I agreed. 'It being a her I could hardly give her a boy's name.'

'Oh!' she said, 'It's so big I thought it was a dog.'

'It is a dog,' I said. 'A German Shepherd bitch.'

'I mean,' she said, 'I thought it was a *dog,* you know—a dog—being so big ...'

'It is a dog. A bitch dog.'

She seemed a little put out at that, and when she was joined by two other women, both of whom patted Lisa, she said to them: 'Her name's Lisa. She's a lady dog.'

'But she's huge,' one of the others said.

'She's certainly a big bitch,' I agreed, 'and I'm sure she doesn't mind being called a lady, for she's just that in the best sense of the word.'

I overheard the first one say later: 'You know, I think David Stephen goes out of his way to be coarse.'

God help me; one of the things that never fails to stir my adrenalin is to hear a bitch called a lady dog.

Another is to be asked whether a kitten is a boy or a girl, because boys and girls are people. And cats, like dogs, have their cat and dog names.

I know a lot of people, who readily speak of someone or other bitching about something, or being bitchy, and who use the word bitchiness, but who call a real bitch a lady dog. The only time they use a female dog's name is to insult it.

Not long ago a woman was telling me how she stopped a youth taking young birds from a nest. 'I gave him h-e-l-l,' she said, spelling it out letter by letter.

'I know how to spell it,' I teased her. 'Why don't you just say it?'

Some years ago I gave a talk on ecology to a certain women's organisation. When I was leaving, after tea, the secretary, who was a friend of mine, said to me: 'I think you shocked some of them tonight.'

'Shocked?' I said, shocked. 'I? In God's name how?'

'Well,' she said, 'when you told of the fox that was shot on the dung midden wall, and when you told them about how you analysed badger dung.'

'Oh! for heaven's sake,' I said. 'Did they expect me to call badger dung manure? What a farmer spreads from the midden is dung; what he buys in bags is manure.'

The euphemism is everything. It seems that, even in this age of sewage, the old honest-to-God words whose respectability is beyond reproach are inadmissable in certain quarters.

I'll give you a further example of this. A minister friend of mine, seeking me out for a lecture he wanted me to do, was told by my wife where he could find me. When he found me I was coming off the moor with a polythene bag filled with fox dung. It was a big bag, and I had it slung over a shoulder.

'What have we in the bag?' he asked.

'Fox dung.' I said.

'Tut, tut,' he replied.

When my daughter was a little girl at school she came home one day and asked me if I would dress a kitten belonging to the aunt of a friend. And I told her to tell her school friend I would, and named a time. Whereupon she asked me:

'What are you going to dress the kitten in, daddy?'

I told her what dressing a cat meant.

Black Wood of Rannoch

The Black Wood of Rannoch is the Dark Wood—memorial to long dead centuries: a remnant of the ancient Caledonian Forest, the great Wood of Caledon that once extended from Glen Lyon and Rannoch to Strathspey and Strathglass, and from Glencoe to Mar; a vast forest of buirdly pines and oaks, of birch and alder, of sunlit cloisters and shadowy aisles, of secret trails and unchancy dens, where dwelt the wolf and the boar, and the bear—the mighty *Magh-Ghamhainn* or *Paw Calf* of the Gael that the Romans knew and caught and crated and exiled to their brutal circuses in Rome.

Caledonia's bears broke the Latin barrier many centuries before the brown bear as a species became styled *Ursus arctos*. The Romans tormented it, and in turn used it to vent its fears and rage and frustrations on crucified malefactors, as witness Martial:

'Nuda Caledonio sic pectora praebuit urso,
Non falsa pendens in cruce, Laureolus.'

But the bears have long gone. And the boars. And some time in the eighteenth century the last wolf padded to inevitable extinction.

The ancient forest is as lost as the Scotland it was part of. But bit-parts of its identity remain as bit-parts of Scotland's remain. The old pines stand sentinel at Rothiemurchus and Abernethy and Glenmore, at Tulla and Mar and elsewhere.

The Dark Wood—the Black Wood of Rannoch—is such a relic, a remnant of aged trees, debilitated by the lesions of history, neither impotent nor infertile, but defeated by environmental abortion, and sorely in need of ecological geriatrics if succession, and therefore survival, has to be achieved.

For centuries the great forest of Caledon suffered from the firestick and if, like the Phoenix, trees grew again from the ashes in the spaciousness of time, they grew up to be burned in their turn. The Vikings were great arsonists. When they couldn't see their victims for the trees they set fire to the trees, and the natives in turn would set the front aflame to delay the enemy's advance.

The Wolf of Badenoch and his followers took the Viking line and plundered in fire and smoke. Bruce burned his share when he was chasing Comyn. The clans did theirs during their internecine wars. (On a larger canvas the Scorched Earth was glorified as a policy in the Second World War, and defoliants were a weapon in the war in Vietnam. Scotland had no monopoly in matches.)

When Monk went on the rampage he took his frustrations out on the trees, and in an order dated 1654 had the woods about Aberfoyle destroyed because they provided hiding places for rebels and mossers. The forest was also a shelter for wolves, and Ritchie tells of the forests at Rannoch, Atholl, Lochaber and Loch Awe being destroyed in the war against them.

But the big destruction, the exploitative destruction, began in the sixteenth century, after Queen Elizabeth called a halt to the devastation of English woodlands for iron smelting. The smelters, as the sheepmen were to do later as a direct result, looked north, and to the north lay the considerable forest areas of the Highlands.

There was now big money in trees, and the big money was made despite the enactment by the Scots Parliament prohibiting anyone 'to tak upoun hand to woork and mak ony issue with wod or tymmer under payne of confiscation of the haill yrne.'

The exploitation went on at increasing pace after the crushing of the first Jacobite Rising, when English companies moved in to work the forfeited estates to the limit. Later on the lairds moved in for the fast make, and Ritchie records that in 1728 Sir James Grant sold 60,000 trees from Strathspey for £7000. In 1786 the Duke of Gordon sold Glenmore Forest for £10,000.

Rothiemurchus, which was so savagely exploited to make ammunition boxes in the twentieth century, provided an income of up to £20,000 a year for many years. In the words of Ritchie: 'The destruction wrought by these later and larger furnaces was irreplaceable.'

When the sheep came to feed on the grazings resulting from the great fellings, they finished the forest clearance of the Highlands, or very nearly so. Hundreds of thousands of acres were burned over, and because of the sheep density put on by the flock-masters regeneration of trees became impossible. The finishing touches were put to the face of that Caledonia stern and wild which some people like to sing about with pride rather than shame.

All of which has left us with a few remnants, including the Dark Wood. So what about the Dark Wood?

Most people would wish to preserve it; but there are those who shrug, why bother? The Forestry Commission has millions of pine trees so why not knock the old wood down like a row of condemned houses and replant a pinewood, instead of mucking about with old trees because they are old and somebody feels sentimental about them? Anyway a pine is a pine.

But is it? The old Scots Pine is a distinctive tree, and a beautiful one; needled in bottle green and red as recently spilt blood in the sun. It is burly, of hardwood girth and byornar muscle, often with a flat crown, and as unlike the well manicured poles of the conventional forest as could be. Its roots reach far back into history. So the Dark Wood is a national monument.

The old forest was a distinctive forest of distinctive trees. The Dark Wood is a distinctive surviving community of those distinctive trees, making a distinctive ecological unit. But this community, like the human communities of the islands, is in decline.

The ageing parents breed but the young are not following on in sufficient numbers to keep it viable. The seeds are shed but few seedlings grow. So the age gap is widening and the trend is likely to continue. The question is, what does one do about this? The Forestry Commission, who took over in 1947, are concerned about the correct management of the wood, and its future.

I am indebted to C.J. Taylor, of Edinburgh University's Forestry Department, for a note on the recent history of the Wood and for his views on the future.

There were no large-scale fellings until the eighteenth century, but this changed after 1745 (what a watershed that date is) and through the Napoleonic period. Damage then occurred from browsing by goats, horses, sheep and cattle, and regeneration was upset.

Ditch and dyke enclosures were made to protect the young trees against browsing, but this policy changed in 1895 when Wentworth took over and established a deer forest. Again the trees suffered from heavy browsing. The Forestry Commission took steps to prevent this after 1947, and some growth improvement was noted after the exclusion of deer.

The problem now facing the Commission is how to keep the Wood in being, and as a continuum. Do they leave it to look after itself or do they manage it as an ecological unit? Twenty years ago they wouldn't have seen a problem; they would have knocked it down and built anew. Those were the days when they could be rightly accused of being fifty years out of date in their methods.

In many contexts they are still embattled, but these battles are out of context here. The Commission has a great potential for good and, like the times, it is changing. How much it has changed is evident from the fact that they are worrying about the Dark Wood at all. What they need here is support and goodwill.

Do they leave the Dark Wood alone or do they manage it? The answer is, I think, that they can do either, depending on the environmental realities of the situation. If a wood is regenerating satisfactorily and browsing is controlled, and all the age groups are thriving, then it would be enough to build a fence around the place and let things go.

But what do you do if regeneration is slow, too slow; or insufficient; or the seedlings, though numerous, are peely-wally, putting on years by the foot? Do you let the place die or do you manage it?

Quick drainage and absence of moss are necessary for strong regeneration of the pines. They can't stand their feet in water and in the wet places the moss takes over. There are such places in the Dark Wood. There are also Sitka spruces that ought not to be there; they are an intrusion and ought to go. Their graves can be the birthplaces of new pines. But, as Taylor says in his note to me, they MUST be seedlings born of the Dark Wood trees, and of no other lineage.

If the young trees need help, or if young trees have to be planted by man, there are many ways of doing this: by felling groups near mother trees to leave space for seedlings; by scarifying or disturbing ground to help regeneration; by cutting ground here and there, or draining here and there. What is done will depend on the actual conditions, and these will have to be expertly assessed. The use of fire as a tool cannot be entirely ruled out, according to certain considered opinions.

The big question, the basic one, surely is: are conditions in the Dark Wood such that, given the time, it could completely rehabilitate itself as a going concern? Taylor believes that help should be given by group selection: clearing bits of ground to help regeneration, planting where there are no mother trees at hand. And using only Dark Wood seed. When you take a fresh look at the Wood it is difficult to find fault with him.

Foxhunting

Routine killing of foxes is general practice, carried on with official approval. That, whether you like it or not, is one of the facts of life. You may, as I do, consider much of this routine killing a waste of time, as achieving anything but control, but it is a fact of life, and we are stuck with it.

Since there are more ways than one of killing a fox, we have to look at more than the man on the horse. He who shoots, traps, poisons, gasses, snares, or sends terriers down holes after foxes is as much a foxhunter as the man who hollers tod in the dewy morning.

You might be forgiven for thinking that if the Foxhunter (I capitalise the man on the horse) got down off his horse, disbanded his hounds, and took up trapping and poisoning, he would become a more respectable citizen.

The Foxhunter hunts and kills for fun, and the honest ones admit it. Others will tell you they help to keep down foxes. Yet others will tell you that the fox would disappear if it weren't for the Hunt.

Foxhunters are the great keepers-down or keepers-up of foxes, depending on the audience they are addressing.

Mainly, they hunt because they like it, and they have to be judged on that.

Personally, I hold no brief for the Foxhunter. The whole horse, hound, and holly-red business is a picturesque survival, and useless.

I can see no justification for it at all.

But I have to admit, bluntly, and unequivocally, that it is not the most cruel way of killing foxes. It is not even the second most cruel way. So where do we go from there?

If the Foxhunter, killing for fun, became a wage earner, killing foxes in more cruel ways as a job of work, would he become a more acceptable type?

Apparently he would. Yet there are some professional fox killers who enjoy their work. I think too much weight can be put on the 'fun' side of all this. You can hardly make it a condition of employment that a man shouldn't like his job.

If Foxhunting were abolished by law tomorrow I would say: Amen. But so long as the gin trap is a legal instrument in Scotland, and so long as strychnine is free, and so long as we have snaring, I shall devote my energies against these.

I would go further and say this: if I were a Scottish fox today I would make my way to an English hunting county and take my chance before foxhounds rather than die a slow death in a gin, or from strychnine poisoning, or in a snare.

In today's context, the worst I can say about Foxhunters is that they are dishonest, because they connive at keeping a stock of foxes, and pay folding money to have dens kept secret; in other words they preserve foxes, so far as they can for sport.

Which seems to me to be indefensible. But theirs is not the greatest cruelty to foxes.

If you say to me that many Foxhunters are hard riders, who risk their necks galloping over the countryside in pursuit of foxes, I will admit that many are hard riders who risk their necks galloping over the countryside in pursuit of foxes.

Which has nothing whatever to do with the rights or wrongs of Foxhunting.

One thing seems to me to be obvious. Much of the opposition to Foxhunting is not really based on concern for foxes, or humane forms of control, but is simply against Sport.

The cruellest weapon used against foxes is, in my view, the gin trap. This is used all over Scotland, whereas there are only a few packs of foxhounds in the whole country. Priorities demand that we go after the gin first. Anyone who can stomach the gin shouldn't find foxhounds hard to take.

That last sentence sums up, more or less, my attitude to Foxhunting.

It is much easier to put one's finger on the acceptable ways of killing, and it has always seemed to me that the professional foxhunter, who goes round the dens with his terriers when vixens are lying up with cubs, does a clean, speedy, unsporting, bloody job of work, which is sanctified by authority, and in keeping with present policy.

Given that foxes have to be killed, this is as good a way as any.

This is, as it happens, the Highland way. Your foxhunter uses a motley crew of small, varminty terriers, hard-bitten and hard as pig-iron—Jack Russells, Borders, Cairns, Fox and all sorts of permutations—to bolt the big hill vixens, who are then shot. The small dogs then kill the cubs. True, many a terrier gets a sore face in a teeth-to-teeth girning match with a vixen, and I know one, indeed, who left his front teeth in a bitch fox's hide.

But that apart, the job gets done, and if some queer characters get fun out of the work, that is just something we have to accept.

For my own part, I think a lot of this is a waste of time, but it is approved work well done.

The fox drive is justified by authority on the grounds of expediency, and the majority of foxes are killed outright, or quickly.

The gin trap and strychnine are abominations.

So what are we to make of it all? To trap, shoot, poison, snare, gas, or use terriers or foxhounds?

It seems to me that the major difference between Foxhunting and the other ways is that the Foxhunter kills for sport, whereas the others kill as a matter of policy, which may or may not be wrong.

The Trap

The darkness was bleary with frost. Through a rent in the smoor a lonely star flickered its feeble light. Other lights—green, blue, amber, yellow, red and white—hardly bigger than acorns, bright and unwinking, made a Pleiades on a tinsel tree at one of the farmhouse windows. The stackyard gloom was damp and greying.

The big, grey-hipped vixen padded through a gap in the stackyard fence, a shadowy shape detaching from the cattle trough. Her thick brush floated in a curve over her hocks. She could hear the cangle of grey geese on the loch, and the whoop of an owl in the spinney, but her ears could listen to several things at once, and

192

were harking ahead for meaningful sounds. Precise and sure-footed, she rounded the first stack, waywise, almost kinaesthetically free of doubt, knowing exactly where she was going.

At the third stack she stopped abruptly, and bellied down. No scrape of pad on brittle straw. No betraying let-out or intake of breath. Just a shadowy shape no longer there. An educated nose, in the right place, at the right height, could have placed her. But there were no noses. The dogs were bedded on straw bales in the barn on the other side of the house.

The lights in the window had halted her. They were new. Perhaps she was aware only of the white ones, but they were enough. She knew what they were, and they puzzled her. And when she was puzzled she pondered. She had to ponder now. So she had bellied down.

She was five and fit; quick to sum up, quick to act; unhurried when in doubt, without doubt when she hurried; cat-patient when pondering. As a ponderer she had what is needed to get a fox to five. The bright lights were new to her, late on in a house of early bedders, off-putting, with nuance of danger, perhaps dangerous.

These were not her thoughts. Her ponderings were a matching of the visual and the odorous and the audible against the fabric of accumulated experience. She was trying to match the lights. With window lights there was usually movement, sound, action, smell or voices. Here were only lights, with nothing within her experience to relate them to. Lights without meaning—detached in space—like the lone star in the sky.

They began to unnerve her, and she edged back into the thicker gloom of the stack, until the cut ends of straw pricked her flank. A half-hour she crouched there, breathing slowly, her brush slack, occasionally opening wide her lean jaws, not yawning but easing tension. When she closed her mouth her breath curled like frayed wool from her warm, polished tusks.

Rats were running in the cart shed behind the next stack. Her ears caught their squeak and scuffle. In a year she had killed twice her weight of rats at this farm. In ten nights, at cub time, she had killed sixty-three. Once, when she was a yearling, she had bitten the head of a broody hen as it poked through the slats of its coop at first light, and had stood by while the body threshed and thumped in the confined space, and the severed artery spurted warm blood in her face.

Now all the hens were in cages; that she didn't know. But the rats were loose. That she knew. She had come for rats.

But this time there were lights where there should be none. The red ones meant nothing to her. The white ones she could see, and maybe she was conscious of brightness in some of the other colours, too.

A white cat with a black tail and a black smudge on one side of its face walked slowly past the front of the house, and passed under the window with the lights. It didn't stop, or hesitate, or run: it walked. It turned the corner of the house, and disappeared.

The vixen rose and stretched a hindleg out the length of her brush. She sighed. She opened and closed her jaws, and this time there was sound—the scrape of pads and the slight click of teeth. She began to stalk towards the cartshed.

She stalked slowly, like a cat, placing each hindfoot with care before lifting a forepaw, without noise despite the crickle of frost in the littered straw. She could hear the rat's feet on trailer and beams, and the patter of a heavy animal on the corrugated roof. With ears cocked, listening, plotting rat feet here, and there, she nosed round the corner upright into the opening and froze. She had to ponder again, and listen, and measure, and look, before making her rush.

And that was when she smelt cat.

He gave her no time to make decisions—whether to out-wait him, or attack—for he shot across the ground, and up the wall, and through the hole in the masonry, into a disused byre. Taken by surprise, the vixen moved too late The rats were fleeing even as the cat was crossing the open. Their pattering ceased. The cartshed was deserted.

Hankering and not unhopeful, the vixen followed her nose along the most exciting runways, but all of them led to holes in the floor or wall. In one corner the smell of rats filled the air like cooking smell in a kitchen. The vixen came back to the entrance, sat down, and stared at the lights with eyes half-slitted. She still had half her mind on them.

The frost was settling; the farmhouse roof greying over; the sky clearing. In the south Orion was shaking off the smoor. The vixen skewed her head as geese passed over, baying in flight, their arrow tip aimed at the North Star. Sighs of chill wind, hardly deep enough to stir the treetops, touched the wet chill of the vixen's nose. There would be cranreuch by daylight, maybe even a scattering of snow to make the morning of Christmas white.

Tonight the wild birds were supposed to be able to speak to each other for an hour. But the birds were silent. Except for the grey geese. And the owl, who was hunting in the spinney, and not conversing, even with his mate. The vixen could still hear his faint war-whoop. She could also hear mice feet on the thatch of a stack. Then her ears pricked to another sound, barely audible—the thump of a dog's foot on straw bale as it scratched its neck. But the dogs were shut in, and out of the way, as she knew.

Cautiously the vixen left the cartshed, moving into the flaffs of wind, mindful to try the other side of the farm. But the lights in the window daunted her. For a moment she stood in the open, with brush down and head high, while the lights touched her red fur and put a highlight in her eye. Then she turned back into the gloom of the stacks. There should be rats there. A quick chop and she would eat. And then away from the unchancy brightness.

Into an alleyway between two stacks she walked—an alleyway criss-crossed with poles shoring up the stacks. Here there was the rustle of feet; rats, not mice. Big rats her ears warned. The vixen lowered her head to stalk under the first leaning prop, the lights forgotten, her mind concentrated on the rustle. She couldn't afford any more misses. She drew her rump clear of the prop, stood tall, plotted the rustle and stepped forward.

Even as she put the foot down she knew there was something wrong, but her weight was following on and she couldn't draw back. She sensed the up movement from the straw, but it was over before she could see it. The jaws of the gin-trap snapped at her paw, and the steel teeth closed with a dull blow, interlocking teeth that gripped an inner and outer toe, and bit into bone and tendon.

The vixen leaped to escape the anguish and was thrown over on her back, held by the anchored chain. She came down heavily on her rump, and there was a second snap of jaws, with steel teeth closing hard, this time on her brush, biting into the bone at docking distance from the root. She fought to rise to her feet, but could do no better than roll on to her side.

Fox-like, uttering no cry, she struggled to free herself. The noise of her struggling went unheard except by the rats and mice in the stacks; the dangerous ears, the ears of dog and man, were too far away to hear. She pulled back to the limit of the chain, and squirmed on to a hip. No thought of concealment now. Or stealth. The lights held no more fears for her. Now she was knowing the great fear—the fear of the unknown—fear of an enemy against which she had no weapons to fight.

Tiring with the struggle, she lay on her side again and rested. The frenzy and the panic were over; raw fear was losing presence to the dull ache in brush and foot. She was now the wounded animal, preoccupied with her wounds. She drew forward against the gin that held her by the brush, and reached out to the trapped forepaw with her tongue. The cold steel stung its moist warmth, and she snapped her teeth at it, bleeding the gum. But there was no rekindling of frenzy, no wild assault, and presently she relaxed again, lying with her flanks heaving, her mouth partly open, and her tongue quivering over her teeth.

The stars were brightening, flashing icy fire. The frost began to tighten its grip, and the vixen could feel its talons. She drew in her tongue when the saliva in her mouth grew cold. The ache in her foot was massive, but deadening with the anaesthetic of frost. She groaned momentarily, and closed her eyes.

The white owl from the farmhouse loft came and pitched on the stack above her, folding his wings and standing tall on long white legs. She didn't hear his flight, but the soft thud of his landing roused her, to wakefulness not movement, with ears seeking and breathing flattened, fearing discovery, for she could not see him, and had no idea what he was.

He was interested in what was down there in the straw between the stack props, and he leaned out from the rim of the thatch to stare wide-eyed and gravely at the fox. Now the vixen could see him, at least the dark eyes and the white, heart-shaped face looking down into her own eyes upstaring at an angle because she was still lying on her side.

The owl danced sideways, right then left, long-legged, with wings half open and dark eyes staring. He tap-danced and bobbed; then leaned out, swaying, making circles with his head, clockwise. He was like a puppet, with the rim of the stack for his stage—a puppet with a voice of his own, but not using it.

Knowing what he was now, the vixen had no more interest in him, but she watched his movements, with her eyes half closed, unable to shut him out completely while he moved. Though disinterested, and unalert, she could not ignore. When he at last lifted away, flapping moth-like on hushed wings, his white underside tinted momentarily by coloured light from the window, she opened wide and closed her eyes in one unhurried blink.

Now and again she would stir, and strain to lick the trapped foot, or struggle on to her other side to ease the stiffness of her hips. The feeling in her toes was dead, and she no longer had any consciousness of them, but above the foot the ache was dull and massive. A similar dull ache was surging up her spine from the deadness at the root of her tail.

Time dragged away slowly from the tightening grip of frost. The farm roofs whitened, and the white thickened to a fur. Rime formed on twigs and buds, and silvered the stack props. Fence wires became sheathed in ice. A dandruff of frost, like a powdering of snow, appeared on the brittle litter of straw, and clubbed the vixen's eyebrows like the antennae of a butterfly. Her breath was white vapour.

195

Round her mouth the spittle formed like foam round a pool.

Deep down she had reserves of strength and will, but only new crisis could have activated them—a visible enemy, recognisable threat, release from the teeth that held her. Her greatest enemy of the moment, the freezing cold, probing her warmth like a skilled surgeon, inexorable as X-rays, she couldn't recognise and had no way of fighting.

She was roused from her torpor by the dingling of fence wires and the trample and thud of hooves. Sheep had burst into the stackyard from the frozen field. The vixen forced herself on to her good foreleg, and with a greater effort got her hindfeet under her too, but she couldn't hold her stance, and quickly collapsed again, to face the sheep seated.

There were six of them, ewes bleared with breath, reeking of warmth, and steaming with fat under their heavy load of wool. They stamped and jostled, not sure of what she was, and not liking her smell. They knew where they wanted to go, and that was through the passage where the vixen was lying.

The vixen struggled on to three legs again, but stiffness, cold, and the anchoring traps, brought her down. This time she fell on to her side. And the sheep were thrown into a panic.

They could have broken left or right, or turned back to the fence; instead they rushed into the gap, and tried to squeeze past the fox, leaping, mounting each other, and knocking down a prop. The prop struck the vixen on the ribs. Miraculously, most of the hooves missed her, but two trampled her belly and two bruised her face.

This threw the vixen into a frenzy. She arched and pulled and tried to leap away. She bit at the gin that held her by the foot and jarred her teeth on the frozen steel. Her breath came in gasps, and the reek of fox lay heavily between the stacks. Tiring, she lay down again on her side, panting, gulping breath, with her eyes closed to moist slits and her hot tongue dripping saliva on to the frosted straw.

Later, she was still quivering, but from cold—the deathly cold that was stealing through the skin of her belly, and into her mouth, and up her trapped leg and her aching spine. Her whiskers were now stiff with rime, her mask hoary, her body fur grey and damp with frost. Her torpor seemed as deep as a hedgehog's in hibernation, yet her eyes slitted questioningly each time a rat scurried past or a mouse moved with whisper of feet in the straw.

The night was far travelled towards cock-crow when a rat peered from under a stack and stayed to ponder her. He was big—big with years and strength—bold and long-toothed, bristled like a dog in anger, venturesome, but wary forbye. He inched out from the bottom of the stack, freezing long between the inches, sniff-sniffing with snout cleft, and twitching his long whiskers. When all of him was out he was a giant, with a tail like a slow-worm to follow.

He moved out, slow-footed, at a tortoise crawl, trailing his slow-worm, well up on his legs, tensed to break back to cover at the first hint of danger. He froze when his snout touched the gin that held the vixen by the foot.

Rasping seconds passed before the vixen stirred to awareness. Her ears heard the sound, like a woodmouse chiselling a hazel nut. The rat was gnawing one of her nails. Not feeling any pain in the anaesthetised foot the vixen lay still until her ears could exactly place the sound. Then she moved her head. The rat moved at the same moment and scurried to his hole under the stack. He didn't look out again.

Unaware of what the rat had been doing, the vixen fell into her cold torpor again. A flurry of snow just before daybreak whitened the fields, and flecked the vixen's fur, but she knew nothing of it. It was the first noises of the day that roused her, and when she awoke, stiff with cold, the lights were on all over the farm.

Now she called on her reserve of strength, but the cold had paralysed her will, and her brief struggle did no more than shake the snow from her brows. She sat on her hip and waited.

She could hear the metallic clatter of milk cans, the rumble of wheels, the clucking of caged hens, the voices of men, and the barking of the dogs in the barn. She pulled back on the gin that held her foot, and tried to flatten into the ground. There was still no movement in the front yard. The door beside the window with the coloured lights was still closed.

It was daylight when the red van appeared in her view and stopped at the door. There was the rattle of bolts being drawn and the door opened. A man appeared at the door—a big man, with curly hair sprouting from the open neck of his shirt. The postman jumped from the van. A black and white collie pup rushed from the house, fussed him, them scampered into the stackyard on its morning round of chasing rat smells.

'Merry Christmas,' the postman said to the farmer.

'The same to you,' the farmer said, taking his mail.

The pup soon found the fox and began barking excitedly. 'That damn pup,' the farmer said to the postman. 'Heather!' he shouted. 'Heather cumeer! Come oot o that Heather d'ye hear! Stupid bloody dug,' he said to the postman.

'She's no very biddable,' the postman ventured.

'Tell the wife that, her wi' her pups in the hoose. Dugs should be kept . . .'

Suddenly he realised where the puppy's barks were coming from. He shoved the mail back at the postman. 'Take this,' he said, 'an haud on a wee.' He ran to the stackyard.

When he came back he was carrying the puppy. He took the mail from the postman and went into the house. A few minutes later he came out again, carrying a gun.

Bit Between the Teeth

There are various ways of losing a dental gold inlay. Some, no doubt, would sound a bit dubious to those who decide nowadays what you shall or shall not have done with your teeth.

You can lose it by chewing tough toffee; which I don't do. You can lose it by biting on a nut in shell; which I don't do. You can lose it by biting on your pipe the wrong way, which I try to avoid. You can lose it by using a too-thick toothpick, which is just silly. Or you can lose it by letting a magpie peck at it, or have a fox pull it out; both of which have happened to me.

I had two magpies at one time, called Jock and Redwing. Redwing got her name when we daubed her white shoulder with lipstick to help us with identification. They were such a pair that their other joint name was 'the terrible twins'.

Redwing was a bird in the classic mould, by which I mean she did all the things of folklore—stealing bright objects, hiding watches, lifting rings—even when they had fingers inside them. She once tweaked out an ear-ring—the kind anchored to a bored ear.

Anyway, it was Redwing who sent me prematurely to the dentist, and earned me from him an old-fashioned look.

It began, I suppose, by my letting her take morsels from my mouth. I used ever smaller morsels, held between my teeth so that she had to peck for them.

Well, one time she was at the pecking, and hit my gold inlay on the edge, and there was I suddenly with the golden artefact on my tongue. I spat it into my hand to gather it carefully for a replacement, but she was in before it hit my palm, and off she went with it in her beak.

Luckily, every door was closed, so all she could do was to skip round the livingroom with it, and eventually I retrieved it from the carpet.

You'd think that no man would be stupid enough to let a similar thing happen again, and I didn't for more than twenty years. Until this fox cub, Sionnach, now four and a half months old, got the tobacco habit and a liking for my pipe.

It was a great joke at first, a cub leaping up and trying to steal a lit pipe, or hanging on to the mouthpiece of a lit pipe as though ettling to smoke it.

And it had to be, and has to be, my favourite pipe. He won't have just any pipe, and when I made him a present of a corn cob he threw it away. He has to have my own favourite, which is battle scarred with vulpine canines.

You can guess the rest. A few days ago I was sitting beside him in his enclosure, saying all the things one says to a tame halflin fox. Eventually I lit my pipe, the one he and I like so much, and, of course, he had to have a puff at it, which I allowed him, being an unselfish sort in these matters.

Well, he grabbed it by the bowl which was glowing and pulled. And I made the mistake of holding on with my teeth. By the time I had pulled him off and taken my pipe in hand I could feel the squint on the sculpture.

Then it clicked back, and I knew the worst was about to happen. It happened later, at dinner, when I had the gold piece on my plate.

I told my dentist, of course—not the old one; he who had grown used to such offbeat capers is now retired—he took it in his stride. After he had put it in and it had been left to set, he did what no dentist can ever resist doing. He did the hook and mirror trick along my other teeth.

Then he said: 'You'd better let your fox know that he's cracked the filling on the other side.' And he promptly hooked it out.

Which means I've to go back again for another going over, and in the meantime I don't smoke anything except cigars (I bought them cheap in Spain) when I'm foxing.

That would be all right, I thought. But the cub is also a cigar addict, and bit the first one in half, going away with the end that smoked.

A Mouse About the House

When our Siamese cat, at the age of thirteen, died a month or so after my German Shepherd Lisa, we decided our cat days were over. Skipper, the svelte, blue-eyed, panthering intellectual, whom the garden birds treated as a piece of harmless decor, was to be the end of the line.

Now, after five cats, then four, then three, then two, then one, then none, we are mouse-less as well as cat-less. And that isn't as strange as you might think, because the one sure way to have a mouse about the house is to keep a cat. This doesn't alter the fact that the best way to have no mice is to keep a cat. These statements aren't as contradictory as they seem.

Skipper, being a country cat with a big reservoir of woodmice to draw on, didn't eat all he caught. Every now and again he would bring a live one into the house, cavort around with it for a bit, then lose it. Then for a week or longer we would have our own private Tom and Jerry show, until the cat caught the mouse and metabolised it.

A good hunting cat will never let the house run short of mice; he will tend, however temporarily, to restock it. Mice can be a damned nuisance, of course; but I share the average person's ambivalence towards them, and must confess that I miss the woodmouse we had around for a while after Skipper died. It was he who unloaded it on us in the first place, and he didn't live long enough to ingest it.

That mouse used to winkle a chocolate sweet out of my wife's box then sit nibbling it in front of the TV set. My wife saw it viewing only occasionally, because she does most of her own viewing while asleep. During her waking spells she found the mouse more interesting than the programme. It's gone now, and I don't know what happened to it.

My late teuchter cat Cream Puff was a devil for letting woodmice loose in the house. He was with us for a very long time and his mice caused us a lot of bother. Like the one that took refuge behind the electric cooker. Later, when the cooker was switched on, there was a blowout and a lot of smoke and bother, and inside we found a fried mouse. That wasn't the only mouse that blew a fuse after escaping from a cat.

One evening I was comfortably seated reading a book when I thought I saw a movement on top of one of the bookshelves. I watched for a bit then saw the mouse perched between the antlers of a model roebuck. At that elevation he could travel three sides of the house without coming to the floor, so he was safe from any cat.

We had that one for a few days, but he ended up right where he had been in the first place; in the cat's jaws. After rubbing shoulders with model roebuck, fox, badger and duck, he was caught by the only living thing in the room; the cat that had turned him loose in the first place.

Another mouse that escaped in that room had begun nibbling at *British Mammals* by L.Harrison Matthews before I noticed him. His researches didn't do him much good because he was caught by the big cat the same evening. And this time he was chewed and swallowed.

I brought a black and white, life-size cloth cat from Ypres (the town's full of them) and planted it on the sitting-room floor to see what a woodmouse would do when it came out on its rounds. I found out. After a bit of suspicious coming

199

forward and drawing back the mouse climbed on to the Ypres cat and fingered its whiskers. Later I substituted Cream Puff and mousie was no more. That was, for the mouse, the wrong night of my ambivalence.

Always you meet the bold mouse, the inquisitive one, with the charmed life and the flair for getting a laugh from the host. We had one such. He used to get into the bottom of the cooker, where books were kept, and rustle like ten mice in an empty box. The cats were almost off their heads trying to get in to him. They didn't. He was caught in a way that none of the book learning had equipped him to guard against. He went out on a short circuit.

There was a mouse at The Hirsel that used to come out every night from the fireplace to say hello and mooch. It even sampled the host's whisky. But one night the host was astonished to see a weasel's face pop from the hole. Mousie was never seen again.

Conversation with a Mouse

It was one of those midwinter days when everybody you meet says to you: 'It's like a spring day'. The sky blue with a gossamer haze; straggles of tinsel on the hawthorns; a cranreuch on the leaf litter; the sun not fit to be looked at.

Blue tits swinging and belling in the birches, and a party of goldcrests making mouse-talk in the pines. On the edge of the birch thicket a blackbird is flicking over curly, frosted sycamore leaves, sometimes lifting and laying them, and now and again picking up and gobbling down whatever.

I come sideways along the middle of the ridge, through the old year's brown ferns, where the badgers have their drag road – the route they take, walking backwards, with fern bedding gathered against their chests. There are bits, and splinters, and flinders of it scattered along the route, for loss is high when you're gathering the frush, withered brown.

The bedding trail leads right to the big badger sett, which has been a sett since I have been me, and in which a hantle of fox cubs have been reared forbye. It was here I placed the birth of *String Lug the Fox* at a time when the badgers had been temporarily wiped out by poisoning. Here I had badger cubs and fox cubs running over my feet in their early, unsophisticated days.

I sit down on one of the big mounds, being a confirmed sitter-down, believing it the best way of doing certain things, although I imagine I walk as much as the next man between times.

I have the wood to myself, I think. I can go for a year without meeting a soul in it, a human soul that is, because one way or another I've met just about everything else.

Two woodpigeons come flying over, in direct flight to cross the T of the ride, but they spot me, as woodpigeons always do, and make a sharp turn away, and up and over the trees, and I settle my back against an oak, and wait.

Before very long a carrion crow flaps across the ride and pitches in the tree right over my head. Now crows have good eyes, and we call them leery, which they are, but that bird is perched there for about ninety seconds before it realises I am down below. That sends it up and away, swearing harshly at being fooled, and when it pitched out of gunshot it keeps up the swear-talk for many minutes.

Maybe it was realising, like the earock in one of Charles Murray's poems, that it's better to be lucky whiles than always to be wyce. If I've got the quote right.

Anyway, I soon forget about the crow, because I see a movement at the nearest badger hole, and I adjust my specs, and look cautiously, and see a woodmouse fiddling among the roots hanging through the roof. It is an alert, bright-eyed, pink-footed, sprightly character, in good nick as they say, and I am a bit surprised because the woodmouse is very much a nocturnal beast, and one going around in daylight is usually a sick mouse.

But, of course, it is dark enough inside the badger hole, and the entrance is facing north, so maybe the mouse thinks it is dusk.

It wriggles and twists and insinuates and upends among the roots, using its long tail at times almost like a fifth hand; then it scurries down into the depths, and back again, and snouts around; then goes back to playing among the roots again.

When its back is turned I unwind my length slowly and manage to get on to my knees without sending it away in fright. I keep my eyes fixed on it, and edge forward each time it runs into the hole or turns its back. In short, I am stalking, if you can imagine something the size of me stalking something the size of a woodmouse.

In the end my face is three feet from its face, and we have a good look at each other. I recognising the mouse but the mouse not recognising me, whereupon the following one-sided conversation ensues.

'Bon jour, mon cher Apodemus,' I say.

The mouse freezes and looks hard at me with its jet eyes. But it says nothing.

I start sending my hand forward, in catching mood, but that it could understand. And it whisked away without a word, down into the badger den, and this time it didn't reappear.

I Marvel at the Mighty Mole

For his size, the mole is one of the strongest animals in the world. He may even be the strongest. I don't know if anyone has worked out the mathematics of it, but I know that a mole can heave and push a stone that would flatten your big toe if it was dropped from a height of three feet.

This barrel-bodied, velvet-coated driver of tunnels has the push of a Shorthorn bull and the pull of a Clydesdale. When it comes to sinking shafts, the marvellous little mole can make mineral borers and miners look like apprentices.

He can, literally, dig himself out of sight while you count up to five . . .

Relatively, he will move as much earth in an eight-hour shift as any mechanical digger. He shifts much more earth than that prodigious digger—the badger. If he were as big as the badger he'd be able to shove a locomotive.

This tremendous tunneler, with the tremendous strength, has, as you might expect, a tremendous appetite. He is one of the world's great gluttons. He is also one of the world's great drinkers. At one sitting he will drink as much water as a laying pullet.

He eats round the clock. After a two-hour fast he's hungry. After three hours he's starving. After four he's a ravenous beast. After six he's almost off his head. And any time after that he's dead.

If he can't get several times his own weight in food every twenty-four hours he's suffering from malnutrition.

Earthworms are his staple food, and the mole will gorge on them as opportunity offers, until he's almost ready to burst. Usually, he begins eating at one end of the worm, passing it into the side of his mouth with his great shovel-like fore-paws, until it disappears like a length of sucked spaghetti.

At other times he'll bite the worm through the middle, and proceed with each section as a separate item. This happens with big fat worms as a rule.

Of course if you want to see the mole eating a worm you have to have him under control. A garden frame, with a solid bottom, from which he can't escape, is a good place. If there's a foot of earth in it, the mole will perform pretty well as he would in a free state.

The last time I let some young people see a mole at work I used raw meat to feed the beast with. That saved me digging up half the garden looking for worms. Moles like fresh raw meat and, if given strips of it, will hide them underground, in a store, as they do with surplus earthworms.

Moles are most likely to be seen above ground in the breeding season April/May, and when at water at any time.

During the breeding season they are most pugnacious and will assault almost anything that gets in their way. Fights are sometimes fierce—in fact they're usually fierce—and cannibalism isn't ruled out where one mole kills another.

Any mole will eat a freshly killed mouse, or a slow-worm, or a cut of the best sirloin. So long as it's meat, the mole will give it a second look.

The Anatomy of a Vole Plague

Last year's plague of mice in Eastern Europe, widely reported in the Press and on television, was not a unique event. Plagues of voles and mice have been recorded since Biblical times.

The late Aristotle described a plague of mice in considerable detail, observing how the population built up to a peak of high numbers then crashed so low that there was hardly a mouse to be seen anywhere. And that was a bit before the Romans came to Rye or out to Severn strode, or saw the voles in the grass at Mons Graupius.

Aristotle's description could be applied, without editing, to a rodent plague today. Cycles of abundance and scarcity, of ups and downs, are characteristic of some small rodents in temperate regions. When there is a breakaway at the peak of a cycle, with the numbers rocketing astronomically, we still use the word plague to describe it. Or outbreak. The cycle is just a cycle.

Classic examples of plagues have been recorded in Scotland. In the Border sheep country the Field Vole, *Microtus agrestis,* reached plague numbers in the years 1875-76 and 1891-92. The voles multiplied to the limit of subsistence and ate out the sheep which had to be hand-fed right through the summer.

Clouds of predatory birds, especially short-eared owls and kestrels, moved in and stuffed themselves with voles; ground hunters like weasels, stoats and foxes fed fat. But you can't stuff more into yourself than you can hold, so the predators hardly chewed more than the whiskers off the plague.

When it subsided, as all plagues do of their own accord, the predators found their main food supply gone. Many of them died; many moved out. The residents remained to make do on what was left. The grass recovered quickly once the vole pressure was off and, in the following year, produced one of the best crops that shepherds had ever known.

This century, as recently as 1953, the voles in the Carron Valley of Stirlingshire broke away to plague level. Plague level can be anything from 300 to 1000 voles to the acre and, with all these teeth gnawing, the habitat is soon beaten up and devastated.

The thick grass, shorn at the base, can be turned back with the foot like a moth-eaten carpet. Tussocks come away in the hand like a wig from a head. Young trees, their roots eaten off, pull out of the ground like garden stakes. The vole population eventually crashes. There is no gradual die-off; the collapse is sudden and spectacular. And they are back to square one.

There is a sort of anatomy about a vole cycle, or plague, which can be drawn or graphed. I raked around in the ruins of the Carron Valley plague, before and after the peak. The Nature Conservancy's workers did a study in depth, with Charles on the voles and Lockie on the short-eared owls that arrived on the scene, as these birds inevitably and predictably do. The pattern was standard; the problems as before.

The Field Vole is a small fat rodent, a blunt-faced near-mouse with a short tail and black boot button eyes. An adult will weigh about an ounce in summer, and a good bit less in winter, but a big male can reach a weight of an ounce and a half. On the whole, voles tend to become heavier northwards on their range.

The breeding season is from March to September or October, and occasionally goes on through the winter. The first young of the year are on their own by May, and can breed at the age of three weeks under ideal conditions. A female can become pregnant while suckling. The vole's potential for increase is therefore great.

Voles, like the related lemmings of the tundra, fluctuate in numbers on a definite cycle. From the low point they increase slowly, then more and more quickly, towards a peak, from which they crash down again to their starting point. It takes just under or over four years to complete a cycle, then the process begins all over again.

The rhythm is only rarely interrupted by explosion to plague height. A peak, with voles scurrying and squeaking everywhere, and even running over your feet, can seem like a plague, but isn't. When there is a plague nobody is in any doubt about it.

There are many ways in which the observer can guess at the stage of a cycle—the point it has reached. If his cat is bringing in two voles a day this year compared with two a week last year, it is obvious that voles are more plentiful this year than last.

On the ground he can judge by the number of voles he sees on foot in a given time; the number of nests he finds with young; the number of burrows in the grass; the number of voles he finds in the nests of predators like owls, kestrels and harriers; or the number he finds on a weasel's doorstep.

At the peak of the cycle he will see voles, vole droppings, and owl pellets all over the place; at the low end he will have to search for all of them.

But for an accurate study he wants to be more mathematically precise than the evidence of a cat's prowess or his own eyes. So he samples by trapping and other means. A general picture might be drawn as follows:

At the beginning of the cycle voles might number from ten to twenty to the acre. After a year or so they may stand at thirty to the acre. Another year may see them at sixty or seventy. After that the rise is steep, and in the third year the number could be 200 or more. If it reaches 300 it is on the threshold of plague.

What triggers off the crash? This is argued about endlessly, and the arguments have become extremely complex, with some observers discounting the obvious.

One obvious explanation—it seems to me a major one if not the only one—is failure of food supply and destruction of cover. The voles eat themselves out of house and home. The grass is their cover and they eat it; they have to eat it because it is their food. At this time they turn to food they normally leave alone—rushes, bark and roots.

In some cases, the crash comes when there are still areas of apparently good grass available. In others one finds small clans of voles eating and living well in small oases of green amidst the general desolation. These are the survivors, and they have to be there or there would be no next cycle. They have enough left of food and cover to see them through.

Disease has been suggested, but there has never been any evidence of epidemic disease among the voles. Stress due to overcrowding? This has been examined with inconclusive results. Predation? This is unlikely; if the predators can't prevent the build-up, it is hard to see how they can cut it to pieces at the peak of the cycle.

In any case, voles lie dead all over the place with no marks of violence on them. It may be, of course, that the voles do something to the habitat in the third year that isn't felt until the fourth when the crash takes place. One obvious thing they do is devastate it.

Even the geneticists have been having their crack at breaking the code, but so far nothing clear-cut has emerged. Stress, despite disproof of it in related species, is still in the running, and it is certainly a fact that at this time voles are less viable.

A vole in the hand is likely to die quickly and unaccountably. Females are more prone to desert their young. It seems reasonable that voles might suffer from the pressure of overcrowding at the peak of the cycle as people do under strain. But what would they live on if they didn't?

The forces at work at the other end of the cycle are also obscure. Why should the initial build-up be so slow?

One would think that the surviving voles, down to say ten or twenty to the acre, would get away to a good start, especially where the grazing hasn't been totally destroyed. But they are under par at this stage and, with little cover, are wide open to predation.

The winged predators take them, and a specialist like the weasel, able to enter the burrows, can put a check on them. At this stage, therefore, it might be argued that predation acts as a brake until the cover comes up again.

Certainly multiplication proceeds faster when there is good cover and food supply. Then the brakes come off altogether and the population runs away from the predators, which can now do no better than live at its expense.

A characteristic of vole peaks, or plagues, is the way the predators react. Most striking of all is the arrival of short-eared owls, nomads which settle wherever voles happen to be abundant. An area that carries a pair of these birds when voles are low, or not high, can have ten pairs at the peak of a cycle, and maybe forty pairs or more during a plague. The denser the voles the bigger the families reared by the owls.

When the voles crash the owls depart. In the interval, ground predators like foxes, cats and stoats kill the owlets crawling in the grass. The short-eared owl is a ground nester whose young wander about long before they can fly.

Resident predators like kestrels, buzzards and harriers, now faced with food in abundance, rear whole families instead of part families. Neighbours also hunt the territory and ferry the prey to their nests on the fringe.

So it is not unusual, when voles are high, to see kestrels and owls sitting around so stuffed with voles that the tails hang out of their beaks. Eagles take a lot of voles at such times, and one can see the living voles mingling with the dead on a ledge beside an eyrie.

Ground predators like the fox find life a bit easier during the peak of the cycle, and kill voles hard. At this time a fox may well get the pound of food he needs each day entirely from voles. Ordinarily he might kill 2000 voles a year on the ground, which is high, but not as high as sixteen or more a day. The fox also takes young owls for a change.

Stoat and weasel on vole ground kill voles almost exclusively. In a peak year the weasel will almost certainly breed more than once, and perhaps even three times. The stoat, a prisoner of his own physiology, can breed only once a year whatever the food supply. When the vole crash comes these related predators are differently affected, because the weasel can go down the burrows that the stoat is too big to enter. The stoat, as a result, suffers first in a crash. My friend Lockie has shown that while the stoat has to get out when voles are down to forty-five to the acre, the weasel can carry on when they are down to eighteen or fewer. Which means it can remain a resident.

Yet, despite all this killing by resident predators, and the influx of additional mouths at the peak of a cycle, the voles have their ups and downs apparently unaffected by either.

The lemmings of North America, Scandinavia and Russia are relatives of the field vole, and also operate on a four-year cycle. They, too, reach a peak. And crash. They have their special predators: foxes, snowy owls, short-eared owls, and a species of skua, depending on the part of the world they live in.

But the peak in lemmings leads to emigration as well as die-off, and this has given rise to a folklore of its own. Emigration of voles is so hidden by die-off that it is hardly noticeable; in the lemming it is spectacular.

Obviously not all lemmings move out of their devastated range. Some remain to begin the new cycle. Many die by predation. But the emigrants have caught the imagination. Their exodus has become a race to suicide or the sea, instead of a search for new range after the destruction of the old.

205

Of course, many are drowned in streams and rivers; many pass over bridges of their own dead; and many reach the sea. But the sea was never their chosen destination, nor death by drowning their aim.

The march of the lemmings illustrates once again man's predilection for attributing decisions to animals involved in some act over which they have no control, and which follows directly on a pattern of behaviour that is a forever strait-jacket.

A White Stoat
in Playful Mood

A white stoat—white, that is, except for her tail tip, and a postage stamp of brown above its root—appeared on the drying green, outside the kitchen window, a week or so before Christmas. I was so pleased that I almost began to believe in Santa Claus again, because, you see, I like stoats, and a lot of stoats is something I don't have where I am nowadays. In fact, in nine years I've seen only two, and both were transients. Plenty of weasels I do have, but that's something else.

White as chalk she was; so whiter than white that she could have crashed a TV commercial. She came out on to the green, whirligiged, sat up, paws to chest, then caterpillared into the laurels, where we could still see her like a travelling highlight in the greenery.

Then she came out again, fiddled about, dug up a piece of meat my tame crow had dibbled into hiding, and went off with it through the wire into the chicken run. Then she bounded westwards into the leeks, and from there into the long grass among the blackcurrant bushes, and was lost.

But she was back next morning, and a sagging smoorey morning it was, and this time she spent about twenty minutes cavorting about the drying green and in and out of the laurels. The birds at the bird table paid no heed to her whatever, and she appeared to have no interest in them.

But she found the rat holes under the fence where the rhubarb grows, and into one of these she went. And out of the other end bolted a big rat, about twelve ounces by my reckoning, and headed for points west. And I said to my audience of three: that should keep her.

This small rat enclave has been bothering me for months, and we're still figuring how to get rid of them (but that's another matter). If the stoat could help to solve the problem she was doubly welcome.

Next day she didn't appear; at least not when anyone was looking. Nor the next day. But on the third day she was back, and in view for half an hour.

I thought of photographing her from a window, but the light was so bad you couldn't have got a reading of half a second at f/0. She worked the rat holes, stopping twice to pick up morsels buried by my crow and magpie.

And that was a caper the birds weren't standing for, although there wasn't much they could do about it. Anyway, at one point they were pitched on posts above her, one on either side, haa-aing and yakkering in their specific swear words.

The stoat paid no attention. When she moved into the leek bed the magpie followed her, pitched a few feet from her, and began to give her an argument. He was near enough to be pounced on so I gave him a holler. He paid no attention to it, but the stoat did, and ran into hiding.

But not for long. She came out on to the green, and entertained us with a display of gymnastics, by nosing around, by standing tall, and by doing about everything one expects a stoat in playful mood to do.

The crow pitched on the window sill to swell the audience; the birds at the bird table went on feeding, a bold robin even daring to feed on the ground a few yards away.

Then the stoat headed for the kitchen window. I put my nose to the glass and could see her eyes. She was climbing the roughcast. The crow sat on, watchful but obviously not afraid. The stoat dropped back to the ground and rippled back to the middle of the drying green, where she began to scrape in a moleheap. She wasn't daft. One of the birds had hidden a piece of meat there.

That day I mounted a sixteen millimetre cine camera at an upstairs window and decided to leave it there for next time. I couldn't get a moment's peace for the crow and the magpie, so I had to close the window.

Next day the stoat was back and the light was good, and I rushed upstairs. Stoat came on after a bit. I began to pick her up in the viewer, and was prepared to run out the whole one hundred feet on her. Then Lisa, my German Shepherd bitch, began to bark, and the stoat took off. Someone had come to the back door.

I waited by the camera until the light had almost gone. But no stoat. We haven't seen her since.

Spilled Oats

Seed-time was late that year. Down in the glen the rosettes of the butterbur shrivelled in the frost; the alders trailed their caterpillar catkins in snow dusty with their pollen. The varnish froze on the chestnut buds. Young rooks were hatching in the Mossrigg elms when the last of the snow still lay in shrunken ridges along the north side of the hedgerows, glassed and wetly gleaming. And, when the swallows came in the frosty morning, seeking the old familiar places, the iron was still in the brown autumn furrows. Gallacher of Mossrigg came out at mid-morning to sniff the air like a setter, standing with hands deep in go-to-hell pockets till his shirt darkened in the rain and the cow-dung was running in a green wash from his horsehide boots. By noon the iron was out of the ground, and when the peewits returned to their eggs, after feeding, they had muddy feet. The rain ceased at darkening; the night was wind-shaken, and in the morning there was no more cranreuch.

Soon the first field was ready for sowing—the long rising field marched by Hackamore Wood on the south and east. At eight o'clock in the morning two tractors were warming up in the farm close, one coupled to the trailer and the other to the newly-painted seed barrow. Gallacher, in ancient out-size battle-dress blouse, was checking the barrow and shouting instructions to Geordie, the cattleman, who was loading the trailer with sacks of oats. Up in Hackamore, three fields away, his shouting was heard by a big fox called String Lug, who was just going to his day-seat after putting down a rabbit for his mate beside a hole under a tree root, where she had five blind, mewing cubs.

The fox, lingering on the wood-edge, saw the movement as Geordie swung the tractor on to the road, and knew it was time for him to go: he guessed where the tractor was coming, for he had heard it roaring in the big Hackamore field for three days in a row. Geordie revved the motor for the steep snap from the close, and swung the wheel hard at the top into the long tree-lined lane leading to the field. The tractor was fifty yards along the lane, lurching in the ruts, spluttering and spouting blue vapour, when Gallacher was manoeuvring the wide seed barrow out of the close.

Its width slowed him down in the lane. Gallacher put the tractor wheels in the ruts and drove looking back over his shoulder. Long briars, and trailing bramble tendrils, clutched at the ends of the barrow and snagged in the wheels; the wheels ripped them off, and spun them, and cast them on the verge. Gallacher wondered at seeing no rabbits when he passed the pasture gates in the lane; he didn't know that String Lug had been that way earlier in the morning. He began to hum *The Muckin' o' Geordie's Byre*. He hummed till he reached the sharp right-hand turn in the lane. Then he swore.

He swore and blasted, colourfully and profanely. He crash-stopped the tractor and jumped down from the driving seat. He knelt down and gathered a handful of clean, shining corn. He let the grain trickle through his calloused fingers. As far as he could see there were little heaps of yellow corn at intervals along the lane. They could only have appeared for one reason. There was a sack with a hole in it.

That was certainly something remarkable, indeed, incredible. For Gallacher was pernickety about his sacks. He was as careful about them as he was with his siller. It was his boast—and it was no idle one—that neither potatoes, oats, nor wheat had ever left Mossrigg in a sack with a hole in it. Holes appeared in sacks at Mossrigg as everywhere else, by mice and magic, but Gallacher was never known to spare the binder twine in strangling them.

'Geordie! Geordie! You've a b-a-g wi a h-o-l-e in it!' Gallacher stopped for breath, and listened. There was no reply. String Lug heard him. The rooks in the Mossrigg elms heard him. The collies, rabbiting away down in the glen, heard him and went skulking home. But not Geordie. His ears were too full of tractor noise, and his thoughts were far away.

So Gallacher had to mount the tractor, and follow the trail. He kept his eyes on the little heaps of grain—some in the ruts; many on the verge—hoping he would see the end of them. But they were there all along the way, right to the gate, and into the field; and beside the stationary tractor there was a greater golden moleheap. Gallacher rushed to the worry without preliminary.

'There's a ton o bliddy corn in the lane! Did ye no ken ye'd a bag wi a hole in it?'
'Aye: A ken.'
'Ye ken? Then whit the . . .'

208

'But A didna ken tae noo! Could ye no hiv shouted tae me?'

What Gallacher said next is best not repeated, though if you'd been within a mile of him at that moment you'd have heard every word quite clearly. His ton of corn turned out to be a twelve-stone bag with four stones short. The oats had spilled from a hole in the middle and the bag's upper half was slack from the loss. Gallacher pulled a length of binder twine from his pocket, made a neck at the hole, and tied it off. 'Let's get started,' he said.

They lined the sacks of oats along the Hackamore fence. They gathered the bags of fertiliser from under a hap in the corner and stood them out too. Then they started to sow.

Rooks left the field as soon as the tractor moved. Cushats whistled overhead, but did not pitch. They were feeding squeakers in flimsy nests in the dark Hackamore spruces. Gallacher swore as he watched them; he knew they would be down in force, poking into the tilth for corn, as soon as the field was sown.

At eleven o'clock the mistress of Mossrigg arrived on a bicycle, with a can of tea and slices of bread and jam in her basket.

'I see the cushies ha'ein a fine feed o corn in the lane doon there,' she said.

'Hole in a bag,' Gallacher mumbled, and started to eat. He was sitting at the field gate into the lane, and presently he saw three cushats flying down to the heaps of spilled corn about a hundred yards away. 'Look at them!' he said to Geordie. 'Ah weel, it's an ill wind . . .'

Round the bend of the lane, out of sight of the men in the field, five other cushats were gobbling grain, pecking it and jumping away with a flick of wings as if afraid of it. They were nervous. After every few pecks they would side-step, and circle their heap, jerkily, with heads bobbing, like marionettes. And sometimes they would flap up into an overhanging branch, and sit there, with their bright far-seeing eyes scanning the fields. Pigeons flying overhead saw pigeons in the lane, and wheeled, and flew down, and soon more than twenty were feeding on Gallacher's spilled oats.

One cushat, with dirt on her beak, flapped up, and circled, then flew high and fast to Hackamore, with one thousand three hundred and sixty grains of oats in her crop. Above the cornfield she sheered away to avoid flying directly over the men, then wheeled round on whistling wings to pitch in the spruce where she had her nest. She flapped down through the smother, with her wing feathers snagging in the fronds, and was greeted by two fat squeakers, with great shovel beaks. She said *Coo-coo* softly, and alighted on the nest.

They thrust their long beaks into her throat, one on each side, forcing her head up and down. And one thousand three hundred and sixty grains of Gallacher's oats passed from her crop to theirs. When she had fed them she was exhausted, panting, with tongue flicking from side to side and her white neck feathers raised. She flew down to the ground to rest, and rub her neck against her flank, and run her beak along the feathers of her wings. Then she flew back to the lane.

A rook had now joined the gathering. The cushat flapped into a lane-side ash, and perched swinging for a moment on an up-pointing twig before flying down. This time she gobbled more quickly, without dance, or side-step, or nervous lifting of her head, for her fear of the new feeding place was gone. When the rook raised the feathers of his crown at her, and tried to scare her off with guttural *caw*, she flicked a wing at him in anger, then went on dab-dab-dabbing at the corn. In the sunlight and twig-tracery of the lane she was all soft grey and salmon-pink, yellow and scarlet and hyacinth blue. With four minutes to live . . .

With hunger sated—this time she was gathering the grain for herself—she hefted from the lane with much clatter of wings and airted for Hackamore, a lone pigeon flying in swift straight flight against the blue of the sky. At every down-stroke, the white bars of her wings could be seen by the men in the cornfield. Gallacher watched her idly, as Geordie emptied a sack of oats into the seed barrow. He saw her falter, and swerve; he saw the she-devil herself flashing to the attack; and he saw pale feathers drifting when the locked birds vanished from his seeing.

Kiki, the sparrowhawk, was the he-devil of Hackamore, and the she-devil was his mate of many seasons. Gallacher knew them well. He loved the she-devil when she was killing cushats; and hated her when she came sneaking after his chicks. But he never in his life pointed a gun at her, nor smashed up her nest, as he did with the nests of the crows. She had once killed thirteen cushats out of Hackamore in a morning.

The she-devil flashed from the top of the tallest spruce as if shot from a gun, and bound on to the cushat before she could crash through the trees. The cushat swerved up and away with one set of yellow talons clenched on her neck and the other reaching for her flank. Instead, they clutched on to the curve of her right wing, where the feathers were white. That made her helpless, unable to control her flight, and she flapped wildly in panic, with one wing beating frantically on the air and the other almost swinging the sparrowhen. Blue, pink and white feathers eddied and drifted above the trees as she lurched and swerved to the far side of the wood. The slim talons tightened on her neck. They had a grip of steel. Her eyes began to blur. Her wild struggling ceased. And at last she was falling. The she-devil released her at tree-height and followed her down.

She hit the grass with a thud, six feet from the Hackamore fence, and the white feathers billowed out from her in a downy cloud. Her yellow and vermilion beak opened and closed spasmodically. Her up-reaching talons clenched and un-clenched, slowly, and she was dead. The she-devil alighted, scissored her wings, and began to tear open her breast.

The hawk had to make her meal on the spot; to carry off such a prey was far beyond her powers. With her sharp, sickle beak, she ripped out pink and grey feathers, which she tossed aside, so that they formed a little ring round the dead cushat. When she had cleared the area of keel and crop she ripped open the skin; she tore away the veined transparent crop tissues and blood welled into the cavity, turning the moist yellow grain to scarlet; then she began to tear strips of dark flesh from the breast. She ate till she had bared the keel, and, when she was finished, she had bloody feet, and white down and morsels of flesh sticking to her beak.

She cleaned her beak on a fence post, whetting till she had rubbed off the last morsel of flesh and the last clinging wisp of down. She ruffled her feathers, pecked at dried blood between the scales of her talons, and whetted her beak again. Then she flicked away in swallow flight through the trees. On the grass lay the dead cushat, with closed eyes and mutilated breast, surrounded by feathers and bloody grain from its own crop, scattered by the she-devil as she fed.

Gallacher of Mossrigg sent Geordie home with the trailer at sunset, saying he would take a turn round Hackamore for ten minutes to see what he could see. He still remembered the assault, above the trees, and he was curious to find out if there was anything to show for it. He had that kind of curiosity.

He cut through on what he thought was the line of flight, with his big feet crackling in the brushwood. The cushats clattered up and ahead of him as he walked, and once he stopped to sniff because he was sure he could smell fox. But he did not wait to investigate; he wanted home with the seed barrow before the light was bad in the lane. The last thing he saw before he reached the far side of the wood was a brown hare running down the inside of the fence.

He legged over the fence thirty yards to the right of the dead cushat, but it was plain to see and he saw it. He turned the body over with his boot and said: 'Uh-huh! A nate enough job at that!' He toed it over again, and, in so doing, emptied most of the corn from its torn crop. Then he climbed back over the fence, with his mind on the she-devil, and walked into a tree.

The cushats came circling down after the tractor had roared into the lane. Roosting crows, alerted by the noise of Gallacher's feet in the wood, turned beaks back into shoulders and slept. Kee-wick, the tawny owl, who had sat up tall on his pine roost, staring wide-eyed, all the time the man was in the wood, shook his wings, gulped, and hooted. *Hoo-hoo-oo-oo-oo-oo!* Like a great moth, he wafted from his perch and flapped silently through the trees; he had a mate and two owlets in the old nest of a crow high up in a pine tree, and this was his hunting hour.

String Lug, the fox, watched the big owl flying from his perch, and rose from his seat under a windfall. He chopped his teeth. He stretched each leg in turn, yawned, and bit at his flank. He, too, had a mate with a family. And this, too, was his hunting hour.

The big fox trotted through the wood to the cornfield fence, and sniffed. He trotted along the entire length of the fence, with the wind on his cheek. He trotted back to his starting point, with the wind on his other cheek. He was checking the wood to make sure that one man had entered and left, and that no other remained. At one point only was there a slight taint of human scent—enough to show a man had been there, but not nearly enough to suggest that he was still there. String Lug ducked under the fence as Kee-wick yelped in the lane. He stopped to listen, but did not go in that direction. His business was on the other side of the wood. And he went there, following right in Gallacher's boot-tracks.

He had no idea what he was looking for—at first; but as soon as the scent of the dead cushat came to his nose, he knew. The smell intrigued him, tempted him, and he wanted that cushat. But he remembered Gallacher. So he inched towards the fence, belly down, brush trailing, with ears and nostrils wide open, wary, suspicious. He knew all that any fox needs to know about men—that they are dangerous. When he saw the cushat, he froze to ponder.

For a long time he crouched there, waiting . . . wondering . . . and who will say he was not thinking? The light was failing fast. Stealthy tread of feet in grass made him turn his head suddenly. A dark shape, like a moving moleheap, came into view, and presently he could hear it snuffling. A hedgehog! String Lug relaxed, but did not rise. He was still wary.

The hedgehog shambled up to the cushat, gathering white down on the quills of his seat. Having none of the fox's doubts, he began to eat. String Lug listened to the cranching teeth for a little while, then he rose, snarling, and rushed bristling to the assault. The fence wire hummed when he hit it with his rump, and the hedgehog was warned. He crouched, with brow quills forward. It was his turn to wait. When he smelt the fox he waited no longer. He rolled himself into an armoured ball.

211

String Lug, ignoring the ball, stalked up to the meat. He wanted to snatch it up, but he didn't. He laid back his ears, and his lips lifted, uncovering his teeth. There was man smell—strong man smell—around the dead cushat, and he didn't like it. He stepped back stiff-legged for ten feet, and sat down to ponder further.

Presently, forgetting about the fox, the hedgehog uncoiled and moved forward. He hadn't forgotten about the cushat. But before he could fairly put his teeth in it, String Lug moved too. There was a great anger in him—born of his own fear, suspicion, puzzlement and frustration; but the hedgehog, blundering in fearlessly where the fox was wary, to poach what the fox considered his own, became the target for it. A nimble black paw reached out and tipped the hedgehog under the snout; it tipped him over on his back and he rolled two feet. But, even as he rolled, he was coiling, and the fox was faced with an armoured ball as before.

This time, however, String Lug did not leave the ball in peace. He rolled it over with light flick of a paw, till he had the beast on his back with the one chink in his armour facing upwards—the place where his hindfeet were tucked into his face. String Lug chopped at the joint, and tried to insert a tusk; but the quills pricked his lips. The hedgehog remained coiled. The fox drew back with uplifted paw, then returned suddenly to the assault; he knew what he was going to do.

Using one forepaw, then the other, he started the ball rolling towards the woodside fence, barely four feet away. Inside the fence was the ditch, running the entire length of the wood, sheer-sided, two feet wide, deep and dark after melting snow and recent rain. The fox pushed the hedgehog under the wire. He snaked through and scooped the ball forward with his paw. He rolled it up the little ridge on the lip of the ditch, made when men were cleaning the bottom with spades, and tipped it into the dark water.

The hedgehog uncoiled instantly. He snorted and snuffled. He sneezed and coughed. He tried to climb out, but could get no grip on the wet, treacherous peat. He swam, snorting, against the slow current, he drifted with it, gyrating. He swam till he could swim no more. He drifted, spreadeagled. He kicked feebly when he bumped against the side of the ditch. He floated, spinning and rolling lazily, for eighty yards. He drowned. He was left stranded on a flat under willows, where kingcups grew, with two wisps of pigeon down wound round the quills of his seat.

String Lug did not follow the hedgehog; he returned to the problem of the cushat. Clearly satisfied now that it could do him no harm, possibly because the hedgehog had bitten at it without mishap, he walked up to it casually, stopped, sniffed, bristled, and snatched it up in his jaws. The instant he had it he jumped away. Outside the ring of feathers he laid the carcase down, sniffed it over critically and touched it with his paw. After an over-the-shoulder glance at the spot where he had found it, he picked it up again, and ran with it into the friendly gloom of Hackamore.

At moonrise a fat vole with boot-button eyes and blunt face scurried from the Hackamore fence tangles and found an ear of corn, the last of a scant trail left by String Lug when he carried his cushat into the wood. The vole nibbled at the icker and liked it. So he picked it up in his forepaws, and sat down to nibble, like a squirrel with a cone. When he had eaten half of it, he whisked back to the fence with the other half in his teeth.

That year the short-tailed field voles teemed in Hackamore. They were thick among the blaeberries, among which they could be seen moving by day. They were in every tussock. Weasels killed them, and foxes. And many died nightly under the talons of Kee-wick, the tawny owl, who carried them to his mate.

Five minutes after his first sortie, the fat vole with the boot-button eyes peeped out a second time from the grass under the fence. His blunt face wrinkled as he smelt the air. Presently he was joined by another vole, which peeped out, whisked out, whisked back, and peeped out again. For perhaps a minute they crouched there, peering bright-eyed from between withered grass blades, six inches apart; then they scurried out boldly, darting this way and that, quivering lips over gnawing teeth, criss-crossing and jumping as if playing a game. Until one of them found another ear of corn.

He gnawed at the ear, and ate it. The second vole darted past him and found one, too. So they fed back along String Lug's outgoing trail till they found the main contents of the cushat's crop scattered inside the ring of feathers. They squeaked their delight and lifted ickers to their mouths, using their forepaws like hands.

Kee-wick, the tawny owl, flew into the wood with a shrew in his talons and pitched in a flat-topped pine twenty yards from his nest. With the prey under a foot, he called: *Kee-wick; Kee-wick; Kee-wick,* and was answered by the softer, less ringing, drowsy *kwik* of his mate, brooding low over her two squirming owlets. The big owl switched the prey to his beak and flew noiselessly to the nest.

When he pitched, he bowed low with his offering. The hen owl, quivering her wings and tail, snatched it from him and dropped it on the edge of the nest. The owlets, sensing food, began to jostle and wriggle, lifting her bodily till their white down was showing in the dark. *Wee-wick; Wee-wick; Wee-wick;* the hen owl sang her querulous song of greeting. Kee-wick turned on the nest, thrust with his legs, and launched out and away. His wild war-whoop was still echoing through the wood when the hen owl pulled the shrew under her breast to tear it for her cheetering chicks.

Three hours later the big owl was flying at tree height, above the cornfield fence, big-headed, front-heavy and moth-like in the full light of the moon. He had killed nothing since delivering the shrew to his mate. His eyes were seeking movement, while his super-sensitive ears, highly selective in their hearing and hidden under the feathers of his head, were strained for the patter of vole or mouse or baby rabbit in the grass. Half-way along the fence, he tilted down, and flapped between the crowded trunks, unerring as a bat, and suddenly he heard and saw the voles feeding on the corn.

Kee-wick; Kee-wick! He swept up, and round, and down, silent as a rag of detached darkness. The voles froze at his cry—momentarily; but in that moment the owl swooped and struck. They saw the falling shadow, and moved. Too late! Kee-wick reached down with clutching talons. He grasped a vole, and squeezed. The pressure of those talons was irresistible. The vole was pinned to the ground, pinned down on many grains of bloody corn, with the owl's weight upon it. It squeaked once, and died.

Kee-wick flapped up with the limp, warm thing in his claws. Three ickers of corn fell to earth before the owl topped the trees, but two remained sticking to the vole's furry flank, and were still there when it was dropped in the nest. They were still there in the morning, long after daylight, when the vole had been eaten, lying on the lining of crumpled pellet-fur in the bottom of the nest. And there was blood on them with dry vole fur sticking to it.

Biting Talk

It began with Cassius, my big dog polecat, as it usually does when you produce a carnivore, however wee or big.

'Does it bite?' asked he-with-the-false-teeth-in-pocket.

'Only folk he disna like,' I replied.

'If I was him I'd be haein a bitin ball right here,' the retired collier said.

Then he reminisced: 'Funny the things that'll go for ye. The gran wean got a white rabbit, an there was me showin him how tae haunle it, an the damp thing but me.'

'I've had it masel,' said I.

He-with-the-false-teeth-in-pocket tapped me on shouther and said: 'A kent a Labrador retriever wance wi only hauf a tail. The ferm boar had bited hauf o it aff.'

The third man broke in, and said to him: 'You'd best be watchin or you'll be bitin a hole in your pocket.'

'Haw Davie,' the collier said to me. 'A'll bet you've been butten mair often than onybody.'

'Often enough,' I agreed. 'But I doubt if it wid mak the *Guinness Book of Records*'.

'Share you've been butten wi a snake.'

'Three times,' I agreed again. 'Bited, but and butten.'

A voice from behind broke in:

'I can hear you're haein bother wi the verb tae bite—even the wan wi nae teeth.' He laughed and, holding his dram to the light, said: 'I bite, I hae bitten, I hae been bitten, I went tae school.'

'You've had a sair time wan way an anither,' the collier said. 'You're the wan that should be in the *Book o Records*. A had a sergeant major that could hae but you in hauf, and he had false teeth.'

They had all come up the hard way, including the scholar wan. He sat down, saying in mock apology:

'I was merely conjugating part o a verb.'

'An is that no a helluva thing tae dae wi ony wee innocent verb,' said he-with-the etcetera.

'If you'll stop grinding your incisors,' said the man. 'I want tae hear aboot this snake bite.'

'So div A,' said the collier, and said to me: 'Wis it a pooshen bite?'

'It was,' I said. 'Adder bite; three times. Once in Spain, twice in Scotland.'

'Where did ye get . . . bit . . . if that's right this time?'.

'Wrong again,' the man said.

'What's the bliddy odds?' the collier complained. 'A bite's a bite.'

'No if it's me that bited ye,' said he-with-the-etcetera.

'I got bitten on the hand,' I told them. 'All three times.'

'But A thocht a bite like that kilt ye,' said he, etc.

'It did,' I agreed. 'Every time.'

I put Cassius on the table and opened his mouth to show his strong canines. 'This one has bitten me when he was in a bad mood.'

'An what else hae ye been but wi?' from the collier.

214

I reeled them off:

'Dog, cat, rabbit, fox, polecat, mink, stoat, weasel, vole, mole, rat, lemming, squirrel, water vole, mouse, hare, horse, foal, girl, crow, magpie, jackdaw, raven, budgie. And been taloned by eaglet, buzzard, kestrel, sparrowhawk, harrier, tawny owl, barn owl and long-eared owl. The last thing I was bitten by was a greylag goose.'

'Fetch the man an anti-tetanus,' the collier called.

Ecology and Brer Rabbit

The ecologist might be affronted by the suggestion that the concept of the Food Chain was plainly stated in nursery rhyme before the word ecology meant more than its derivation; but it is true.

'This is the Cow . . . that tossed the Dog, that worried the Cat, that killed the Rat, that ate the Malt, that lay in the House that Jack built'.

Nobody ever said it plainer than that. That man Jack built the habitat (house) and the food supply (malt) that fed the primary item (rats) to feed the first line carnivore (cat) that was assaulted by the second line carnivore (dog). Fox in place of dog would have been better because fox would have eaten cat. Cow is a fortuitous follow-on—a herbivore booting a carnivore for reasons best known to herself.

If the author of the rhyme had stated the quantity of malt stored, and related it to the barley crop and acreage, then given the number of rats feeding on it, the rhyme could have been expressed in true Pyramid of Numbers and Energy Flow, and maybe used in the universities as well as the nurseries. But alas! he stuck to the dramatis personae, demanding only representational illustration.

But all modern ecologists are not conservative. Only the other day my friend Lockie, a lecturer in Ecology at Edinburgh University, drew to my attention a quote from the immortal Uncle Remus, that could be used as a sort of Hippocratic oath for scientific tearaways. It was:

'I aint disputin about it, but I aint seed em, an I don take no chances dees days on dat what I don see—an dat what I do see I got to examine mighty close . . .'

Was Joel Chandler Harris an ecologist? He had to be before he could make Uncle Remus one. And you've just read Uncle Remus's own words. I confess to being shocked that in the hundred times I have read that darling book I missed the glitter of this jewel.

If one defines ecology as the study of animals and plants, their relationships to each other, and to their total environment, one quickly appreciates the admirable scientific caution of Uncle Remus, who obviously realised that things aren't always what they seem.

It now seems to me, indeed, that if Uncle Remus is reread as a study in ecology, a new dimension is added to it. The tilts between Brer Fox and Brer Rabbit are excellent psychology—do anything you like with me, even eat me, but don't throw me yonder to a fate worse than death, as it were—but the ecology is there trying to get out.

215

The fox would surely do what the rabbit feared most, as he thought. But what he did at the end of the day was to throw Brer Rabbit into COVER. And COVER, as the ecologist would tell you, is an important aspect in predator-prey relationships. It makes it tougher for Brer Fox to catch Brer Rabbit.

Mink-Wrapped Against the Cold

Frush with cranreuch the tussocks crackled and snapped under my feet. Every twig and branch was ermine wrapped. The vapour breath of Shona, my young German Shepherd bitch, drifted past me and trailed away, and when I sat her down and walked ahead to face her from the front, I had the sudden feeling that this was how the first puffs of the gas cloud at Ypres must have looked to the soldiers holding the tip of the Salient of terrible memory.

A hare snaked away in front of me, running low with ears laid back, then high with ears up, and I called the bitch to heel. We headed for the big badger sett on the knowe to await moonrise, stirring up clouds of frore on the way. When we got there we both had a dandruff of frost powder.

The cold was painful; it clawed at my ears like the talons of an angry bitch sparrowhawk. I could almost hear it, like bacon sizzling in a frying pan. I can usually wear cold as easily as an old coat; but this was getting to me. I thrust my gloved hands under my parka, and waited . . .

Syne the moon surfaced, monstrous and red-faced, like a great inflated balloon. When it was a hand's breadth above the trees it silvered, and a gelid breeze began to torment me where the sparrowhen had already been at work. Glow-worms of scintillating fire sparkled in the heather snugging the badger sett. My near-side damaged big toe said quit. So I quit . . .

I took the path for easier walking. The ponies came to the field gate, warm-lipped, padded with fat and winter overcoats, radiating comfort, snorting like dragons, not fire but frore. I moved into the bunch, and knew central heating, and warmed my numb fingers in the stallion's mouth. The bitch rolled in the ground ermine.

The snow hadn't come to much, so it was no more than inch deep, and frozen, and the walking was easy. On our left we passed the wa-gang tracks of a hare, followed the neat prints of a walking fox, then stopped where two roe deer had crossed our front. But there was no sign of them.

I veered off down to the frozen burn. In the burnside trees a tawny owl was calling, but there was nothing moving in the spectral radiance. Until the black shape appeared from under the washed-out roots of a beech tree . . .

I sat down with the dog, holding her close to me for warmth as well as to restrain her in case she had the notion to investigate. The black shape appeared from a clump of white-furred thistles, ferret-like, not cat-like. It bellied low across the ice on the burn, then bounded, weasel-like, back to the beech tree. The moon leered frozenly down; the black fur sheened. The beast was mink-wrapped against the cold. It was a mink.

It appeared again, came towards us, then turned right and recrossed the burn. My toe was still giving me orders to go, but old habits die hard, and the black beast, and the star glitter, and the sardonic moon ordered a little stronger, so I tholed ...

The minutes froze on, and the temperature dived still further below zero; and the stars winked mirthlessly, and the moon laughed in its sleeve, and the bitch poked my arm with her muzzle, and my big toe sent waves of warning up my tibia and fibula. I was nearing the point where sitting still would be about as easy as not breathing. Then the mink reappeared.

It was going up the bank backwards, like a badger dragging in bedding. But it wasn't dragging bedding. It was dragging a body, a body as big as itself. And the body was a rabbit. It dragged the rabbit under the beech roots. I listened to my toe and rose. I walked, with the bitch at heel, along the middle ice of the burn and stopped opposite the beech tree.

The dog knew as well as I where the mink was, and I had to *wheesht* her to stay still. I could see nothing, and I could hear nothing. But my nose suddenly owned a smell I could recognise, the sort of camphory funk scent of a mink. So it had got the dog smell, and my smell, and was letting off.

I left it at that, and walked home on feet I couldn't feel, and with fingers I didn't want to own.

Death in the Afternoon

A stickit ermine (I've just invented the term) is a stoat wearing only part of its winter buskins. The one I'm going to tell you about had white gloves, a white splash on the face, and the white on its flanks a bit higher up than in summer.

Which was just as well, there being no snow, so it wasn't all that obvious on grass verge and tarmac, although the extra white was flashing signals enough to draw the kenning eye.

He was big, too, at a guess within woodmouse weight of 12oz., and he had a fine muscular tail, longer than many I've seen, with a well bushed black tip. He was the first of his kind I'd seen on the ground for many a long day, so I was pleased to see him, and almost stood the van on its nose to stop in case I frightened him away.

He sat up tall when I stopped, looking left, right, and round, with his ear trumpets harking for translatable noise. With the glasses I could see the bright, lustrous, alert eyes of him. He was, as stoats usually are, quickly reassured and dropped to all fours, a sinewy bundle of nervous energy who had to be doing something.

The first thing he did was a massacre of white crocuses, but there was no savagery about it; he did it like a puppy slaughtering balls of wool or bits of paper.

He liked the look of them, and spent some acrobatic seconds ca'in the heids aff them and rolling on them in a kind of ecstasy. Then he flicked the heads around with nimble right and left handers from his forepaws.

Next he flashed along the verge, stopped, back somersaulted, and tossed a

217

damp leaf into the air, biting at it as it came down. But that was merely something he had met on the road to where he was going, and the place he was going to was the wire netting surrounding a grey mink.

Up, down and across the wire he climbed, expert as any squirrel, while the mink inside went up, down and across inside. Then the mink squalled, and the stoat spat, and both came down to earth, eyeing each other through the mesh.

That was all he wanted of the mink it seemed, for he whisked across the tarmac, along the green, and up into a tree from which he peered down at me. He swung on a branch, came down hind end first, then round the trunk, and up again, then down head first, then up again, then down to earth, and round and round the maypole.

I've told it quickly, but not as quickly as he did it. You have to see a stoat at high-speed play to know what high-speed play is.

Then he did something I've seen stoats do more times than I can remember. He leaped at the trunk, and down, leaped at it and down, in a figure of eight movement that a camera wouldn't have stopped at one thousandth of a second.

I'm sure he did the figure of eight acrobatic dance fifty times before he dropped down, looked around, and went back to his crocus slaughtering. And at that point I started the engine. He looked up for a moment, then carried on. I began to reverse slowly, without disturbing him. At fifty yards I turned and headed for the house to collect a camera.

They're always telling me I should carry a camera about with me, but I never do. This time I could be there and back in plenty of time, because once a stoat starts on a play session he keeps at it for a long time.

I grabbed a 35mm, loaded it with fast film, and ran out to the van, noticing without particularly noting that two cars had just passed the place where the stoat was.

When I got round to where he had been I had reason to think again about the two cars because one of them had taken more than I was aiming at—his life.

There he was on the edge of the verge with his face bashed and the hind end of him flattened. 'Will you chaps never learn?' I said to him as I picked up the supple, warm body and threw it as far into the field as I could. It was fit for only the crows now anyway.

Research into Rats

The day my wife came to me and said there was a hedgehog on the meal-store shelf, I didn't believe her, because there was no way the beast could have got up a smooth wall to a perch as high as I am tall. But I went out anyway, because the something had to be something that looked like a hedgehog.

It did, hunched there on the shelf, with hair bristling like a hedgehog's quills. But hedgehogs don't have long scaley tails, nor the face that turned to greet me, with buck teeth threatening. It was a rat I was looking at, an outsize one, and when I approached within its toleration distance it took the shortest way down: via my face.

I side-stepped it, and it bumped my shoulder before taking cover among the packed corn and meal bags on the floor. I closed the door and went into the house for Nip the terrier, about whom a poacher friend of mine once said: you couldny keep up wi her shovelin rats for her. I told her there was a rat and that got the cutting edge on her before I took her into the meal-store and shut myself in with her.

But she was impatient and began thrusting her nose between the sacks instead of waiting for me to move them. She got bitten in the face for that, and the bite started her fizzing. I began to move the sacks. The rat kept dodging behind the next one and the next, but in the end it had to break cover, and the terrier went in for the chop.

It bit her again before she killed it; and it bit her hard, marking her for life. She did something with that rat she had never been known to do: she went back to it and bit its spine after it was dead.

The rat weighed 1¾lb., and measured 18½in. in overall length. And it was a buck; not a female heavy with young.

Anyway, the point is it had a go at me, then a go at the terrier, which is the kind of thing one would expect from a big rat with no escape route. People often speak pejoratively of the rat as 'cowardly'. In fact, brave and cowardly are terms that really have no place in this context, although all of us tend to use them.

What is certain is that a big rat will face man, dog or cat when cornered; and there are rats that will face man, dog or cat when not cornered, or if they even think they're cornered. I knew one that chased my son for 100yds., jumping at the seat of his trousers all the way. And that was simply because we had burst into her breeding nest.

Besides being aggressive and aggressively defensive, the brown rat is also resourceful. It takes a lot to keep a rat out when it wants to be in, or keep it in when it wants to be out. If there's food available it will try everything to get to it. My wife watched one trying to get at a nut dispenser put up for blue-tits. A rat came but couldn't reach it. So it climbed on to the paling, about 18in. away, and waited for the wind to blow the dispenser closer. Then it leaped on to it, stuffed itself with peanuts, and dropped to the ground. It was back the same evening.

Although the rat is wasteful when it comes to damaging or fouling what man wants for himself, it is provident when it comes to looking after itself. When I had my small farm I spent a lot of time watching the few rats I allowed to remain for a time in my poultry house. I worked at night, using infra-red lighting. I would put a heap of whole oats on the floor, then watch the rats ferrying away the ickers to a hole in the wall about 8in. from the floor. They always used the same route.

They did the same with eggs, moving them by nosing them along the run, then hefting them to the hole with nose and fore-feet. They did this often, without ever breaking an egg. It was the only way I ever saw them moving eggs.

Like rats before and since then (like rats now indeed) they showed the 'new object' reaction as it is now called. If I put an upturned bucket on their route they would approach it suspiciously, then go round or over it.

I used to leave the bucket, or other object, for two or three nights then remove it. If the rats had been going round it they would continue to do so for at least 24hrs. after I had removed it. Thus, they were hidebound. But they soon reverted to the old route once the object was away for good.

At the end of my stint I introduced the rats to the terrier. I threw her in one night when there were three rats on the floor. One made the hole in the wall. The next night she killed it. I then plugged the hole in the wall, and that was the end of rats in the henhouse. It was a kind of treason, but I'd given them an extra month of life.

The Antithetical Rat

Some years ago, during a discussion on rats, when everybody said the worst they could say about them, which was about the worst that anybody could say about anything, I made the point that I found the rat an intriguing animal, not without some admirable traits. Incredulity was what I was met with.

Not long afterwards, when asked to sum up the rat in a few words, I said, and wrote, something like: 'The rat is tough, resourceful, cunning and aggressive; bold and shy; wary and reckless; clean and dirty; wasteful and provident; highly adaptable but hidebound; quick to learn and slow to forget'. My opinion hasn't changed. I said, and wrote it; I say and write it now.

A big rat, 'with its wild up', as The Wafer would say, will face man or dog or cat, even when it has some freedom of action. Cornered, it will face anything. And fight. It isn't a nice thing to have a big rat leaping at your face, or hanging on to the seat of your trousers, but you have to admire the beast just the same.

I doubt if anyone who knows anything about rats, apart from just killing them, as I do most of the time, would deny that they are tough, resourceful, cunning and aggressive. But let's look at my group of antitheses.

Bold and shy: Just watch a rat figuring out how to reach a bait in the open, which is a problem I often set them when I'm in studious, not slaughterous, mood. It will fiddle about looking for an undercover, or concealed approach; that's its shyness. But its boldness will keep driving it until it finds a way. At best it will find concealment where there is none; at worst it will make the quick dash out, the quick pick-up, and the quick return. If its suspicions aren't aroused it will work away at a puzzle in the open, when there is no other way.

Clean and dirty: The rat is clean in its personal habits, forever grooming itself. And you'd be hard pushed to find anything more sartorially correct than a rat in a corn stack. It gets dirty in dirty places, as pigs are forced to do. And it spreads the dirt everywhere it goes, until it finds time to clean itself again. Droppings are something else. The rat isn't particular where it leaves them.

Adaptable and hidebound: Its adaptability is legendary, which is why it has conquered the world and most of its habitats. But in behaviour it can be hidebound e.g. skulking warily round a strange object in its path, then going round it less and less warily when it becomes used to it, then continuing to go round it after the object has been removed.

Wasteful and provident: The rat doesn't waste the food it gathers for itself; it doesn't, in fact, waste food at all. But it wastes a lot of food designed for human consumption merely by contact with it: nibbling, treading, fouling, contaminating, or otherwise rendering it unfit for people to eat.

We call it waste when rats make meal of corn ickers; but the corn isn't wasted for the rat. The rat eats it. We lose it. Rats do enormous damage to human foodstuffs, but the loss is all ours. The rats do all right.

Quick to learn and slow to forget: how true this is. Give a rat a fright once, and it will remember for a long time the place of fear. Frightened on its normal runways, it will react by changing them. Caught once in a cage trap and released it won't be caught again. In fact, some rats will become cage-wary by seeing other rats in one.

I kill rats in a variety of ways, and my terrier kills rats in the only way she knows. As far as rats are concerned my terrier has one attitude; as for myself, I can make varying decisions about them. But, because of the damage they do and the threat they can be to health, and for other reasons, I kill them most of the time.

Of Mice and Men

Men have always, more or less, taken mice for granted; women have opted, more or less to be scared of them. And, as far as I can see, Women's Lib hasn't changed that much, if at all.

Like the brown rat and the ship rat the house mouse came from abroad, but it has been here so long—since the pre-Roman Iron Age—that it has acquired full citizenship. Ever since its arrival from the steppe zone of the southern Palaearctic region is has lived largely at the expense of man (a commensal), eating his food, living in his buildings, gnawing his books and his clothing, fouling everything it walked over, and leaving an unpleasant smell.

Yet there has long been an element of whimsy in man's attitude to the mouse, a refreshing ambivalence that is totally lacking in his attitude to the rat. It has figured prominently in poem, story, cartoon and cinema; and Disney made it immortal. We can't get away from the cartoon theme of cat and mouse; a predatory-prey relationship where man again shows his ambivalent attitude—employing the predator while having sympathy with the victim.

Even in more serious affairs the cat and mouse theme creeps in, as when England released political hunger-striking prisoners from Mountjoy jail, so that they could eat at home then be brought back again; and thus gave rise to the Cat and Mouse Act of immortal memory.

Wherever man is the house mouse will try to be. It is most abundant about farms, houses, and warehouses, but it is also found in hedgerows near farms. It was the common mouse of the old-style, old-days, corn stacks, where it throve exceedingly, making meal of the ickers. Such places were the local reserve depot for the outdoor populations.

But it is as a commensal of man that it is best known. When the St Kildans were evacuated, leaving the house mice in house and byre, the field mice moved in, and eventually the house mice disappeared completely. So St Kilda today has field mice but no house mice.

The reverse happened on Lunga of the Treshnish Isles, where there were and are no field mice, but where there were and are house mice. Eighty years after the island became uninhabited the house mice were still there, living the free life afield as their ancestors did on the eastern steppes. Yet when Fraser Darling moved into a bell tent on Lunga in 1937, to do a job of work, the house mice moved in and began work on the stores.

Two conclusions can be drawn from these documented happenings. The first is that when man leaves a house the field mice can take over from the house mouse. The second is that house mice that have lived afield for eighty years, will latch on to man and his goodies as soon as he reappears.

The house mouse is extremely adaptable. It can live in refrigerated warehouses where meat is stored at a temperature of -10 °C. These mice, although living in darkness, and on a strict diet of flesh, have a higher breeding rate than mice in houses with all mod. cons., and produce six or seven litters a year. In its place of origin the house mouse is still found on dry steppe feeding on the seeds of grasses. When it associates with man its habits change, and it develops new types, with longer tails and darker bodies. Such is the type we have with us.

Indoors, in houses or corn stacks, mice breed throughout the year, but there is a winter halt, or cut-back, among mice living afield. In corn stacks, where they had everything, the mice produced as many as ten litters a year. In houses, where they are more disturbed, they might produce five or six litters a year.

Cats and barn owls are the main predators on house mice, except in corn stacks where weasels can do great execution. Weasels will often occupy an infested stack, but it is unusual to find more than one weasel, or a bitch with kits, in a stackyard, regardless of the number of mice. The weasel is territorial and doesn't like to share with its cater-cousins.

High Hopes—For a Vole in One

The pool mirrored blue sky and clouds, and the rusting rowans that crouched on the bank with their backs to the weathery west. Sycamore leaves came sailing down where the water ran deep and glassy under the high bank, and were carried away round the corner, buoyantly birling, to be lost.

The morning was soft, blue and bleary-eyed, but the tractor was out on the high field, with the digger in tow, and the voices of the assembling tattie howkers could be heard clearly at the pool.

I had spoken to some of them as I left the road to walk the burnside, and now I could hear my name being shouted by one to another, from which I gathered that they would rather have been with me than getting back-ache lifting cold, wet, glaury tatties.

I sat down behind a gnurly rowan and lit my pipe, and it wasn't long before Frank arrived. I heard him before he appeared muttering 'frank', as he flapped down to the pool on wings like blankets, with a couple of crows narking swear words in his ear. Every time the heron came down to that flat he had the crows threatening down his snakelength of neck, and sometimes almost striking him in flight.

The heron pitched on the pebbly flat opposite the high bank and stood with the water dancing over his long toes. He gathered in his neck until his dagger beak was bedded on his craw, blinked his jewel eyes, and stood, just stood. He began to look like a grey stump. He was hearing every word of the tattie howkers that I was hearing and paying as little attention as I was now.

The heron stood, and stood, and the digger had turned up four drills before he moved his beak, and that was only to tilt it slightly for a moment, open it a finger thickness, and swallow saliva, or so it seemed.

It was then that the vole appeared from a hole in the bank, upstream from the bird, clear of the pool, about a dozen heron strides away, and crouched down on a sandy peninsula about the size of a saucer, to nibble at green shoots hanging down to eye level. The voices from the field were loud, and the vole stopped frequently as though hearkening, but he stayed where he was and nibbled between listens.

So there I was with the heron twenty yards from me on the water's edge, to my left, and the water vole in front of me, not fifteen feet away.

He was a big vole, an old-man vole, the kind a farmer friend of mine would have described as 'a water rat as big as a cat' when he meant a water vole as big as a big brown rat. All rats, and big voles, become as big as cats when seen through country eyes.

Alas for the poor water vole, alias Water Rat. He is not a rat but he suffers for the likeness such as it is. He is more beaver, more porcupine-like, more coypu even, than rat-like. He is an altogether different beast from big scaley-tail with the long snout, but the diagnosis is usually made after the thunder of the twelvebore, or after the terrier has been sent in to chop. The trouble is, of course, that brown rats often occupy water vole burrows in summer and are frequently seen in the water.

Anyway, this big vole began to patter and paddle downstream.

I knew there would be fun in a minute. For he was headed for the pool where the heron was standing like a lifeless figure on a Japanese screen.

The vole splattered through arm's-length rapids and over slippery stones, reaching the pool close to the high bank where the water swished deep. Fortunately he sailed with it close to the bank, so that he was well out of the heron's reach.

The watching bird unlimbered his neck, a little bit, twice, in readiness, and slowly turned his head to keep track of the vole crossing his front. But he would have needed an extension to his neck to reach it, so all he could do was wait for it to come close enough for a stab. But the vole held to the bank, and turned the corner in the wake of the sycamore leaves.

There was a flicker of the heron's eye and nothing more, and that, I felt sure, was that. But I was wrong.

Presently the vole appeared on my side of the burn, below the heron. Then he began to scurry in bursts towards me—that is towards the heron. It looked as though he meant to give himself up. In fact, he came right on until he was a yard behind the bird. The heron knew he was there, because I noticed the half turn of the dagger beak. But how does a bird, standing in the water, strike behind at a shore-based target?

The heron tried it. There was a great mix up of legs, and a flurry of wings, and a drunken lurch, but the vole saw the first move and was into cover and under the banking while the heron was still trying to disentangle his fankled anatomy.

And that, this time, was that.

Weasel's Weakness Leads
to the Death Trap

A snake-ripple of red fur in seven segments—a weasel with six kits, most of them as big as herself—drew its length across the lane.

It looped round a mossy stone, then disappeared into the dog-roses and creepers in the ditch, whereupon a yellowhammer and a pair of blue tits flipped on to a spray to voice unfavourable comment.

There was no commotion in the ditch, no shaking of vegetation, and for all a man could see there might have been nothing there.

Then from my side of the lane a small, sharp face with dark eyes peered from the drain, withdrew, peered, withdrew, peered, then pushed out on to the ridge of turf on the verge. The whittret was brown, with white waistcoat and white gloves, nose glinting like a ripe blaeberry, and a tag of a tail twitching.

Behind her emerged the other six segments, with some clucking and mouse squeaks. The weasel turned left down the ditch, and the segments fell into line, running in her slots, with rumps up, on their way to the stackyard.

But before they reached it, they disappeared again, under the lane and out of the drain on the opposite side. Now they ran this ditch, having changed sides without crossing the open.

I lost them in the drystane dyke at the stackyard.

During the next few days, however, we saw weasels in twos or threes, running into cover with mice in their jaws, dragging or chasing rats. They were completely oblivious of spectators so long as there was no interference.

After a week or so the group disappeared, and from then on we saw only a single weasel from time to time.

One day my wife did a high jump at the back door when she saw a rat diving into a drain followed by a weasel about the size of a sausage. The weasel emerged on another part of the ground, after negotiating the drainage system, and had the rat in tow: a rat that must have been four times its weight.

Weasels are great subterranean geographers. Being territorial they know every foot of their beat, every mousehole, vole creep and mole run, and they can follow mouse, vole and mole along their own highways.

Surprise a weasel, and it will disappear in a flash into the nearest hole. It will disappear in front of your nose and be out again looking at your back before you can turn around.

But curiosity is their great weakness. Having gained the safety of a hole, or dyke, they must immediately poke their faces out again to see what is going on.

Persistance is another of their characteristics. Put one off a prey, and it will be back to retrieve it almost under your feet.

Both these habits make it an easy beast to destroy.

There are times when you might actually think the weasel went out of its way to get into a trap. And the surest way of luring one there is to bait it with the dead body of another.

The other day I was perched on an old drystane dyke watching a squirrel in a tree, when out of a hole in the crumbling stonework a weasel poked its face.

I moved along to the spot and poked my nose into the hole. The weasel chirp-squeaked and drew back. When I stepped away the face appeared again, the bright eyes taking me all in, curious rather than afraid.

I repeated the gambit several times, with the same result, then walked away. But every now and again, when I peered over my shoulder, there was the weasel on the dyke, or looking from a hole in it, not far behind me.

The beast's curiosity—probably on free rein because it was well fed and had nothing else to do—was driving it on to keep in touch with me.

Anyway, it followed me to the end of the dyke before parting company with me.

Mole as Prey and Predator

The moudie poked its pig-face out of a spoil-heap on the washing green and crumbles of soil trickled down the slopes of the heap. It stuck out its shovel-feet then withdrew below ground just as my Jack Russell rushed to pounce. She dug after it but, of course, hadn't a hope of catching it. She dug out about a cubic foot of Planet Earth. We let her dig because she needed the exercise.

Last year she did kill a mole in exactly the same circumstances. Half the mole was out of the heap when she pounced, and it didn't get back in: the other half came out in the dog's jaws. She killed it but didn't eat it. She played with it. Young predatory mammals do that too. Fox cubs are specialists at it, and I've found several mummified moles at fox dens.

I had a Labrador retriever bitch who was a specialist at scooping surfacing moles from their heaps. But she didn't kill them: she didn't kill anything. She would drop the moudie, lift it and lay it again, and again, but when she was finished with it you'd hardly have guessed it had ever been mouthed.

My Siamese cat managed to catch one out of a heap. He played with it for a while, as cats often do; then he killed it. But he didn't eat it. He ate the other things he caught: voles, mice and young rabbits. But no amount of coaxing would persuade him to eat mole.

I was at a kestrel's nest one day when the cock bird delivered a mole in his beak. I checked it later and found that the bird had killed it all right. But the hen didn't use it. It lay on the nest until it rotted and dried out. Yet the birds were feeding shrews as well as voles to their young. One wonders why a bird that will eat shrew should turn up its beak at mole.

The same thing happened at a buzzard's nest. I checked the mole and found that it was heavily taloned, so the buzzard had killed it. Yet it was never used as food. I wondered just how the bird had killed the mole. Had it been watching moleheaps like my terrier? I think I found the answer on Mull when I watched a buzzard gathering earthworms in a field among moleheaps. A mole poking its face out then would have been easy prey.

Despite these instances, and the fact that I've had foxes, wildcats and polecats that refused to eat mole, I'm sure that they must do so at times as, for example, when the alternatives are mole or hunger. In every case I've mentioned there was never any scarcity of other prey. All one can say is that the mole is not a favoured item with these species.

The mole itself is a specialist predator on earthworms, which it eats in prodigious quantities; but it will eat other soil animals when it comes across them in its burrowing. In captivity it will eat strips of raw beef, with a little oatmeal, and do well enough on such a diet for a time. But earthworms are its main food.

A hungry mole will eat worms by the foot while you watch. And if there's a surplus he will store them, biting to disable them, so that they remain fresh in the cache. I found one such cache in a cold frame in the garden. It was more than a foot down. The worms were crammed together like spaghetti, and would have filled a pint milk bottle.

Round the clock, and throughout the year, is the mole's hunting pattern. He is soon hungry after a complete gorge; a glutton who sleeps between bouts of eating, an eight-hour fast would be the death of him. When you keep a mole you either dig up half the garden to keep him going on worms, or you switch to raw meat and oatmeal. I'd rather purvey for eight weasel kits than for one adult mole.

Nearly everybody knows the mole on sight, yet nearly everybody hasn't seen a wild one alive. This is natural enough, for the beast spends most of its life underground and is seldom seen above ground by day. Mole heaps are something else; they don't go away. So most people walking in the country become familiar with them and know whose they are.

The popular belief is that the mole is blind. It isn't. It has pinhead eyes and it can see with them; but the beast might as well be blind for all the use they are to it in the kind of life it leads. If you part the fur on a mole's face you will see the dark little dots of its eyes without any difficulty.

No Respite for the Shrew

A shrew's life is a hectic life, round the year and round the clock, in the leaf litter in summer and under the snow in winter, a bustling, nervous, hither and thither scramble, with orgies of gorging between bouts of sleeping.

A few hours in the shrew's day can mean the difference between life and death; if they are sleeping hours or feeding hours the shrew lives; if they are waking hours or sleeping hours without food the shrew is dead. The short fast ends up in the long sleep. The shrew must eat every few hours or die.

When the hedgehog and the bats are in the death-like sleep of hibernation, and the badgers are lying in for the weekend, and the squirrels are too lazy to stir, the shrews are active, hounded by a metabolism that leaves no margin for over-sleeping or poor hunting following on bad weather or disablement. Only the fit survive and only if they eat their own body-weight or more of food each 24hrs.

All the shrews are small. The common species weighs about half an ounce; the pygmy is smaller still, maybe a fifth of an ounce or less. The water shrew is the biggest, and that isn't very big, perhaps body of 3½in. and 2½in. of tail. And that is a giant.

Shrews aren't blind. They have eyes, pinhead eyes, big enough to be seen and show a highlight, but not much use for seeing with. It is highly unlikely that the shrew ever uses them for seeing with, or ever sees anything with them.

If you watch a shrew rummaging in the ground litter, which is it's way of hunting, you'll see it probing with its sensitive, flexible, mobile, trunk-like snout. This wiggles and feels around like a finger. The shrew doesn't seem to recognise a prey until it literally falls over it, or makes contact with it. It will walk past a prey two inches to one side unless it hears it wriggling.

It has been suggested that perhaps they find food by echo location as bats do, and as dolphins do. Shrews are noisy at all times, but they may utter ultra-sonic sounds that reflect back and guide the shrew by echoes.

The water shrew lives by water, on burn and river banks, and it needs clean water. It is an expert swimmer and diver, and can prowl the bottom turning over pebbles with its forepaws. It has swimming hairs on its feet and uses its tail as an oar. Underwater, it can look silvery because of air bubbles trapped in its fur.

On the surface the water shrew will swim after floating insects, or come up underneath them and swallow them in the way he himself is sometimes taken by the pike. Underwater, the water shrew swims like a newt. Ashore it can bound like a ferret, and there it will eat much the same prey as the common shrew—earthworms, spiders, woodlice and other small creatures.

Common shrews and pygmies hunt the ground litter, and can be seen moving under it, heaving it as they go. When either finds an earthworm it bites it along the back, pricking it a hundred times and disabling it. The shrew usually drags the worm into cover where it begins to chew it, making a great pig-in-a-trough noise as it does so.

It is now known that the saliva of shrews contains a poison that can disable very small mammals if given in big doses. This may explain its ability to tackle prey bigger than itself. But it is harmless to man, and a shrew bite is less noticeable and less important than the prick of a two-day chin-stubble.

Shrews are common enough, and any ear that can hear will hear them in the grass or hedgebottom if it sits down and pays heed. You will soon hear the squeaks and mini-shrieks and the shouting of odds, and it isn't all that difficult to locate the shrews themselves. They are easy enough to catch, and quite harmless to handle.

Shrews have a powerful musk gland that leaves a distinctive and unpleasant smell. You will smell a shrew in the house quicker than you'll smell a dog, however dirty. As a result, apparently, few mammal predators eat shrew unless the alternatives are shrew or nothing. But hawks and owls eat them. Weasels and foxes will do it in emergency. So will cats, but cats usually bock them up again. So the stink isn't much of a defence. What is the use of it if it prevents you from being eaten but not from being killed?

The White Thing

Overhead, the moon was in a clearway of thrush-egg blue, but the sun was having a struggle breaking clear of the muffle of bruised clouds at my back. I had already been sitting under a big beech on the bank of the burn for half an hour, waiting for the sun to break out and beam some colour on to the steel and pewter water of the burn.

It came out in the end, reddening the sky, but it was the usual false promise, another of the too-bright-too-early mornings that end up in a sunless day, usually with rain.

But at the moment the light was welcome, for I could now see the brown hares grazing near the pool, and a cock pheasant parading with a hen and halfgrown poults. Then there was the roebuck downstream, with his fore-hooves in the water, while he rubbed his dirks in the low branches of a willow. There were crows and curlews, and a flock of starlings to my right.

And there was the brown rat—a big one—who appeared suddenly in front of me, straight out of the water. He didn't see me. His fur was sleeked with his swim, and he took time to shake himself like a dog, before running along the bank with his rump up, and his scaley tail following like a slow-worm.

He turned back to the water a few yards downstream, swam across, then began to prospect along the opposite bank under the beeches. Presently he disappeared into a water-vole burrow.

Now a vole is a vole and a rat is a rat, and there should never be any doubt about which is which. The rat with his sharp face, trumpet ears and long, naked tail bears little real resemblance to the water-vole, with his porcupine face, musquash coat, short ears and furred tail. But here was the classic situation, of a rat on water-vole territory for the summer, masquerading as a vole, thus causing confusion among the uninitiated.

Well, he had hardly disappeared when two water-voles popped out of the ground almost under my nose, and began fighting. They would grapple, then draw away to sit up and threaten each other, then fly to the clinch again. They rolled over and over, making plenty of vocal noise the while, and the racket was enough to make the roebuck throw up his head to listen.

After a few minutes one vole dived into the water and swam downstream. The other drew back to the burrow from which it had emerged and began to nibble the grass beside it. All round the hole, the herbage was close-cropped; voles like to eat with their back half still in the burrow.

The grazing vole ate, then withdrew, ate, then withdrew, and eventually withdrew completely.

Then the rat appeared again, coming almost from behind me, between me and the water. And as I froze to watch him, I became aware of a brief flash of white in the grass forward on my left, about twenty feet away. And that was where the rat was headed.

While he was working his way forward-leftward, the water-vole appeared at the burrow mouth again, but I was now more interested in the rat. I could see no more white flashes, but there was something there, and I just couldn't figure what

228

it was. After all, how many white things are there going around?

I was just reckoning that the rat was going to flush out the white thing at any moment, when the white thing sprang from cover right on top of the rat and there was a scuffle, and the rat was squealing, but not for long, and the white one was suddenly standing there with the limp thing in his jaws.

'Skipper!' I called, and he put up his tail, but he wouldn't come to me in case I might take his rat from him. I followed him home, knowing I would get the rat when he deposited it in the garage. The white thing was, of course, my red-point Siamese cat. But the cat that would eat his rat would be my wild-cat Teuchter.

Why I Believe in Murphy's Law

One of the laws to which there seems to be no exception is Murphy's Law: if it can happen it will happen. So-called nuclear accidents confirm this. My three-ounce tame bitch weasel, Teen, made the same point a wheen years ago.

STV wanted her on a live programme—a programme going out live, that is. I would rather have had her as a filmed insert, not because she would be any problem to me—her tameness was known to one and all as they say—but because I am a great believer in Murphy's Law.

We agreed that I would bring her in a big glass-fronted box, in which she would have plenty of room, and into which the camera could pry with ease. It was the editor, I think, who remarked that if she was kept in the box nothing could go wrong. The famous words . . .

Well, I was seated in the studio off-centre, with the big showcase at right angles to my feet, and Teen was happy chasing my finger up, down, and along the glass front. My misgivings, I began to think were unfounded. Then somebody wanted to speak to me for a moment. I walked maybe ten feet to be spoken to. And somebody put their foot through the glass front.

That was Teen out. And there was panic (weasels being what they are) and a lot of people discovered that they could perform acrobatic feats they had never dreamed of.

Of course, Teen got into a panic too. But I managed to grab her before she escaped into the studio hinterland, and very naturally she bit me before I could pacify her. By the time she was pacified the blood was running down the back and front of my hand from twin pairs of needle holes in the tip of my right forefinger.

Murphy's Law took time off from there on in. The programme went on, and everybody thought what a wee stotter she was. I was glad to get her back to the car, but getting home was something of a pantomime. I had to put her in my side pocket, but she kept poking her face out, and I kept pushing it back. I daren't leave her loose in case she got among my feet, and maybe got trampled to death.

Syne I reached the village near where I lived, by which time I was feeling that a drop of the malt was a good idea, so I parked the car and entered the small hostelry, with my right hand in my right pocket to restrain my three ounces of bitch weasel. The joint, as they say in polite circles, was stowed. And I was just deciding I had better leave without the malt when this wee man with one eye hailed me:

'Haw, Davie! I seen ye on the telly wi' your wheesel. Ha'e ye got it awa hame?'

I said: 'No. I hae her in my pooch.' And I withdrew my hand from my pocket and put Teen on the bar. And almost cleared the pub.

Now there were only three of us at the bar: Teen, the wee man, and myself. I made my apologies to the boys and reached to pick Teen up, but the wee man said: 'Leave her be. She's daein fine.' Then he proceeded to knock off the untended drams to right and left. And nobody said a word to him.

I've never been able to understand this terrible fear of small weasels that is as rife today as it was when I was a boy. The beast shares with the wolf a reputation that is in no way related to the facts.

Teen's mate was a big nine-ounce weasel named Tammas, and they lived together in a small upstairs room specially fitted out for them. Mind you, they often came downstairs to play and chase each other, much to the alarm of visitors and to the disgust of the dogs, who didn't like their pads nipped.

One day my wife said to me: 'I wonder what all these people on the road out there are getting to stare at?' I went outside to find out. Tammas and Teen were entertaining them doing their famous curtain climbing act. A man said to me: 'Thae twa should be on television.' I could have told him . . .

A young student called Carolyn King, doing a doctorate on weasels at Oxford, wrote to ask if she could come to meet my weasels, and I said yes. She came, and stayed two days, spending most of the time sitting in the wee room upstairs with Tammas and Teen running all over her. She was charmed by them, and they got a mention in her thesis.

Those Hybrids That Never Existed

Nowadays, the term 'hybrid' means something to everybody. The gardener, the poultryman, and the cage-bird fancier are all familiar with it. There are many forms of hybrid; there are thousands of hybrid strains of this and that; but this article is not about them. It is about the hybrids that never were . . .

The question of hybrids between this and that often crops up. It met me again the other day when I was talking to a business man, who had just met someone, who had told him . . . Such conversations usually begin something like this:

'Have you ever seen a cross between a rat and a rabbit? I was speaking to a man . . .'

I say 'Um,' or variants of it, and listen. I learn that at such and such a place there are rabbits (scarce now, but still about) that are half-rat. There's no question about it. The man had almost seen the mating take place.

As a matter of fact, I am told, you could spot them at a glance—those ratty rabbits; and see the rats running in and out of the rabbit burrows.

The real fact is that rats and rabbits don't interbreed, anywhere. Some rabbits become lean; some that live in dirty places become dirty. But they don't interbreed with rats.

Rats running from rabbit burrows means no more than that rats are occupying the burrows. In summer, rats frequently do so, and breed there, with their own kind.

It is very strange that one hears a lot about ratty rabbits, but never anything about rabbity rats!

You will hear, quite frequently, the same story about foxes and dogs. I often meet a man who knows a man who met a man who has seen a genuine dog-fox hybrid. But I've never met anyone who has actually seen one, or anyone who can tell me where to see one. And I have never seen one myself.

Not that I haven't tried to produce one. I have. But while I've had dogs and foxes become the best of friends, both sexes with both, I've never seen anything approaching an attempt at mating.

Yet, anatomically and physiologically, there is nothing to prevent such a mating. Dogs will mate with wolves, either way. But nobody has produced a dog-fox hybrid that has been proved.

The wildcat will mate with the domestic cat under certain circumstances. Lion and tiger can be crossed, and have been crossed, but the hybrids are sterile. The cross is either a Tigon or a Liger, depending on how the cross was made.

I once received a water vole from a correspondent. He described it as a cross between a water vole and a mole. The beast had the velvety coat of the mole, and was quite black. But it came from an area where black water voles with glossy coats are the rule. It was not a hybrid; it was a pure vole of a geographical type.

Hybrids do occur in nature. I have already mentioned the wildcat. Among certain game birds, often in predictable circumstances, genuine hybrids occur.

A few years ago I ran a cock pheasant with some bantams, and set some of the eggs. They were fertile, and I reared the hybrids. I have since seen crosses between a ring-neck pheasant cock and Rhode Island Red bantam hens.

The mule is a hybrid. (Any interspecific cross is usually referred to as a 'mule', and the term is common among cage-bird fanciers.) But the mule in the best-known sense is a cross between the Spanish jack donkey and the horse mare. The cross works the other way round, too, but has never been popular, and so is not produced.

Carrion crow and hoodie crow interbreed freely, and produce fertile offspring. But these birds are not really separate species; they are rather geographical races of the same species.

But in any consideration of the problem of strange matings one has to be very clear about one thing. It is that the mating urge can drive an animal to attempt the movements without mating at all.

Dogs come readily to mind in this context. I had a tame roebuck who pestered my dogs, my goats and myself in August. I have known an old roebuck, a wild one, who harried sheep and heifers in the same way. So it isn't always right to say people have not seen what they say they have seen; it is simply that they haven't seen as much as they thought, or would like to believe.

Henry Tegner, who has stalked roe all over Europe, and knows the roe deer as well as any man in this country, once wrote very strongly about a farmer who had seen a buck trying to cover a ewe. The ewe's family were supposed to have been hybrids, and Tegner rightly ridiculed this. But he was wrong to insist that the farmer had not even seen an attempt at mating. He probably had. I have seen this ritual myself.

The Pine Marten and the Mull Polecat

Behold the pine marten—the cat-size tree weasel, with fox face, bushy tail and throat patch of creamy white or orange! See him on a freezing morning, after a snowstorm has blanketed the hills, when the white hares and ptarmigan are down, and the red grouse are running in dozens on the ruffled, roadside snow.

Small birds are scolding in the trees, in one of which is the marten, alert, with head up, tail down, and rump arched. He leaps from branch to branch, climbs down the trunk backwards, and bounds across the snow with bushed tail twitching, slotting his footprints deep for any man to follow.

That's as much as you're likely to see of the pine marten, even when you know what you're doing and where to look. Yet if you wait long enough, and often enough, overlooking a known trail in the ancient forest, at near sunset, you may prove to yourself that interception is an art of the possible, and see him as he curves round the base of a tree, snake-like, pale-fronted and bright-eyed, stalking low through the glare and shadow. But give him a whiff of your scent and he'll vanish as though blasted from the spot.

Although a haunter of mature coniferous forest, the pine marten has shown itself to be extremely adaptable. When the felling of the forests, and intensive game preservation, forced it into the wildest country of rock and scree, often treeless, it contrived to survive into the twentieth century, when tree planting on a big scale helped to tip the balance in its favour again.

Now it is turning up in many areas in the Highlands far from its strongholds in Wester Ross, and has been reported in Ayrshire. Border country martens are almost certainly beasts from the North of England, itself colonised from the Lake District.

Male martens are bigger than females, measuring about 2½ft. in length, of which about 10in. or so is tail. But bigger males have been recorded. Weight ranges from 2lb. – 3½lb., according to sex, age and time of year. Colour varies from chestnut-brown to near black, with pale edges to the ears, and a throat patch of creamy white. This patch is of variable size.

Like the stoat the marten breeds once a year, and there is delayed implantation. Mating is in July and August, and the young—average litter is three—are born in late March or April of the following year. Implantation is delayed until mid-January. Almost all that's known about the marten's breeding is from studies of animals in captivity.

Martens prey on small rodents and small birds, the field vole being the commonest mammal prey. Small hares and rabbits are also killed. But the prey range is wide and catholic, and includes beetles, caterpillars, cocoons, deer carrion, wasp grubs, fish, and berries of many kinds.

Man is the main predator on the marten in this country, and the beast is still caught in illegal gin traps set for foxes. Foxes have been known to kill martens, and eat them, but it is unlikely that such predation amounts to much.

Pine martens will rest in, den up in, and even breed in the old nests of birds of prey, including the eyrie of the golden eagle. They will also store food in such places. Like so many other animals the marten sets scent on rocks, trees, stumps

232

and branches, and this probably performs the double function of marking territory and bringing the sexes together.

The polecat, often called the foulmarten or foumart, is officially extinct in Scotland, although there are plenty of feral polecat-ferrets around that closely resemble the real thing. Such animals have been established on the island of Mull for a very long time, and some of these are really indistinguishable from the genuine polecat. In addition, it is known that real polecats have been released in parts of Scotland in recent years. So you could be right or wrong in your identification, just as the animal could be right or wrong.

Polecats measure about 23in. overall, of which 7in. or so is tail. Weight is about 2¾lb. In the right polecat the outer coat is long and dark, with a purplish sheen, and parts readily to reveal the buff underfur. The ears are rounded, with white edges. There is a mask of dark fur enclosing the eyes and extending over the muzzle. This mask is framed in pale fur, almost white, giving the impression that the animal is wearing spectacles. Lots of Mull polecats conform to this specification.

Anyway, right polecat or wrong polecat, the argument becomes academic if one happens to break into your henhouse. Real or ersatz, the result is the same.

Hedgehogs in Hibernation

Hibernation is much more than sleep, and hibernation from November to March is more than a long sleep. If a hedgehog were to fall asleep at the beginning of November, and simply sleep on in the ordinary way of sleep, it would quickly waste away and would never wake up.

But, of course, it cannot do this any more than you or I could, unless it was concussed or suffering from some other brain injury. It would die if not nourished in some way.

Hibernation is a bodily state that enables the hedgehog to remain 'asleep' for four or five calender months, and the state is brought about by a combination of low temperature and physiological readiness in which insulin plays a large part.

The hedgehog is a warm-blooded animal that can hold its temperature steady under varying conditions of heat and cold, although it overheats at a temperature of 34 °C. and becomes distressed, just as it does when subjected to extreme cold, below zero, while awake.

When hibernating the hedgehog relinquishes its thermostatic control and all its bodily activities slow down—its respiration, heartbeat and expenditure of energy. It hibernates in a place insulated against the outside cold, well wrapped up in blankets of leaves and grass. The critical temperature for the hedgehog has been shown to be between 17 and 15 °C. in the immediate surroundings, which means that the outside temperature may be as low as 10 °C.

The hibernating hedgehog sleeps in a tight ball, not relaxed. If a thermometer is pushed into the middle of the ball it will register the temperature of the surroundings. Breathing is slow and regular, as against 15 or 20 breaths per minute in ordinary summer sleep, and there are prolonged periods when there is no obvious breathing at all. Although extreme cold will rouse a hibernating hedgehog, causing it to warm up to compensate, the awakening in spring comes about by warmth.

The process is slow, with much snuffling and nose-running, and quick breathing, and it takes some hours for complete arousal, and before the animal can walk steadily and confidently. It is then hungry, and thirsty, and I have noticed with my own animals that fluid was as much wanted as food at the start of activity.

When the hedgehog emerges from its hibernaculum it often trails some of its bedding out with it, leaves, grass and fibre. This material is gathered by the animal in its mouth when the hibernating nest is being made up, not carried in on the quills. The clinging of such material to the quills at awakening is accidental. It needs only a little direct observation to make these things clear. The hedgehog no more carries nesting material on its back than it does apples.

From time to time, during hibernation, the hedgehog will exhale air in great snorts. It will also do so when touched during hibernation. This is a sound that can be heard, even in a wild animal, if it is not too deeply buried, and I once found a hibernating hedgehog by sitting on the tree root under which it was sleeping.

The funniest incident I can remember about a hibernating hedgehog was when a rook was confronted by one. The bird had been dibbling and leaf-turning along a banking on the edge of the rookery when, suddenly, a ball of grass began to roll clear. It rolled a few feet, sending the bird flapping up with a startled *caw*. Then the grass ball came apart and a hedgehog sprawled out.

The rook had obviously arrived at the moment of awakening, and perhaps expedited it by accidentally pecking the padding over the hedgehog. Or the bird may even have heard snoring and pushed in a prospecting beak.

It used to be quite a usual thing in my neighbourhood (when frogs were more common) to find hedgehogs eating frogs in the early days after awakening. The frogs were available at the right time, and in large numbers. I once found a hedgehog asleep with a frog clamped in its jaws, and another eating a frog that was still alive, and eating it from a hindfoot forward at that.

Birth Control and the Rabbit

To breed like rabbits is a pejorative term when applied to people. One doesn't hear it so much nowadays, but it was common enough in the grim 'thirties, when it was mainly directed at such social units as coal miners, immigrant Irish, and the poor in receipt of parish relief, and mostly used by the comfortably off with a small family and supposed economic security. I believe it is not unheard of in other parts of the kingdom, in other contexts, today.

The term is meant to suggest indecent fecundity and an almost non-stop breeding cycle. Presumably those who made such accusations against their fellow citizens expected the human lagomorphs to practise some form of birth control. The joke is that if the human-rabbits had done so they would have been acting more like the real rabbits, and less like people.

It is a matter of biological fact that rabbits practise birth control—not by decision, or with knowledge of it, but in response to the existing environmental conditions, which include the size of the rabbit population at the time. It is also a fact that rabbits do not breed throughout the year; there is a peak season and an off season.

Birth control in the rabbit is a built-in mechanism, which operates with greatest effect when rabbit numbers are high, and with least effect when they are low. In other words, the control varies according to the density of the rabbit population on the ground. It is, therefore, a sort of stabiliser.

What happens is that a percentage of litters is lost in the uterus. Does conceive, and the young develop, but instead of going on to full time, and being born, they die in the uterus and are reabsorbed into the mother's tissues. Reabsorption begins about the twelfth day of pregnancy and is complete in about two days.

Before myxomatosis Professor Brambell established that about 60 per cent of all litters conceived were reabsorbed. Despite such losses the rabbit population remained high before myxomatosis, and between 60 million and 100 million a year were killed for the market.

In the rabbit there is a well defined breeding season from January to June, but there is sporadic breeding in all other months of the year. In the period January to June there is a pregnancy rate of about 95 per cent.

From all this it is quite obvious that there must be few, if any, human beings that breed like rabbits, in the sense of restricting their breeding season to six months, and suffering intra-uterine mortality of 60-plus per cent.

In any case, in total output of young the rabbit is a low producer compared with many other animals. A queen bee produces thousands of young. A queen wasp is the mother of every wasp in the bike. A single female frog produces a regiment of tadpoles. A hen salmon lays more eggs at spawning time than a dozen high grade fowls together in two years; an 18lb. fish will lay close on 8000 eggs. The rabbit's average of five young per litter is a tiddle by comparison.

In the brown hare, as in the rabbit, there is considerable intra-uterine loss, especially in the autumn, and this has been reckoned as high as 25 per cent by Kolosov for the hares of the Caucasus. Hares give birth to three or four litters in a season, but litter-size varies with the time of the year; spring litters are the biggest on average and winter litters the smallest. Young hares do not breed until eight months old.

There is also the phenomenon in the hare, the reverse of reabsorption. It is superfoetation. It is not known how often, or in what circumstances, this occurs; some zoologists consider it normal, others consider it exceptional and unimportant in terms of population. Superfoetation means becoming pregnant while already pregnant, conceiving a second litter before the first has been born.

It is a remarkable fact that Pliny the Elder, born A.D. 23, knew of superfoetation in the hare. In his fascinating, and usually fanciful, *Natural History* he wrote:

'The hare, which is born to be all creatures' prey, is the only animal beside the shaggy-footed rabbit that practises superfoetation, rearing one leveret while at the same time carrying in the womb another clothed with hair and another bald and another still in embryo . . .'

Maybe he was overloading the conveyor belt, but he knew it was there, and it was a long time being rediscovered.

Daylight Aerobatics

On the second day of March, with the afternoon sun shining and the snowdrifts crusting in the frost, there was a wavering flyer among the trees, like maybe an avian drunk, but when it came butterflying into the open, I could see it was a bauckie bird, about the length of my hand across the wings, looking black against the snow until it caught the sun when it became bank-vole brown.

The coldest night of the year, they had said it was the night before, yet here was a pipistrelle out in daylight, in full sun indeed, when icicles were hanging from the roadside dyke and the glass was on the stones in the burn.

It takes a man a moment or two to collect his wits in such a situation, although I had seen bauckies out in snow before, and recall especially the two or three my cousin and I watched at Loch Tulla on a February day, when Rannoch was blanketed and the drifts were six feet deep.

On that occasion there was insect prey to catch and the bats were snatching them from the clouds above the drifts. They were in view for a minute or two then all of them disappeared.

Today there wasn't a sign of anything likely to attract a bauckie bird, although I suddenly remembered that my wife had found a bluebottle wandering around the kitchen sink a few hours earlier. So maybe there were other things out.

The waverer did a Links-of-Forth flight along the top of the hedge, fluttered out to the middle of the road, wavered back towards the hedge, dipped and touched the snow, making a little puff, then came right at me and lifted drunkenly over my head—a fine carry-on for a snowy March afternoon.

I wheeled, almost falling, to keep the bauckie in view. It was now on the other side of the road, but it crossed to my side, and again lifted over my head.

Its echo-sounding equipment must be deeving it, I thought, if the waves were stotting back off me, yet there it was, almost hitting me before taking avoiding action. Maybe the racket was so much like an orchestra that it couldn't separate one sound from another.

After a second close look at my not inconsiderable thatch of hair the pipistrelle stachered away over the hedge, and kept going towards the old toom hoose on the brae-face. I watched it do a couple of turns round the front, then it disappeared. And at that point I went visiting where nobody lives any more.

I searched every square foot of the old house, which smelled of damp and decay despite the frost and snow. I knocked down flaking plaster and started sneezing when I disturbed the old lime and rubble, but no sign of bat or bats could I find. I came to the conclusion that my winter sport was up the chimney, which would be well rain-washed and free of soot.

The old house is near the place where I used to live, and in the old days we were accustomed to watching the bauckie birds from spring to autumn, but I had never seen one so early in March, even without the snow.

In those days I had a very nice bird bath in the garden, and I've seen the bats on a summer night coming down for a drink. I've even watched one in the bath, apparently having a swim, certainly a paddle, because there's no doubt at all that the act was deliberate. But try as we might we could never catch one at the water. Only twice in over nine years did we get one indoors, one in the sittingroom which managed to get out again, and one in the front porch which injured itself when it was grounded.

When we had the tawny owl, he used to go for 'walks' with us after dusk, and many a time he amused himself hunting the bauckie birds, although we never knew him to catch one. But tawny owls do catch them, and the birds that nested down by the Luggie used to bring one occasionally to the nest.

My snow bauckie, like the ones at Tulla, had been flying mute (not calling that is; its echo-sounding system would be working) yet the pipistrelle is one of the most vocal of bats under normal circumstances, as anyone with ears to hear can hear.

Her Place Was on the Fist of a King

Time was when the place of the peregrine falcon was on the fist of a king: herself one of the lordly ones, she was the falcon of the high and the mighty, and woe betide the common man who tinkered with the royal birds.

Falconers of old, not to mention one William Shakespeare, almost always referred to the peregrine as She, because the female was the one most in demand. Bigger and more powerful than her opposite number, she was the Falcon; the lesser bird, the dashing male, was the Tiercel. The names still stand today. In the literature of falconry the peregrine has pride of place.

But the heyday of the lordly birds came to an end. The nineteenth century saw them reduced to the status of vermin. The new generation of sportsmen had no use for the peregrine; they wanted to fill game bags as a coalman fills coal bags. So they killed everything with a hooked beak on principle.

The peregrine's place was no longer on the fist of royalty; she was more likely to be found nailed by the neck to some keeper's vermin board. The nineteenth and twentieth centuries were the killing time for falcons, and it seemed at one time that the peregrine must go the way of the kite and the osprey and the goshawk. Instead, it contrived to hold on, mainly because of protection in the Deer Forests.

Then, in 1954, the birds were given total protection by law, at all times, from all people, in all places. There are no exceptions. No one has any right to kill a peregrine today.

What the law says is one thing; how people obey it is another. There are still a lot of people who destroy peregrines as a routine; there are others who scrupulously observe the law. At the same time there has been a slight revival of the ancient art of falconry, and permission is granted to bona fide falconers to take young birds from the nest for training.

Long, long ago I got around to calling the peregrine the 'gentleman in blue', a fit title for this blue-plumed, ermine-trimmed marauder. The late Mortimer Batten coined the name, I think; at least he was the first man I heard using it. It is pleasant to hear boys who have walked the hill with me calling the peregrine by this title.

The peregrine is a brilliant flyer, in my view the king of all the falcons, fit to top sixty miles an hour in level flight, and to double that speed when hurtling down on his quarry from the sky. The stoop—the high speed power-dive—is the most spectacular assault made by any bird of prey.

I can recall the morning in Argyll when a peregrine tiercel struck a cuckoo out of the air above the corrie. The peregrine struck, sheered away, then threw up at dizzy speed while the cuckoo fell to earth a twisted wreckage of feathers.

And there was the morning of the hoodie crows. The greybacks were loitering about the corrie where the falcons had their nest, when down came herself in her power-dive—a feathered missile, a rushing thunderbolt of feathers, and thud! a crow crumpled and fell, struck dead from the grand stoop on a snell morning and not recovered.

But the peregrine doesn't always kill in this way. She may bind on her prey in the air, if it is small enough, and I have seen this happen when the quarry was a wheatear. She will take the prey from the ground from time to time, prey like young rabbits and waders. And she may pursue prey on the ground as in the case of a grouse hiding in deep heather. It is amusing to watch a falcon walking about flat-footed trying to stampede a grouse from cover.

Peregrines kill a variety of food, and take whatever is readiest to hand. Birds on the coast kill seabirds to a great extent; others kill a lot of jackdaws and rooks; others kill a lot of pigeons. On a grouse moor they take grouse—as they have been doing for thousands of years, before man ever got around to messing up the grouse and everything else by greed and mismanagement.

On the coast the peregrine nests on rock ledges, using no nesting material; or she may choose the old nest of a raven. Inland, the birds use the old nests of raven or buzzard, or may lay on the bare rock of a ledge.

Peregrines are noisy when their nesting territory is invaded by man, and that is when you will see some wonderful flying exhibitions.

When the chicks are small the tiercel provides all the food. Later on, both birds hunt and carry food to the nest.

A Poor Man's Ode
to a Six-Letter Word

Every British bird of prey is protected by law, which doesn't prevent every British bird of prey from being killed by somebody, somewhere at some time, and of course there was the man who shot an osprey because he thought it was a gull, or something.

But the great frustration must be for the man who gives a bird of prey sanctuary on his side of the fence only to see it blasted by his neighbour on the other.

It happens more often than you might think, because there are still a lot of people, a very great number indeed, who think that the law in favour of hooked beaks should be a law against hooked beaks. Such people make up their own laws as they go along. They are the faceless outlaws, for whom the law so often retains a surprising near-affection.

In their attempted rearrangement of immemorially established wildlife relationships, these people use the word 'vermin' to describe anything they don't like.

In their canon the following are 'vermin'—eagle, peregrine, buzzard, harrier, merlin, kestrel, sparrowhawk, owl, raven, crow, gull, heron, goosander, cormorant, fox, wildcat, marten, stoat, weasel, otter and badger.

Because of this concept which they believe as others believe in God, these people will flout every protection law, break every section of the law, and yet still look upon themselves as pillars of law and order. This dualism never ceases to amaze me.

Fortunately, there are keepers, foresters and farmers up and down the country who have rejected or never accepted the 'vermin' concept. They are the leaven in this mass, and they are not faceless men. They say their piece openly.

Some of them, like myself, use the word 'vermin' all the time in a spirit of ridicule, in mental quotation marks, in the hope that it will become a word to amuse, and the concept kidded out of existence.

Thus I said to the head forester:

'You'll soon be up to the neck in vermin if you don't do something about all these harriers, and buzzards, and kestrels, and things.'

And in the true spirit of him who catches on, he agreed—in quotation marks. So every harrier and buzzard and kestrel was as safe as tomorrow.

Each time we looked round all the 'vermin' characters we made jokes on the subject. Others joined in and it was nice to hear the six-letter word used so often by people who wouldn't have let anyone put a finger on harrier, buzzard, kestrel or sparrowhawk. I offer the idea as a way of combating the concept in the future.

We ended up trying to make ridiculous little rhymes about the 'vermin', even in other languages, and one day I found myself scraping the barrel with mongrelese like this on the harrier:

When first we saw der vermin hawken
Der vermin hen was hard a-clockin'.
Den in cam der vermin cocken
To deeve us wi his craiks an croakin'.
Der vermin cocken kills der burdens
For to feed his clockin' hennin.
Der vermin young wans soon is hatchin'
And den der cocken goes a-butchin'.
Der vermin hennin feeds der young yins
Wi' der prey der cocken brings in.
Dem harriers a' is vermin hawken
Killin' lots o' grousen chicken
An murderin' a' dem singing burden
For to feed der vermin cleckin'.
Der keepers canna sleep ein winken
Thinkin' o' dem vermin stinken.

Swallow Tale

The mist was so thick this morning that we were beginning to wish we were bats, using echo-location to find each other, and envying the geese and the gulls flying above it all, across the wan disc of the morning sun towards a loch they couldn't see. And there was a lark up there too, with no view of the ground but maybe one of heaven's gate.

'It's X-ray eyes it'll need tae get back doon,' one said.

'All it has to do is fold its wings, and drop, with a delayed opening,' I said. 'Gravity and all that.'

And we went on to talk of migrant birds, and how they found their way, and I was saying how they navigated by the astral bodies and all that. And the man said: 'They'd be pushed tae see ony astral bodies in that soup.' Well, it led, as it always does, to someone saying the swallows'll soon be here; and so they will—round about the twentieth of the month, going by past experience. They arrive more predictably than they leave.

When I had my farm, I always began watching for the swallows on the eighteenth and the wife of one of my farm neighbours, who had, and has, a great interest in the comings and goings of swallows, always phoned me when hers arrived.

Many a year I used to phone her on the seventeenth or the eighteenth to say my birds had arrived, and she would come traipsing down to see them. And, of course, they hadn't.

Over the years we broke even in having swallows first, the nineteenth being the usual date of arrival. She had a lot more swallows than my five pairs, but I knew mine better because I ringed them. This is how I knew that one of my home-bred birds had made the journey to South Africa and back five times.

I checked the birds when they arrived, climbing the rafters in the dark, and blinding them with a torch so that I could catch them up and look at their legs.

Ringing helped in another way. It enabled me to prove to myself that the birds came back not only to the same place, but to the same building, and the same rafter. Even to the same spot on the same rafter.

I had no way of finding out how my five-times-back bird fared in 1958 or subsequently, because my place was pulled down to make way for a factory. Anyway, up till then, she had created no record, because I recall reading about one that had made the journey nine times and was killed by a cat in the end.

Swallows show great persistence in getting into the building they know. If you shut a door, as a friend of mine once did, they'll squeeze under it if there's an inch of a space. And that's the way they'll get in when they're feeding young. Last year we had a pair in a new house, in which lived a new transformer, and their way in and out was via the mouse-size hole below the door.

A farmer friend of mine, who had just bought a braw, new, expensive car, shut his garage when the swallows arrived. They weren't going to be allowed to polka-dot his new machine. But he discovered that they'd squeezed in at ground level and had begun a nest. So he nailed sacking under the rafters and opened the door for them. His comment: you can't win. But, of course, like most of us, he had a soft spot for swallows.

Departure time varies from year to year, much influenced by weather and temperature. I've had birds feeding young the last week in September. My farm wife friend had them feeding young in the second week of October, and the pair stayed on into November, long after all the others had gone. We thought they might leave their young in the nest when the frosts came. But they waited on.

'That shortens the winter a wee bit,' was my friend's comment.

The Biggest Grouse in the World

The forest is dark at half past two in the morning, and a fine smir of rain, like gossamer, cobwebs one's clothing and makes the eyebrows itch. Here, in olden times, marched the Marquis of Montrose and Viscount Dundee.

But I am not out at that hour seeking the ghosts of the two Grahams, Great Marquis, or history's Bloody Clavers.

I am out seeking a great bird that disappeared from Scotland the following century, in 1760 or thereabouts, and which remained extinct until 1837, when it was successfully reintroduced from Sweden; the capercaillie.

Now, there's a broth of a bird for you, the *cabhar-coille* of the Gael, the biggest bird of its kind in the world, beaked like an eagle and booted like a grouse, which in Scotland is, and in America and England is not, although both countries ettle to have it if they can.

In these days of small Scots it is nice to know we have a big grouse a yard long, who can attract the ornithological tourists.

Anyway, there I was in the smirring rain, and the pine gloom, being put into hiding by one of the most knowledgeable and observant keepers in all Scotland.

The forest was silent. Nothing, it seemed, was stirring. And then, faintly, came a sound like the ticking of a watch.

Tick! Tick! Tick! Tick!

I strained my eyes, and, suddenly, there he was in the gloom, parading round his pitch, with his tail fanned, his head erect, and his beard aggressively on end; the great cock capercaillie, most magnificent of grouse, of eagle bulk, making a noise like a halfpenny watch.

He strutted and slow-walked, ticking and clicking his beak. Every now and again he would leap into the air like a blackcock, and a man could hear the catsquech and pop of him. But, in fact, a blue tit could have made more noise than the giant cock of the woods.

There were other cocks parading round the knoll, posturing and challenging, but none came to blows that morning. Indeed, the display of the capercaillie is largely formalised, like the opening of Parliament. Fierce fighting does, however, sometimes occur.

The capercaillie is one of the birds which, occasionally, makes news. Once in a while a displaying cock becomes so territory conscious that he will assault anything that intrudes, including you, me, Land Rovers, sheep, keepers and postmen.

Then he gets the headlines, although he isn't a tough guy trying to be a predator

241

on men, but a slightly mixed-up bird who doesn't know what he's doing.

As in the black grouse, the sexes in the capercaillie are quite dis-similar: the cock all blues, and browns and beetle-green, the hen dappled in buff, black and white, with a rufous breast.

Hen capercaillies usually lay their eggs in a scrape of a nest near the base of a forest tree, or stump, although some will do so among brushwood and others among thin scrub or heather outside the main tree cover.

An untended nest is open and usually easy to see, but when the bird is on she is ill to pick out, and will sit until almost trodden upon. I should think that the majority of nests are found by accident; by flushing the sitting bird from her eggs.

A fortnight ago, I was looking at a nest with eight eggs, which had been deserted. A collie, gathering sheep, had flushed the bird accidentally and she didn't come back.

Her eggs were only two or three days clocked. Had she been sitting longer, she might have made the return journey and a nest would have been saved. But this is one of the working risks that ground-nesting birds have to run.

It is, nevertheless, remarkable how seldom such a thing happens. A man covering the same ground every day will tell you that he sees a dozen hen capercaillies for every nest he finds, and that his dog may put a hen off only once in every two seasons or so. I am sure that this must be about the way it is with wild predators like the fox or the badger.

The broody hen capercaillie is a ponderous slow-moving, couthy bird, who can sit out the daylight with hardly a movement, except when she stirs to turn her eggs.

When her chicks hatch, she leads them away into the forest after their down has dried. They are no bigger than day-old poultry chicks.

The Farmer's Tale

This retired 75-year-old farmer was visiting a non-retired 70-year-old farmer friend of mine while I was visiting my 70-year-old farmer friend, so we met for the first time, although my 70-year-old farmer friend seemed to think I should have known the 75-year-old because he had originated in the neighbourlikehood, which he had left nearly fifty years ago while I was still sweating over Highers at school, and I had to tell the 75-year-old that I had no recollection of him whatever.

My 70-year-old farmer friend said: 'This is David Stephen who is running a country park near here, ye ken, a naturalist that likes wildlife an that, but he keeps cattle beasts and horses forbye, an ferms enough tae feed the cattle and horses and deer an things.'

'Is that a fact?' the 75-year-old said. 'Man I've read some things you wrote in some paper or ither (I like the way *The Scotsman* was referred to as some paper or ither) an ye ken I've aye wanted to see a badger, an I'm livin among the bliddy things an have never seen wan.' Or words like that.

So I invited him up to see our Spick, who is one of the only two daylight badgers in the UK.

242

And then he was telling me of the day he went out to look at his coos on his Essex farm, and found a youth lying in a patch of deadly nightshade, and how he touched the lad with his foot saying: 'Come on laddie, oot o that,' and how he suddenly realised that the boy was dead from eating the berries, and after that there was the police . . .

From that he came on to saying that he was perturbed about the way that wildlife was disappearing from some farms, and he had some sharp things to say about the cowboys who shoot anything and everything throughout the year. I told him we had them too, but that the disappearance of wildlife from farms was a complex thing, attributable as much to modern farming as to cowboys. Farmers are the victims of a system that forces them to adopt practices inimicable to wildlife of all kinds.

And then the human being in the modern farmer sat up and spoke to me in the way that makes me, the supreme pessimist, become a temporary optimist.

'Weel,' he said, 'Afore I retired there was this day I was looking at the hay, bendin doon ye ken, an suddenly there was this twa shotes fae a twelve-bore, an two pairtritches was doon—wan o the cowboys fae a neeborin ferm.

'So I lookit aboot, and picked up maybe ten or sae wee cheepers, nae mair'n a day or twa auld'(when he wasn't speaking Essex his Scots was as good as any) 'an I took them hame in ma bunnet, and my dochter Margaret took them and pit them in aside some rabbits and things she had in a run, an she fed them, and they cam on good, and syne they were sittin on the windae ledge wi my wife tryin tae shoosh them awa.

'But they widna gang, and they bydit on and grew and flew, gey pairtritches by ony standart. An they stayed on and next year some o them were there, and they bred there and were gaein aroon wi their cheepers the next year, fair settlet doon. An it was a gran sicht, and made us a feel guid.

'Ken,' he said, 'I just thocht that the world wid na be the same if a thur things were gane.'

And I said to him: 'Man, when I was in Dublin's fair city eighteen months ago, speaking to schoolchildren, a Dubliner said to me: 'Wouldn't the world be a hell of a place if it was filled with nothing but people?'

Same thing, whether in Dublin or Essex. Same thing, from China to Peru. And the old man almost said Amen.

Much Ado About Rabies

The man, who had just seen a fox near my house, said to me: 'Should you not be killing foxes now, with all this rabies scare?'

And I asked him: 'What rabies scare?'

Whereupon he looked at me, in genuine surprise, as though I had an extra arm growing out of my ear, and said something like: 'But surely you of all people know about the rabies scare?'

A few weeks ago, when I was in France, where there is rabies, the only people who brought the subject up with me were UK tourists. The French discussed it with me when I brought it up. Yet, when I came home, every second person seemed to be talking about it, and getting all het up about the menace of foxes.

And how's that for getting carts before horses?

Certain of the media have done a good job about getting the horses in this position, and succeeded, intentionally or not, in focusing attention on the fox. Highland sheepmen could not wish for a more favourable climate of opinion in which to push for action against their historic scapegoat—the fox.

The fact is, of course, that no fox in Britain has rabies, and there is no rabies in the country. If it arrives, as I am sure it will because of the number of stupid people there are around, the blame will be people's. And the victims will be legion: foxes, dogs, cats, voles and whatever.

The permutations are intriguing, apart from being frightening. Siamese cat attacks big rabid rat, and gets bitten before it makes a kill; later on Siamese cat bites owner. Or imagine baby weasel mistiming its pounce on a big vole and getting a bite that is rabid.

I often think of those MPs who guffawed so loudly when a member suggested imprisoning people who tried to smuggle in pets. They should be sent into the wilderness to grow up. One shudders to think that people like these are part of our legislative defence against rabies.

In my view the time has come to make prison sentences mandatory in the case of pet smugglers. The time has also come to spell out to the public that we don't have rabies, that it can only arrive by human agency, and that nobody at this moment is under threat from fox, dog, cat or anything.

Just before writing this—indeed what decided me to do so now rather than later—I had a phone call from one of the management of a big industrial complex, and the conversation went something like this.

'We have a dog fox, a vixen and five cubs in an old shelter in the middle of the yard. Are you interested in having them before we set about exterminating them?'

'I don't want them,' I said. 'But why do you wish to kill them off? Are they causing some upset to your work or your production line?'

'No, not really,' he said. 'But you know it's this business of rabies, and I understand that foxes are a danger.'

'There is no rabies in this country,' I pointed out. 'So foxes are no threat in your place or anywhere else. Have you a lot of rats?'

'Yes,' he admitted, 'we have.'

'Well, that's probably why the foxes are there.'

'But children climb in, at night, illegally I admit.'

'I see no problem there. The foxes will kill rats, not children.'

'Yes, but the foxes are also killing cats.'

'That,' I said, 'is a habit they have. What's wrong with their killing cats in a place like yours.'

'Well,' he said, 'the workmen feed the cats, and rear them, and the foxes are killing them.'

'Now,' I said, 'you've given a valid reason for killing foxes. The men don't want their cats killed. Fair enough. But why bring in rabies?'

'Thanks for your help,' he said.

'Not at all,' I said.

244

Something to Declare

When I came off the Bilbao-Southampton boat I joined the red sticker queue, announcing to the world and HM Customs that I was in a declaratory mood. I've said queue, but there was no queue. Only one car. Mine.

'Will you come round here, sir?' said the big Customs chap, an affable six feet or so. 'You've got a red sticker up. What is it you wish to declare?'

'Well,' I said, 'if you know what quail's eggs are I've got eighty-two of them.'

'. . . no!' he said, but riding it well, and wearing a half smile.

'. . . yes!' I assured him. 'That's why I'm wearing the red sticker . . .'

So I told him that I'd been with a friend of mine in the mountains of northern Spain, one Armando Pousada, who breeds quails for the gourmet's market, and when I say breeds quails I mean quails in thousands, and that the idea of taking a setting of eggs back to Scotland had matured while he was showing me around.

So, I added, I said I'd take a setting of eggs, laid on the day I was due to leave, and that I'd put them under a bantam if I was allowed in with them, and I could see no reason why I shouldn't be—the quail being a British bird, though extremely rare and on the First Schedule of the Protection of Birds Act, which means I could be clobbered if found in possession of them.

Anyway, I went on, he brought the eggs round late on the evening before my departure, and I said to him: 'That's a gey box for twelve wee quail eggs.' And he said to me: 'There's eighty-two in it.' And I began wondering where I would find a hen with a cluck-seat big enough to cover six times twelve plus ten. 'And that's it,' I said to the nice big man in the uniform.

'But you didn't get an import licence.'

'I never knew I was going to import quail eggs, but I'm in this queue to find out whether they've to be taken or allowed in. These are First Schedule eggs, and if I get them in I want it to be clear that they *were* brought in.

And I told them I would have them hatched in an incubator, hens being too wee, and would then have the birds on display at Cumbernauld, if any hatched. At which he said he'd have a talk with his colleague. And presently another nice man came over and looked at the eggs, and said he hadn't realised quail eggs were so wee, and said he'd have to phone the Min. of Ag. and Fish., which is the way he said it, and being a bright lad myself I knew what he meant.

Presently he came back and said he was waiting for a phone call, so we talked about quails and capercaillies, and Scotland, and there they were, the two of them, hoping the eggs would be allowed in.

A good while later, as we say, the phone call came through.

'You can take your eggs, I'm happy to say,' the second nice man said. 'But in future you should contact this address for an import licence. New regulations are apparently imminent. And I wish you every success with the eggs. I hope they aren't spoiled by the journey.'

They were such a change, these two born public relations chaps, from some of the hard-heads at Dover with whom I've waged guerilla war, on and off, for forty years. And so thorough and with it withal.

I was packing the eggs safely when the senior chap came over to say *bon voyage*. As I was stepping into the car he said: 'Well, say hello for me to the little rascals when they hatch.'

I freely acknowledge that these two blokes made my day.

A Skinful of Make-Believe

The tractor was straining at the throttle and metaphorically tossing its head, its horses raring to go to shear the furrows and hap up the tatties we had planted by hand—thirteen drills, forty yards long, with the Golden Wonders bedded down for the long God's-time parturition of a murphy that could look Raleigh straight in the eye; the murphy that technology cannot make better, but has certainly made poorer.

The farmer gathered his mechanical horses by the wheel, and you could almost see them rearing on their monstrous wheels while they snorted extra puff.

'Are ye no for gie'in them a wee pickle fertiliser?' he laughed.

'That's a dirty word around here,' I said.

'Aye, of coorse,' he laughed again. 'You like them kinofa natural.'

'Kinofa,' I agreed.

Aha! you might say; we're on to the muck and mysticism again. But it was the man who used the word *natural* not I. And what he was really saying, seriously facetious, was that the commercial article was unnatural. But he meant more than that. He was also saying that the natural product was better. And he is right.

The modern potato is an imposter, a blob, a squid, a skinful of make-believe, a beefed-up, bowdlerised beast whose lack of savour and texture is not improved by the maltreatment it gets at so many hands after it is born. Any man who argues that the modern breed is a true son of its father would, in Mann's phrase, argue that 'a table is a fishpond'.

Those who knew not Joseph accept the modern heresy of the potato. Those who did will ask such questions as: 'Why is there no such thing as a really good potato today?' Or refer, nostalgically, to the way mother used to cook them. A good potato can be spoiled in the cooking. The modern one is spoiled before it sees the light of day. Mother couldn't do with the new what she did with the old. Bad cooking is the great leveller; but it takes more than a good cook to mask the modern murphy.

W.H. Hudson, that great, and now somewhat demode, writer of English prose, grew up in the homeland of the potato, where the wild tubers were wee as bools, and the cultivated ones as good as any in the world. Hudson reckoned they were the best in the world, and wrote of them:

'They were beautifully white and mealy, with that crystalline sparkle of the properly cooked potato in them which one rarely sees in this country (England). In Ireland and Scotland I found that the potato was usually cooked in the proper way by people of the peasant class.'

Modern field chemistry has done its worst for the potato, but one doubts if the end result would be much different if it did its best. Hudson would be hard put to it today to find the peasants, or their handiwork. But one knows what he meant.

Writing of his first meal of potatoes in England, he asked his hostess:

246

'Is this the way potatoes are cooked in this country?'

'Why, yes,' was the answer. 'How else would you have them cooked?'

He described the dish as 'a sodden mass of flavourless starch and water that made him almost sick, a mass that looked like a boiled baby in the dish, boiled to a rag.' To this extent the English Farmers' Union is right in saying that people mishandle the potato. But they should remember that what they are giving these people is a replica, and a poor one at that, one that goes black in the face if you turn your back to answer the door bell.

Hudson had a weak digestion, and was warned off potatoes by his doctor. Yet when he had an attack of indigestion he ate nothing but potatoes for a few days, and cured himself. He cooked them in their jackets and ate them with pepper, salt and butter. Potatoes only. Nothing else. And, for him, it worked.

He considered that the potato, a native of America, was one of the two greatest food plants in the world. And it is a great food plant. Or was. It was the food of the aborigines of America North and South before Raleigh ever heard of them. The Peruvians, who built temples to the sun god, and worshipped the rainbow, also worshipped the Potato Mother, and prayed to her when the seed were committed to the earth.

The Potato Mother of today, whoever she is, deserves to have the murphies thrown at her.

Preservation and the Gun-Toters

There's a kind of magic about the word 'Game' when used of animals that are, and even of some that are not. Away back a farmer friend of mine once said explosively:

'Gem! Gem! Every pauper atween here and hell, withoot wan penny tae rub aff another, has got gem on the brain!'

I knew what he was referring to. There was an old man who lived in a house with an earthen floor and a gable of uncemented bricks, who hardly knew where his next meal was coming from, yet he hardly slept a wink if he saw a fox heading in the direction of the pheasant covert at dusk.

A few years back a young man from my home town, who had just taken up shooting—he might as easily have taken up golf, or fishing, but he chose shooting—came to see me with his braw new gun, and while we were talking a kestrel appeared and hovered above the field across the road. He asked what bird it was, and when I told him he said: 'That's vermin isn't it?' The ink on his gun licence was hardly dry but he was soaking up the traditional attitudes quickly.

That is part of the magic of Game. Like God and Devil, Black and White, it has its opposite, and its name is *Vermin*. And vermin is practically anything you like to dislike.

At a fox drive just after the war a policeman shot a tawny owl, and when remonstrated with said it was sair on the young pheasants. He didn't have any pheasants, of course, and the only ground he owned was what stuck to his bootsoles on a wet day, but he was bemused by the magic word.

A keeper I knew, who emigrated to Canada, used to say I went over the score, as he put it, on the subject. He was meticulous himself in observing the bird protection regulations and tended to think of himself as typical. He came home on holiday and, wanting to get the heather under his feet, hired out as a beater on a Perthshire grouse moor. He packed in on the first day because two short-eared owls were clobbered in an afternoon drive.

The most valid criticism that can be directed at many of those who preserve or shoot game is that they shoot other things other people wish left alone, and very often protected species at that. This may not be—is not indeed—as generally true as it used to be, but it is still only too true of many places. It is equally true that the ancient fortress of prejudice has been thrown open to reason.

What impressed me at the first Game Fair at Blair Drummond, where I was invited to lecture on predators, was the space given to modern research on game and on predator/prey relationships. The Nature Conservancy's scientists had a free run with their research on grouse, and there was an impressive stand on the role of the weasel and stoat in the countryside. The Forestry Commission took space to show what it was doing in the field of deer management.

But, of course, there are those who prefer the old ways. At the last Fair I listened to some of the comments at the weasel and stoat stand. 'I don't believe that.' Better still: 'I winna hae that at a!' And again: 'They're still vermin tae me.' But these are the incorrigibles, like the keeper who once said to me apropos eagles: 'A' I ken aboot eagles is that they're nae guid!' And how's that for profundity hey?

Bodies like the Wildfowlers and the Nature Conservancy have got together and worked out management plans that meet the needs of both. This was one of the great achievements. One can see a similar trend in certain quarters concerned with game birds, but there is a lot of slack to take up. Here the British Field Sports Society has a role and a great responsibility. Occasions like the Game Fair provide a great opportunity for education.

It seems to me that if sportsmen, and especially their game managers, once recognise that they are subject to the laws of the land like everybody else, they would be in an unassailable position. The only battle they would then have on their hands would be with those who want to abolish field sports altogether.

Personally I can see nothing wrong with shooting grouse, or partridges or pheasants or woodpigeons or wildfowl which, after all are food species. I have reservations about other kinds of sports. But there are certain things I would make obligatory; that shooting men should have good dogs, that they should have a shooting test, and that they should undergo a test to establish that they could recognise all the protected birds of prey.

The sorry truth is that the old brigade's worst faults are being copied by the new legions of free-ranging gun-toters, responsible to nobody. Democratisation of sport hasn't changed anything.

Malice Through the Looking Glass

The most dangerous, destructive, wasteful, irresponsible and unteachable animal on earth is the one we see when we view our collective kissers in a keekin glass. Yet we expend effort, emotion and a lot of money preening on our egocentric Olympus, blaming everything else within sight, and riding ruthlessly with spiked feet over all the facts in pursuit of the fashionable woe of the moment.

A man, a person rather, could give a hundred thousand million examples to prove the thesis that a person is the most dangerous, destructive, wasteful, irresponsible, unteachable animal on earth, but if a person accepted that she/he was the most dangerous, destructive, wasteful, irresponsible, unteachable animal on earth, she/he would have to give up blaming everything else or stop being the most dangerous, destructive, wasteful, irresponsible and unteachable animal on earth. And that is unthinkable.

A person said to me the other day: 'Isn't it dreadful about the poor orang-outangs in Sumatra and Borneo? Why must these people go ahead and wilfully exterminate such a harmless and gorgeous creature?' This person is paraphrasing what the sayer person said.

And this person said to the other person: 'Take a gander at Mull, where persons with twelve couples of otterhounds, or thereby, come up from somewhere in England's brown and polluted land, where persons, including themselves, have killed the otters or made the land unfit for them to live in, to hunt the otters of Mull. And if the thick-headed teuchter peasants of Mull object whassamatter? What do Mullachs know about what's what. What?

Persons, this person continued, will always find some sort of justification for anything, often with legal and so-called scientific approval. The bunch of chinless wonders who invade Mull with blameless hounds would probably argue that hunting is legal and, anyway, they are UK otters, what? There's no troops along the Border yet, what? It might not be a bad idea to have the Argylls at Craignure, though, this person said. What?

We all ken there's no cruelty to otters; they like being tossed and chopped (and for God's sake don't any chinless wonder hit me with semantics and say I'm using the wrong jargon). And maybe the chinless wonders will say the otter is a menace to river stocks; if they believe that they'll believe in the flat earth or leprechauns. But don't blame the poor hounds; they're only dogs, and nice dogs, blame the persons who lay it on.

This person then said to the other person: 'You know, when Farley Mowat was about to start out on his wolf stint in the Barren Lands, the Establishment chief said to him that hunters were complaining about the scarcity of deer and blaming the wolves. So the Establishment (note that) were set for a wolf killing campaign. The fact was that there were five person hunters for every deer! And how did this come about? Because of earlier hunter persons.

But the wolves were handy for blaming. The Eskimos knew better when they said, in their primitive, no-degree-at-Oxford-way: wolf is good for caribou. A person might well ask them what they thought of the person packs.

Pests have been the in-thing for a long time now, and if only we could control

249

the pests we could feed a hungry world, and a hungrier world, so they say. America probably holds the world record for use of pesticides, every one a magic formula for salvation, and a lot of people have been in on the act of polluting the US of A. It creates jobs, though, you might say.

Well, here's what the US Department of Agriculture says: it says that the proportion of food crops eaten by pests has increased since the large-scale use of pesticides. The proportion eaten between 1941 and 1951 was 31.4 per cent. It is now running at 33.6 per cent. So where have all the dollars gone? And what price the Silent Spring?

This is the achievement of the face in the keekin glass.

The Cow Wore a Gate as a Collar

There was this pigeon with the bread collar, which made it a collared dove of a sort, and what had happened was that it had been pecking and flipping a slice of bread until it had made a hole in it, then the thing had accidentally slipped over its head, like a jersey, and there it was, all Elizabethan pan-loaf frill, and wondering how to get rid of it. It eventually managed to do so by some more head shaking.

As I was saying to The Wafer afterwards, I used to keep braw wee doos called Oriental Frills, which passed on their frills to some of my Tumblers' offspring through miscegenation, but this was the first Bread Frill I had ever heard of, complete with farinaceous collar.

He smirked, and said: 'D'ye mind o Tammy at the railway cottage who had the greyhounds? Well, he had a bitch that was broken doon an he put her in a shed.

'The shed was maistly auld planks or corrugated, but the wan end was ply-wid. Well, the bitch scarted an scarted until she had a hole tore in the ply-wid, then she gied the bit brienge, and was awa wi her head in a widden frame.

'She got stuck in the door an they had tae turn her sideways tae get her intae the hoose.'

'That's nothin,' I said, borrowing a favourite expression of The Wafer. 'I once met a coo wi a gate round its neck, an iron gate at that. She had been clawin her neck an got stuck, and by the time my wife and I arrived she had broken wan o her horns, and had the gate aff its hinges.

'I had on a new suit, and by the time I had her loose I was slaigert wi blood and spittle, and had tae send the suit tae the cleaners right away.

'But I think the funniest thing I ever saw was a roe deer wi its face stuck in a can. It had been joukin aboot a dump on the edge of the wid, an got tae lickin the dregs oot o a syrup tin. It shoved that hard it got its face stuck, and there it was runnin awa wi a tin muzzle wi Syrup printed on it.'

Later I recalled the number of times I had removed hedgehogs from snares set for rabbits. In fact, I remembered freeing the same hedgehog from the same snare night after night, so that I was forced to ask the rabbit catcher to remove that one altogether.

The beast changed course and walked into the next one! An insurance man would call the hedgehog accident prone.

A guinea pig my daughter had when she was a very little girl was a great one for being framed in wool; every time my daughter let the beast up inside her jersey it ate its way through the front then sat there looking out, framed in green, scarlet or dove-grey.

Once it tried to make a hole for its head in the dog's skin but the dog refused to co-operate.

Nowadays, wild animals with collars are not an unfamiliar sight in some parts of the world, including this country. These collars are coloured, and are for identification purposes during field studies. But the ear tag is a much better idea.

At the moment I have a pair of weasels, in adjoining enclosures, with a hole slightly over an inch in diameter providing the only entrance from one to the other.

I am trying to breed weasels, but I have to be sure that the female can get out of the male's way if she has to, in case he might kill her.

She can get through the hole into his section, but when he tries to get through to hers he finds himself in a hardboard collar, jammed.

Death by Poisoning

Many years ago, on one of the western islands, I picked up three rabbits which, as I suspected, had been doctored with strychnine. I put a buzzard off one of them. The three had been planted within shouting distance of an occupied eagle's eyrie, and it doesn't take a genius to guess why they had been put there.

The Royal Society for the Protection of Birds has been publicising again the fact that birds of prey are still being illegally poisoned by way of illegal baits laid out for them, and I am pessimistic enough to believe that this will go on as long as poisons are available.

Over a very long period of time I have learned one thing: that people who didn't poison birds still don't poison them, and that people who did still do. The 1954 Act didn't change anything very much; nor has subsequent legislation. The little bit of change brought about is that there are penalties now, if you can bring a prosecution. And that, as it always has been, is the problem.

Strychnine was, and is, a favourite with the poisoners, although there are more sophisticated methods nowadays. The use of strychnine is, of course, illegal except for moles. Once you've got it you can use it for anything that takes your fancy, including foxes and other people's dogs, as well as eagles and other birds of prey.

A dead sheep was a vehicle for poisoning in the old, and not so old, days. Doctoring the carcase was illegal, as was leaving it unburied; but that didn't stop anybody who wanted to use such a vehicle, and the foxes and eagles died the death of clenched teeth and talons.

Another way with the dead sheep was to ring it with gin traps, which are also illegal but still used. You might catch anything from an eagle or a fox or a dog, to a human being. I got my boot in one once. Then you could poison the same carcase as a bonus.

Another way of using poison was to doctor an egg and set it up on a peat hummock in a waterhole. That got the crows and the ravens, or a fox that was prepared to jump or swim. The drowning set for foxes was used in such places. It is presently illegal but the pressures still go on to have it back again.

Sheep farmers naturally defend themselves against all accusations that they use poison. I know a great many sheep farmers, none of whom would dream of using poison; but a lot of others do. Many of these will say they want to kill only foxes, but eagles fall to the baits too, and in any case bait is unconditionally illegal.

Keepers are in the same boat. There are those who never shot or trapped a bird of prey in their lives, and there are those who do it all the time.

Eagles and harriers are still killed on grouse moors, despite the fact that predation has been clearly shown to be an insignificant factor in grouse numbers. But the bird of prey is there; it kills grouse; therefore it must be taking grouse that would otherwise be available for man. And it's a fallacy.

I don't know what the answer is. Passing laws does very little except achieve an occasional prosecution; otherwise things go on very much as they were.

I like to remember a sheep farmer friend of mine who discovered one year that he had a pair of nesting eagles on his ground. When the eaglets hatched his shepherd found two dead lambs in the eyrie. My friend phoned me and I went down to have a look. We found that the lambs had died of pulpy kidney (certified) and when I skinned them there was not the slightest sign of predation.

I was once asked to look at an eyrie to see if there was any sign of lambs in it. There wasn't. But when I went in beside the eaglet I found myself up to the knees in dead hoodie crows. Conclusion? Nothing: except that this pair of eagles happened to be killing hoodie crows. Just as other pairs sometimes kill fox cubs.

Triumphal Progress

Motor cars are my greatest non-subject of conversation, and you couldn't pay me enough to get me to a motor show. If somebody asks me the name of a car I have to look at the nose or tail or wherever the manufacturer sticks his label.

But I'll still talk motor bikes for half a shift, especially the great breed of Nortons that used to clean up Europe with such monotony that we used to wonder why anybody bothered to hold races at all. I rode Nortons for twenty-two years, and I've been driving all told for forty-six years. And for a long time I was as often on two wheels as four.

When some of today's two-wheelers talk about doing the ton I modestly remind them I was doing said ton before they were born. And every now and again one of them offers me a whirl; and every now and again I accept. And enjoy. Reckon I'll aye be a two wheel man at heart.

Oddly enough the bike I am liable to talk most about was a 1919 belt-drive Triumph with rim brakes. On that old thing I covered most of Scotland. I forked it over some of the roughest ground in the country, and there was a lot more of it then. I eagled, foxed, hawked and badgered with it. And laughed.

It was ten years old when it began to take me over much of Scotland and I took it over the rest. It went over the old road to the Isles without a complaint, and still had a kick left in it when I traded it in for twenty-five shillings against a Norton.

I had the nerve to put a sidecar on it, and it had the nerve to pull it, although it tired every ten miles or so and had to be given a rest, just like a thing of flesh and blood.

I stabled bike and sidecar in a bottling shed, where operated a builder of bicycles, a former naval engineer and then beer bottler, and one of the great jokers and engineers of this or any other century.

When I took the combination out, and gathered my two passengers, the builder of bikes would say:

'That's right. Tak plenty wi ye. If ye don't need them gaun ye'll need them comin back.'

And he was right. We pushed that outfit over half of Scotland.

Most of the trouble was the belt drive. When the weather was fine the outfit burped along, but when the rain fell we did a great imitation of moving while standing still. The belt slipped and was the only thing that moved.

One wet day we managed to keep up a crawl, and I remember a horse passing us, drawing a trolley with a ruck of hay aboard. It beat us by half a length to the top of the hill, and the farmer said: 'Try the whaup.'

One day the gear lever went on strike. It began to jump the ratchet and every few yards we were unwillingly lapsing into neutral, with the engine revving and the outfit dying to a stop.

My sidecar passenger came up with an idea. He tied a length of string to the lever, passed it under the tank, and held the end in his hand. That way he held us in gear, and we got miles at a time before the cylinder head burned the string through. From that day the outfit was known as the string change Triumph.

That trip we burned our way through two balls of ham twine before we made home, and never again did we set out without plenty of length.

The last joke that bike played on me was when I was weekending at a farm in the Blane Valley, where I used to do some field work. The combination was put in a shed over the weekend, and on the Sunday the cat had her kittens in the sidecar.

The check questions after that were: 'Petrol? Oil? String? Kittens?'

This Harvest of Ugliness

In the birch thicket the sun had polished the silver. Tits were churring in the wine-red shadows; there was a lark up singing in the newly washed sky; two peewits were whooping in bat-flight above the field.

The air tasted very sweet. I sniffed it appreciatively, like a setter; preed it as though it were twenty-five years old malt; drew it down into my lungs with uninhibited addiction; stepped into the wood and took the Lord's name in vain.

In vain is right; and I found myself echoing Charles Murray's 'Gin I was God':

'To some clood-edge I'd daunder furth, an feth,
Look ower an watch hoo things were gyaun aneth.
Syne, gin I saw hoo men I'd made mysel

253

Had startit in to pooshan, sheet an fell,
To reive an rape, an fairly mak a hell
O my braw birlin Earth—a hale week's wark—
I'd cast my coat again, rowe up my sark
An, or they'd time to lench a second ark,
Tak bak my word an sen anither spate,
Droon oot the hale hypothec, dicht the sklate,
Own my mistak, an, aince I'd cleared the brod,
Start a'thing ower again, gin I was God.'

And, if it came to pass, I'd echo my late father, whose invariable comment on just retribution was 'Hell Mend Us'.

For, just on the edge of the birch thicket, somebody had begun one of those do-it-yourself rubbish dumps that are now as thick as fernytickles on the face of the countryside—created, not by people throwing things over their back wall but by people in motor cars who drive deliberately to some quiet, attractive spot, like this wood of the roe deer and the badgers, and there bring forth the symbols of their moronism.

Here were the bottom of a baby's pram and all of a doll's pram; a suitcase, a briefcase, and briefs (or should it be scanties, or panties? I can never be sure nowadays); a can of creosote, a cardboard box, a pair of slippers, and a hotchpotch of paper, tin, old iron, plastic, polythene, old carpet, bits of linoleum, sundry cloots, a bit of a ladder and a child's chair.

Of these by-products of the so-called affluent society, the small chair and the doll's pram would have delighted many a small girl of the thirties, and no one would have been ashamed to present a child with either. They represented waste as well as disfigurement, the mindless arrogance of surpluses.

But why? Why this sowing of consumer's surplus to reap a harvest of ugliness? I am told over and over again that any local authority will collect special items or large rubbish by arrangement. So there is no excuse at all. Yet these vandals will transport their rubbish for miles, at their own cost, to dump it where it most disfigures. Why?

Looking at such dumps one feels like rewriting the lip-service catchwords of Conservation Year in Orwellian style—Ugliness is Beauty, Muck-up is Make-up, Amenity is Amort, Conservation is Floccinaucinihilipilification.

The creators of this kind of countryside dermatitis are not from any particular stratum of society, although all are wheel borne—they range from the 3½ litre to the jalopy with lesions of rust. I have seen paper and fish boxes spewed from a fish-van, linoleum from a limousine, and a bed roll from a beetle polished like one.

Litterbuggers may be class conscious but they belong to no one class. Planters of broken bottles are not confined to the carrier of screwtops. Our rookery has now more crockery than rooks; there's a wood with more rocking chairs than roe deer; a corner with more records than song birds. Every other field has glass teeth. The bigger the barricade you build the higher the heap grows inside. No maginot line can hold off the litterbugger on the offensive.

I must confess to a wry, unwilling smile when I came back for a second look at the spring sowing beside the birch thicket. There was a blue tit on the hand-grip of the doll's pram, right way up then downside up, and repeat, like a plastic bird falling upside down in a cage.

One can't win, I'm sure. When I got back home I had to gather a scattering of the paper linings from tomato boxes from the verge opposite my front gate. It's happening round every other corner.

Wildlife Under Fire

March fires are part of the calendar; the nose expects, and would miss, the smell of them. In the dusk the dry grass makes a braw lowe and nobody bothers overmuch. These are small fires, often the arson of small boys. The big fires, by law approved, the muirburn, are a tool of land-use, to hold vegetation at a fire climax, as on a grouse moor.

The trouble with the big fires is when they take place long after the allowed date, sometimes by the vandalism of the match-happy who can't see a bit of dry stuff without wanting to light it, and sometimes by the damn-the-consequence occupiers of the ground who often feel the same way.

A March burn-over, if not controlled, can be serious enough and a threat to forest or property. Later fires cause great destruction of wildlife and sometimes of domestic stock as well. One man, helping fight such a blaze, forgot about the terrier he had with him and it was roasted in the fire.

Some years back I was standing talking to the firemen during a big June blaze that was threatening the cottages by the roadside. There were three fire engines out but no water and the heat was too great to face.

At one point the flames parted and out bounded a roe deer doe. She stopped clear of the smoke and looked back over her shoulder, obviously expecting a follower, and I said to the firemen, 'She'll have fawns in there.' But no fawns appeared and the doe, to my surprise, bounded back into the fire through the break that had let her out. She didn't reappear.

I found her, and a fawn, the following day, burned to death.

On another, and much bigger moor, where there had been a much bigger fire, I found a charred roe fawn whose mother was also keeping tryst. She ran in circles when I approached and was still there, keeping tryst with her dead, when I left.

This big moor gets an occasional touch of the fire still, but has had two major, devastating burn-outs in the past thirty years. During one of them I caught fire while speaking to one of the fire officers; the hot peat would suddenly flame into life while a man was standing there.

That year there was a pair of merlins nesting on the moor—they were the last—and I was anxious to save them if I could. The birds were nesting in tall heather, and the five eggs were near hatching; the chicks had been knocking at the door the previous day. By the time I reached the nest the fire was a few yards beyond it, and the chicks were roasted in their shells. The two merlins were flying about, calling in alarm, blinded by smoke and soot.

There was a kestrel in a tree, sitting on eggs, but she was safe from the fire, because there was a boggy buffer between her tree and the fire. But I could hear her sneezing when I climbed into the next tree, where I had a watching place.

255

I went to look at a curlew nest, expecting to find the eggs cooked like the merlin's. Instead the curlew was still sitting on them, her feathers singed to the quills! She had stuck with them. She went off calling when I came up, but she was unable to fly properly, and raced round me instead. Her eggs were unharmed. A ring of green round the nest had saved it from being burned over. The curlew returned to her eggs when I left and a week later hatched out four chicks.

I found a grouse crouched in the heather, cooked to brown. Beneath her she had small cheepers, also dead, but not cooked. That was the only time I have seen such a thing.

The moor that year was a graveyard of burned birds, and the crows and foxes spent some days afterwards scavenging the dead. Nothing that couldn't run or fly escaped, and even some of the runners and flyers were caught—like the curlew that stayed to singe and the grouse that stayed to die.

During the other big fire on that ground many animals gathered as refugees on a damp area of rough pasture that became close-curtained with smoke but did not catch alight. I counted them during one clear moment—seven hares, one fox, three roe deer, and a mixed parcel of curlews, grouse and peewits.

Woodpigeons and Rooks
Out in the Springtime Air

It's here at last, I think, I hope—the spring that is—because the rooks up there say so, behave so, sitting on eggs in wildly tossing nests, while March's lion, ignorant of folklore, rages on in non-stop activity rhythm, determined it seems to have the last word, and go out with more than a wee curmurring.

Ah! the bold rooks, in plumage sunned to beetle purple and shot ebony, rocking round the clock on precarious footholds, or, in their nests, like the wee man in his wee boat bucking the Horn. Deeved by the lion's roar, their loudest caws reach grounded ears as small-talk.

On the ground woodpigeons seek dropped twigs for their nesting, leaping aside every now and again with flick of wing when the leaf-litter stirs to the tread of March's waukrife lion.

In the field, clear of the trees, half a dozen woodpigeons are feeding among the molehills, dancing in unwilling circles as the wind tips them under a wing. They fan their tails, which act as sails, and they are tilted forward on to their faces. In whatever position they stand, their feathers curl, bend or quiver.

There are rooks down in the field, too. It is a small field, touzy and wet. Some rooks are dibbling, gathering food and pouching it; others are tearing up small sods which they will carry away for nest lining. The birds leave scars, as on a golf course where players forget to replace the turf.

A sudden squall of sleet drives the birds away. I move into the tree shelter myself, and come out again with the sun, as a collier appears on the road, walking with his bike, side on to the wind.

'It's a tempestous kinna day!' he greets, slowing down and coming on to the verge. A grand word that, I think, a word to cut coal with.

'It is that,' was the best I could do in reply.

'I see an egg broke on the road back there, wi a green kinna shell. That's what got me aff the bike, in fact. Wid a craw drap that fleein?'

That's the way to get messages, I'm saying to myself. Economy of words, as if they were costing him at telegram rates, but clearer than any telegram. And the craw, for rook, was right zoologically.

'If it's a craw's egg a craw drapped it,' I agreed. 'It can happen.'

'How?'

There it was, the three letter word, how for why, a quarter-inch of question demanding yards of answer, like the child's 'What Makes an Engine Go?'

'Caught short, I expect,' I said to him. 'Flushed from the nest on point of lay' (that was the poultryman talking) 'or on point of lay wi nae nest ready, or mebbe just in bother wan way or anither. It can happen.'

'Aw weel,' he said mounting his bike, 'it's wild wather, whit . . .'

'It is that,' I agreed, waving to him.

He would be no more than a quarter of a mile away when from his direction, the fox came dawdling along the edge of the rookery, probably disturbed by him in the hedge-bottom at the roadside. A step to one side put a tree between me and the reynard, and I was able to watch him walking past.

Walking he was, with his ears up, not scared, and his brush in a nice curve. He crossed the open field, scaring off some woodpigeons that had returned, and disappeared into the spruces. And that was the last of him. I wondered if the collier had seen him. If he had, I thought, he would be back in a hurry to tell me. So I waited. But he didn't appear.

So I was saved from his probable question, which I asked of myself: 'What's a fox daein joukin aboot at this time o the day?'

And didn't give myself an answer.

The Poisoned Landscape

Everybody talks about pollution these days, and you don't have to work your palate to death or stick your nose out very far to understand why.

There's water so polluted that you could almost set it on fire, or walk on it without committing a miracle. The soil we walk on, that generations of farmers left better than they found it, is now a drug addict, needing an annual shot under the skin to keep it working.

We're all Borgias these days. We buy garden seed by the ounce and poison by the pound. Today's well manicured, shampooed gardens are death traps for the birds we lure into them.

But gaily we go on spreading the Borgia bree—slug slaughterer, worm-wiper-outer, spider spiflicator, beetle basher, butterfly belter, sawfly slayer, earwig eliminator, caterpillar crucifier . . .

Aha! But look at the hygiene. So look at it. What hygiene we have: the kind that tells you how much mercury you can thole without dying, or that makes mother's milk a menace to her child.

But food was never so clean. So it wasn't. It comes glossed and glittering, waxed and varnished like the plush picture on the packet, mummified in indestructible plastic.

The taste has gone from the tasty; and to the tasteless has been given taste. And what a taste! Cosmetics is all. What matter if an apple is saccharine slush, so long as it has a pretty face.

So roll on you broiler chickens, choking in the ammonia fumes of *haute* stink.

Hail to you little bobby calf, wobbling thick-kneed and anaemic in your technological coffin.

How's it gaun you steaming pigs, sweating, packed like sausages, in your antibiotic stupor?

It's all in a good cause. If man himself is prepared to live like a battery hen, what right has any hen to walk on grass, or pig to root in the sod?

Our Father, that used to be in Heaven, why don't you do as the poet once said and get oot to some clood edge and tak a gander at your braw bricht birlin Earth, a hale week's wark, and see what the miracle mongers are doing to it? Maybe anither wan o yon heavy shooers wid help clean up the mess.

A while back I was talking to an old Irish priest, telling him about the speed-up in beef production, the barley beef, the baby beef, the whatever you like to call the stuff beef that lost all as it grew at double pace. And he said to me, he said:

'Well, you know, where I come from they're all little farming folks—crofters I suppose you'd call them—and they weren't all that educated, you know—in the academic sense that is—indeed they were mostly ignorant you could say without offending anybody very much—and they didn't know anything at all about these new-fangled short-cuts.

'But there's one thing they did know—it takes tree years to make a tree-year-old . . .'

For me that says about it all, and how I wish I had thought of it first.

Ah, well, we go from bad to worse, but where the bad is considered good the worse isn't hard to take. If there's a dollar in muck let rip the reek even if we bock up half a lung in the meanwhile.

A Confirmed Brambler

What is a bramble? it's a plant. What is a blackberry? It's the fruit of the bramble. That's the orthodoxy of it; but to me a bramble is a berry—jet as a blackcock's feather, sheened like the nose of a roe fawn, totally luscious and delectable when metamorphosed into unshoogley jelly the colour of dark, glowing port.

Blackberry hell!

I used to be a confirmed brambler, i.e., one who brambles (and if you say there's no such thing as a verb 'to bramble' well, there is now). I still am a confirmed brambler, an addict even, so long as someone else is doing the picking. I work on the jelly.

Brambling has been in my blood since I was a small boy carrying ripe fruit, always black, home in a hankiepoke. At bramble time all my white hankies became purple, and I gave my mother more work undyeing them than I ever gave her making jelly.

Brambling is, at one and the same time, a disease and one of the most democratic of all pursuits. It is also one of the most uneconomic, judged by economics. If you don't believe me just take a look at some of the bramble brigade, anywhere.

Somebody drives six or seven miles in a big car, using maybe a pound's worth of petrol out and back, then out come two old dears with legs sheathed in nylon, carrying a small basket apiece. They gather perhaps six pounds of the tempting fruit between them, and go home with hands bleeding and nylons wrecked. Uneconomic, yes; but think of the end product that not even Jove would have turned his nose up at.

The bug bites some people so hard that they will actually go out, getting themselves cut and hacked, to gather brambles for somebody else. I have always drawn the line at that.

Observe my two old dears as they wrax into a bramble thicket. Game as they come, they reach into the depths where two berries are hiding coquettishly, or strain high to reach luscious fruits almost out of reach, defying arthritis, rheumatism, doctor's orders and unreachable roe-noses. Aye; it's aye the tap pickle that causes the bother.

One of them, snagging a foot in clawed tendrils, sits down suddenly and unintentionally, ensuring that sitting down later will be painful. Or one will fall on her face, to rise with a thorn in a nostril. They lacerate wrists and leg-gear, but keep on overstretching and picking.

It has been said that women make the best bramblers, because they can stick longer at it. This isn't true. The best bramblers I know are men, and the best swearers I know are bramblers. If you're ever doubtful about the bona fides of any non-swearer, take him brambling. If he sweareth not when he trips and coups ten pounds of brambles back into the thicket he's a non-swearer.

Apart from men, my wife is the best brambler and the most dedicated I know. With her it's a disease. In fact if all the Established Religions in the world were to declare brambling a mortal sin she'd be first in the queue at the Gates of Hell.

Well, she might tie for first place with one of the most elegant bramblers of all—the roe deer. I claim to know something about brambles and bramblers, and the roe is more fascinating to watch than any of the hominids, however garbed and however mighty in prowess.

I love watching a roe deer brambling. Mind you, you won't see one at it if you're a raker, that is one who keeps on moving about. You have to wait for the roe. You have to sit up, and out, for him or her. You have to do it late and early. And you've to be used to disappointments. Often the roe won't appear, but often it will. Then you'll see the most pernickety and accomplished and dainty brambler in the business.

'Fall Out' for Potato Drill

After much flooding, the low-lying cornfield is a loch, and the ungathered stooks are islets, each with its pair or trio of baggy-trousered rooks, which remind one of the familiar cartoon castaways. While the rooks peck hungrily at greening ickers of wasting corn, the herring gulls sail along the canals between the potato drills in the field over the fence.

One day the tattie howkers are out in force, at the most back-breaking job in all the world; the next the field is being lashed by rain, and the pheasants, out gleaning chats, walk sideways, with twisted tails, against the wind.

The topsoil becomes like glue, and a fourteen stone man goes in ankle deep where yesterday the tractor left a spoor as clean as a jelly mould.

Tattie howkers are notable for their earthy wit and uninhibited ribaldry, but this year the trend is towards astral physics and the terminology of nuclear research. There is talk, maybe not surprisingly, about ten, twenty and thirty megaton bombs.

When one picks up an outsize potato—of the type suitable for the chip trade, but a ghastly object to the man who likes a good tattie—he no longer calls it a football, or a turnip, or like a fermer's heid: he calls it a sputnik. Then somebody calls out: 'Open it up, then, and let the monkeys oot.'

A sudden squall of hail, stinging the face with merciless skite, is no longer compared with hazel nuts, or bools, or moth balls, or doo eggs. Its modern name is atomic fall-out.

When a man on the headrigg stops to pluck a bramble sting from his hand, he is asked: 'Is it a thorn?' And the reply is likely to be: 'Naw, it's wan o thur wee copper needles the Yanks put up.'

The man at the bottom of the field has a transistor set perched on the drystane dyke. It is giving out hot music. When farmer comes down the field on the tractor, he sees the man jiving while he waits for more potatoes to be thrown out for him.

As the potatoes fly sideways farmer says: 'Noo ye can get your back bent and stop the jivin'.'

'I'm no jivin',' the man says. 'I'm radioactive.'

I sit on the old dyke at the bottom of the field after the tattie howkers have gone, and watch the gulls gobble bool potatoes. A few crows come down too, and walk about with swaggering gait. Presently all the birds are looking for leftovers.

The crows leave first. At dusk the cushats rise with loud slap of wings and wheel into the wood. Then, when the light is so poor that a crow would be invisible, the gulls cascade into the air and fly off towards the loch. The field is now open to the mice and the voles, and the owls which will come later.

Then I hear the hound music of geese.

Here, in the potato field that has become a loch, I find geese, perhaps the birds I saw arriving the night before. They are swimming out to slumped stooks and pulling out sowpit heads of corn. Their heads go up as I let the car drift slowly by, but they don't panic.

The rain is driving hard. On the moor road I stop to make room to let another vehicle past. It is The Wafer himself, and he gives me a blow.

'Lookin' for the geese?' he asks.

'That's right,' I say.

'They're away ayont the rigg at W's ferm. In the tattie park!'

'The tatties must be good the year,' I say.

'It's a good job something likes them,' The Wafer says. 'It's no me that'll be pittin' them in my mouth . . .'

Four-Footing the New Year

At four minutes past the howdumdeid on Hogmanay the door dinger donged peremptorily, and I hastened to let in my first foot knowing whom I was expecting, but the whom I was expecting turned out to be a whom of another Order: mammalian all right, but instead of a biped with a lump of coal she turned out to be a coal-black Labrador retriever carrying in her jaws an outsize cock pheasant jewelled like a pasha.

A man can be forgiven for being a little anthropomorphic at four minutes into the New Year, so when I say that the black bitch looked puffed up with pride, and full of *joie de vivre,* and *bonhomie* (or should it be *bonfemmie*?) you'll know what I mean. I gave her the pheasant back and told her to take it through to mother, whereupon she breinged through to the sitting room and presented it to my wife.

In came my daughter and her husband grinning like cleft neeps. It was they who had donged the dinger, then sent the bitch ahead, and they were feeling about as pleased with themselves as though they had sent her to put the usual fiver on number two at Ayr. What did they expect from a bitch trained by *moi meme*? You may recall her yourselves: she took up a lot of *The Scotsman* space when I was schooling her.

Anybody prepared to bet that mine wasn't the only house to be first-footed by a Labrador retriever carrying a pheasant?

The most bizarre first-footer I ever saw (and he wasn't first-footing me) was a horse: a big horse pushing 16½ hands. My wife and I were calling on this bachelor farmer, who had passed out before the bells and left the door open for visitors. The horse was in the lobby when we arrived, and walked into the sitting room when we opened the door.

I roused the farmer, who was as fu as a dirigible, whereupon he reached out a hand and shook the horse by a forehoof, saying: 'Gie us a bit shake o your haun auld frien! An a guid New Year!' The horse made no reply . . .

When I had my own wee farm I also had a pet raven named Pruk, who was a broth of a boy, and whose main aim in life was scalping strange women. On Hogmanay my first-feet were a husband and wife. They didn't know the raven was sleeping in the porch: nor did I. When I opened the door the raven first-footed me, then tried his scalping act on the woman. I had a hell of a job getting him out of her hair. You can't train ravens the way you can Labrador retrievers.

Hogmanay 1981 was quiet for us, with my wife out of action for several days with busted intercostals. I busted them, when lifting her gently out of my way in

261

the playful manner I have of half-killing people. We talked of 1981, as people do, and I realised that, for me, it had been a dog year.

First there was, and is, Shona, replacement for my old German Shepherd Lisa, who died in November 1980. Miss Moncreiffe of Moncreiffe very kindly sold me Shona, whom she had intended keeping for herself, when I told her that Lisa had died. The day after her death I presented my wife with Shona. And in due time I was schooling another young German Shepherd bitch.

I worked on her during the spring, day after day, and it can be a boring routine, but as week follows week and you see what you've got the boredom wears off. Maybe boredom is the wrong word, but those who have worked on a puppy from first to last, without falling down on the job, will know what I mean. I had a target for Shona, and its name was Lisa. She is doing very well and maybe she'll make it.

Well, I'm more or less through with Shona, when these good friends of mine decide that they'll get their number three son a G.S. bitch from the same place if I'll school her. I say yes. Boy wants a bitch like Lisa. I point out that Lisas don't come in gift packets, but I'll do my best. And I have been doing just that. Since 16 November, when I started on her, she's had maybe seven days off school.

Her name is Elsa. She's gentle if not yet genteel, and she has more than a distance piece between her ears. She is strong, a thruster, handsome and big: all the things I like. She has found her place, happily, in the peck order. She treats Shona like a mother, and Shona treats her as her own puppy. I'm enjoying myself.

Lisa would have approved.

Turn of the Year Tales

It's almost a ritual at this time of the year to be asked to dredge up from one's memory the funniest or most memorable thing that happened to one at this time of year. And it seems that the older one gets the more frequently this ritual takes place at this time of the year. If you have bother with that sentence you can blame it on the time of year.

Well, there was this bullock that was a great one for jumping fences, and a farmer had slung a fence post from its neck to stop it from jumping fences. Yet there it was on the road, having just jumped a fence. We were footing it on a first-footing jaunt when we saw it make the road. It came lolloping towards us, shied, skidding on the ice, and went sailing past us on its backside with its forefeet in the air! I thought it was one of the funniest things I had ever seen: I still think so.

Then there was the old shepherd who had his Hogmanay joke with a snowman, or rather snowoman. He built it near a roadside pool where a body had been found a long time ago (not as long ago as the days of giants and fairies but before my time). Then he draped it in some old white curtains. And in the sma' hours he waited by in the wood to watch the homing revellers. In those days they mostly walked, there being far fewer cars around.

As he told me afterwards he never knew so many people could break the 100yds. and ¼-mile records. As they came down from the village singing lustily, or not so lustily, one and all turned about and went haring back whence they came.

And next day, in the hostelry, they were all very humdrum, not wishing to speak about it.

Aye: but whit aboot the drama? Everybody's hot on the dramatic bit. Well, I suppose it all depends on what you mean by drama. Let's try this one, which I've been telling just about every year, because it's the only real one I have for the ... yes ... for this time of year ...

Well, there's the loch there, and tonight it is wind-ruffled, with an egg-shaped moon beaming down, and there's a tawny owl calling, and grey geese baying somewhere. It's the same loch as it was more than half a lifetime ago, yet not the same.

It was frozen that winter; everything was frozen and crackling; the moon was ice, and my dogs came home with me with icicles hanging from the hair of their bellies.

I had a little hiding place on the shore of the loch, and it was like an ice-box. Out on the loch, at the tip of a long shadow cast by the moon, were some mallard ducks. Near the middle were some grey geese, and a few other odds and ends. Then the otter came along.

He came weaseling across my front, and along to the high bank. There I lost him, and I didn't see him again until he came out at the tip of the long shadow, slap into the parcel of dozing mallards. Then there was a tushkarue! Up went the ducks, up went the geese, and the sky was full of noise, and the moon laughing in its sleeve. And the loch was deserted, except for one otter, straddling a mallard duck.

I said to myself: 'Well, well!' Then I said 'Well, well, well,' when two foxes came in on a converging course on the otter. There was a brulzie, a bit of leaping and snapping, and the otter was dour. But in the end he had to give way, leaving the foxes in possession. They trotted off, one leading with duck in jaws, the other following in its slots.

So?

Well, next day The Wafer said to me: 'You should have been at the loch last night. Two foxes and an otter had a bit argument ower a duck, and the foxes won.'

'Good heavens!' said I: 'I was there. Were you there as weel?'

'Naw,' he said. 'I only read aboot it in the snaw this mornin.'